ON THE REVOLUTIONS OF THE INTERNAL SPHERES

A NEW THEORY OF MATTER AND THE TRANSMISSION OF LIGHT

SECOND EDITION

K TROY

authorHOUSE®

AuthorHouse™ UK
1663 Liberty Drive
Bloomington, IN 47403 USA
www.authorhouse.co.uk
Phone: 0800.197.4150

Published by AuthorHouse 01/24/2018

ISBN: 978-1-5462-8761-2 (sc)
ISBN: 978-1-5462-8762-9 (e)

Print information available on the last page.

This book is printed on acid-free paper.

Dedicated to my father, Edward Troy

Contents

Prologue

"Realism is dead. . . Its death was hastened by the debates over the interpretation of quantum theory, where Bohr's non-realist philosophy was seen to win out over Einstein's passionate realism. Its death was certified, finally, as the last two generations of physical scientists turned their backs on realism and have managed, nevertheless, to do science successfully without it. . ."

Arthur Fine in ***The Shaky Game***

Imagine the following situation. A physicist in his laboratory is trying to understand the strange phenomenon that is unfolding before his eyes. He has set up an electrode in front of a diaphragm with two apertures. The electrode, he has good reason to believe, is emitting *particles*, and some of them are making it through to a phosphorescent screen located at some distance behind the diaphragm. The puzzling thing is that the pattern building up on the final screen is an "interference" pattern, composed of alternate bands of many and few particle impacts. Such a pattern is typical of a *wave* motion. If we pass a water wave through two apertures (under the right conditions), two trains of waves will issue from each aperture and give rise to alternate areas of greater or lesser disturbance in the water, depending on the degree to which the crests and troughs of one wave interfere constructively or destructively with the crests and troughs of the other.

But our physicist is convinced that his electrode is shooting particles, not waves. How can particles give rise to an interference pattern? He decides to reduce the output of the emitter to just a single particle at a time, on this occasion leaving both apertures open. The particle, as expected, lands on a particular spot of the final screen, confirming that it must indeed be a particle, occupying as it does a localised area of space. Now the physicist leaves the electrode on, so that it emits individual particles, one at a time, every few seconds. To his consternation, an unmistakeable interference pattern builds up again over time on the final screen! How can this be? Is there some kind of wave *guiding* the particle to its destination? Each particle is going through a single aperture but is there an associated wave that goes through both?

Of course, this situation is not imaginary at all. It is the situation that the physicist has found himself in for one hundred years now. How is he to account for the unparticlelike behaviour of the particles that (he believes) populate his atomic experiments? This difficulty – and it remains a difficulty - has not held up the advance of physics from the point of view of its growing practical ability to manipulate the physical world. Whilst our *understanding* of what is happening at this level has remained quite primitive, our capacity to *predict* the outcome of such empirical situations has developed in a way that is unprecedented in the history of science. But the very sophistication of our theoretical apparatus for describing the behaviour of atomic particles has obscured our discernment of what is actually happening on the concrete level. The behaviour of causal players in atomic experiments is now couched in the language of complex mathematics, and it becomes ever more difficult to discern what this language might correspond to in the real world.

Careful and rigorous variations of the two-path experiment described above apparently demonstrate that systems which seem to have waves moving in them suddenly and discontinuously *collapse* to particles when certain sorts of interactions take place. Let us allude to just two of the perplexing conclusions that physicists have been driven to make on the basis of empirical data such as this. The first is that causal entities can change their character and become particles or waves depending on the empirical conditions they find themselves in. Among other things, this means that they can go from having no localised position at all in space to occupying a pin-point position on a screen. Secondly, the transition of a system from having a wavelike character to taking on a particlelike character occurs *discontinuously*. The transition is not governed by deterministic laws, but can only be described in a statistical fashion. Before the collapse, the causal entity does not have a definite value for an observable such as position, for example. Which value of position the entity takes on as a result of the collapse is considered to be a matter of pure chance. There is *no objective state of affairs* that determines the value into which an observable collapses. If this claim is true – and refined laboratory arrangements seem to indicate powerfully that it *is* true – then any hope for a comprehensive explanatory story of what is happening in the quantum domain is effectively quenched.

In very simple terms, our most advanced description of the constitution of the material world holds that reality is populated by ghostly entities that, much of the time, have no definite values for observables such as position. Physical systems are believed to evolve in ways that are fundamentally indeterministic. The age-old dream of explaining the deepest causes of natural phenomena - *why* things happen as they do – has lost ground to a descriptive account that makes do with predicting *what* will happen.

But what if the physicist at the beginning of our story had some of his fundamental presuppositions wrong? What if the causal influence evolving in his system was neither a wave nor a particle? Would the evidence still compel him to conclude that the influence evolving in the system had properties of a paradoxical sort? Would he still be led to think that something not strictly causal was happening in the system? It is perhaps sobering to consider that all of the unsatisfactory aspects of modern physics – the indeterminism, the contradictory behaviour of causal entities, the semi-mystical descriptions of the role of observational conditions in collapsing reality into a tangible form – all of these implausible features derive from our assertion that *particles and waves* are moving in the system. The vexing empirical data suggests that the hypothetical *particle* sometimes exhibits properties that should not be possible for a particle, and the hypothetical *wave* sometimes ceases to be a wave for reasons that cannot be expressed fully in causal terms.

What if we were to re-evaluate that vexing empirical data and consider if *another* type of causal influence altogether were evolving in the system, an influence that is neither a wave nor a particle? The task of this book is to do just that. The basic foundational evidence for the standard theory of the atom is reinterpreted in a natural and plausible manner. We aim to show that the picture of a positively charged nucleus surrounded by a hive of frantic electrons is not at all compelled by the evidence. Better pictures are possible. A simpler model can be formulated that will explain the data in more self-consistent terms. Will this new model of the atom involve an implausible forcing of the evidence? What sort of unlikely story will the unsuspecting reader be asked to accept? On the contrary, what follows in these pages is easier to digest than the unlikely claims that three generations of students have been expected to swallow. All that is asked for is an open mind, and a willingness to evaluate critically the views of the status quo.

This book accepts the numbers and mathematical formulae that have arisen out of atomic physics, but it reinterprets them so that they become the basis of an orderly account of the world. All of the fundamental quantities of physics – the atomic number that Moseley realized pointed to an essential constituent of matter, the valences of the various elements, the length of the different rows in the atomic table, the numbers associated with the Pauli exclusion principle, the magnitudes of frequencies and wavelengths of light, the startling simplicity of the Rydberg formula – all of these numbers will be shown to have their basis in solid objects with determinate properties exerting causal influences that evolve deterministically.

In the pages that follow, we can promise no sublime mathematics nor sophisticated analyses of the theories of others. What we have focussed on are the simple and classic experiments that have been interpreted in certain ways but could have been interpreted differently: Young's observation of the interference of light; Weber and Kohlrausch's comparison of the electromagnetic and electrostatic units of current; the null result of the Michelson-Morley experiment; Thomson's measurement of the mass/charge ratio of the electron; and the mysterious patterns of the Stern-Gerlach results. The fact that the most recent of these experiments dates from the 1920s does not mean that their significance is obsolete. We hope to show how an original viewpoint on their results reveals a picture of reality than is comprehensible, ordered and one.

Author's note regarding the second edition: This second edition clarifies and corrects some of the ideas contained in the first. A different account of the nature of the "neutron" is presented that is more consistent with the evidence. Reflection on the composition of deuterium effectively motivated these changes. The account of the way that groups of atoms align during the absorption of light has been modified. A section has been added dedicated to Maxwell's theoretical derivation of the velocity of light. In order to make this work more interesting for a general audience, some of the philosophical material contained in the first edition on the nature of space and the development of theory has either been placed in the appendix or eliminated altogether.

Chapter One

THE HYDROGEN ATOM, ATOMIC BONDING AND THE PERIODIC TABLE

"If, in some cataclysm, all of scientific knowledge were to be destroyed, and only one sentence passed on to the next generations of creatures, what statement would contain the most information in the fewest words? I believe it is the atomic hypothesis . . . that all things are made of atoms — little particles that move around in perpetual motion, attracting each other when they are a little distance apart, but repelling upon being squeezed into one another. In that one sentence, you will see, there is an enormous amount of information about the world, if just a little imagination and thinking are applied."

Richard Feynman in ***Lectures in Physics***

Overview of this chapter and its principal claims
1. A brief account of the development of atomic theory.
2. The presentation of a new model of the hydrogen atom. The "genesis-unit" has no protons, neutrons or electrons inside it, but

it can account for all the empirical evidence usually explained in terms of these particles. The unit is constituted by a fundamental B-V polarity.

3. Moseley's work demonstrates beyond doubt that the atomic number of an element corresponds to some fundamental quantity within the atom. The standard view considers this quantity to be the number of protons in the nucleus. We interpret the quantity to represent the number of genesis-units in the complex atom whose electrostatic influences are not counterbalanced by other units within the structure. Beginning with helium, we show how each successive element can be accounted for in terms of the progressive fusion of one or more genesis-units to the original structure. Fusion requires the presence of "binding genesis-units" – units whose electrostatic poles hold other genesis-units in place. These binding units are usually called "neutrons". Whilst it is true that genesis-units in general have no net electrostatic charge, in normal circumstances their poles *do* give rise to noticeable empirical effects. However, in the case of binding units, these effects are counterbalanced because of the position they occupy within the complex structure. Thus, for all the world, they behave like a "neutron" – a particle with no charge.

4. A few principles are set down for determining how many binding units are required for each successive fusion. Using a simple framework, we show how a significant number of the lighter elements in the table are formed. Each successive fusion confers greater or lesser equilibrium on the composite structure. The basic feature that increases equilibrium is the *symmetry* of the arrangement of the genesis-units in the atom. Symmetry entails that residual electrostatic influences are better reciprocated. The noble gases are characterised by perfectly symmetrical configurations of genesis-units.

5. We show how symmetry can explain the numbers in the periodic table, such as the amount of elements in each row, and the shared characteristics of elements in the same group. The Pauli exclusion principle can be described as being underpinned by the question of

equilibrium/symmetry. We show how each fused genesis-unit has a unique role in reciprocating the residual electrostatic influences of the rest of the structure in order to achieve maximum equilibrium. Four structural features of the position of the genesis-unit in the atom correspond to the four numbers of the exclusion principle. They confer a unique role on each genesis-unit in the stability of the whole.

6. We see how chemical bonding has a corresponding tendency to increase the equilibrium of atoms. All stable chemical bonds lead to composite structures that are more symmetrical (and hence more electrostatically stable) than their constituent atoms. Thus, using the simple causal principle that there is a natural tendency towards symmetry of structure (because it increases electrostatic equilibrium), we explain the arrangement of the periodic table, the ontological basis of the Pauli exclusion principle, the characteristic properties of groups of elements and the bonding patterns of atoms.

1.1 A brief account of the early development of atomic theory

When the atoms of any particular element are stimulated in the right way, they emit characteristic "wavelengths" of light (the term "wavelength" is more appropriately used for a motion of *matter* in space, but the term will be used throughout this book for convenience). Atoms of an element also tend to absorb light of the same wavelengths. The characteristic band of wavelengths emitted by an atom of a particular element is known as the "emission spectrum" of that element. This spectrum is unique to that element and can be used to identify which elements are present in a sample of matter that is being spectroscopically analysed. During the nineteenth century, the spectra of elements were systematically studied. Analysis of the typical wavelengths of solar light revealed the presence of certain elements in the sun, including ones that had not yet been discovered on earth.

The ability of an atom of a particular element to emit and absorb light of certain specific wavelengths was taken as a sure indication that the atom had an internal structure of some sort. Scientists speculated that these characteristic spectra must be produced by components of the atom in states of vibration. Balmer's discovery that the hydrogen spectrum could be described by a simple mathematical formula reinforced the conviction that such spectra were generated by structures in the atom whose operation was governed by precise physical laws. By the end of the nineteenth century, significant empirical clues were discovered that seemed to shed the first light on that internal structure. The most important development was related to the study of the nature of what were known as "cathode rays." When an evacuated glass tube was fitted inside with electrodes (at either end) and a voltage applied, the glass at the anode (positive electrode) end of the tube began to emit a fluorescent glow. This seemed to be due to rays being emitted from the cathode towards the anode, which flew past the anode and struck the glass, giving off light. Physicists asked themselves if these rays were particles of matter in motion, or if they were some sort of immaterial process in the invisible ether that was believed to permeate all of space. If it were the case that the cathode rays were a process in the ether, then, it was assumed, they could not be electrically charged in themselves, since such a disturbance in the medium could not in itself carry a charge. The charge, then, would have to be some sort of by-product of the transmission of the cathode rays.

J.J. Thomson set himself the task of investigating the nature of these rays to see if they were a form of immaterial radiation or if they consisted in a transmission of charged particles. In an impressive series of experiments, he established that the rays were deflected by a magnetic field, and that the amount of this deflection was equal regardless of which material was used to make the anode, or which gas was used in the tube. This showed that the rays were identical in nature regardless of their origin. Another experiment supported the claim that the rays did not produce charge as a mere by-product of their transmission, but were electrically charged in themselves.

Two other experiments were to have the furthest-reaching influence on the subsequent history of atomic physics. The major obstacle that remained to establishing the claim that cathode radiation consisted in a steam of negatively charged *particles* was the fact that no-one had ever succeeded in *deflecting* the rays in an electric field. Indeed, the fact that Hertz could not detect any noticeable deflection in the rays convinced him that the radiation could not consist of charged particles but must be of an immaterial sort. Thomson suspected that the traces of gas left in the discharge tube might be responsible for preventing the beam from bending in the field, so he evacuated the gas in a more careful and systematic manner than had been done previously. When Thomson performed the experiment using a well-evacuated tube, he found that the beam was indeed deflected by the electric field. If the upper plate was connected to the negative terminal of the battery, the beam was bent downwards. When the polarity was reversed, the opposite occurred. He concluded that cathode rays consisted in a motion of negatively charged particles.

In his most celebrated experiment, Thomson calculated the ratio of mass to charge of the hypothetical particle. We will examine this experiment in more detail in Chapter Four, but consider for a moment the significance of this finding. A value had been calculated for the ratio of mass to charge of a particle that until then had only a dubious claim to existence. Thomson called his particle the "corpuscle" but this soon gave way to the term "electron" coined some years earlier by George Stoney for the fundamental unit of electric charge. In the 1897 paper in which Thomson describes his experimental work, he hypothesized that all matter is made up of aggregates of the particle, the mass to charge ratio of which he had just calculated. The consequences were enormous for the history of science. It is no exaggeration to say that this was the defining moment in the development of the fledgling atomic physics. For the first time, convincing evidence had been found for the existence of a subatomic particle, a universal building block of matter. And something even more significant had been achieved. It wasn't just that the *existence* of the electron had been "demonstrated": a measurement had been made of the most fundamental property that we associate with a physical object - the quantity of its matter.

It took some time for Thomson's discovery to be accepted by the scientific community, but once it was accepted, atomic physics never looked back. A pattern developed throughout the twentieth century where a new particle would be "discovered", its properties quantified, and its role in the functioning of the atom established. No-one today can question Thomson's logic in asserting that the evidence surrounding cathode rays indicated the existence of a subatomic particle. The series of experiments that he carried out pointed to no other conclusion. And the additional claim that atoms were *composed* of electrons seemed an obvious one. Similarly, when radioactive atoms were observed to emit a form of radiation known as "beta particles," it was natural to assume that these supposed particles had once been *inside* the atom, rather than produced spontaneously by the process of radioactive decay.

Later evidence from experiments involving "electrons', however, cannot be reconciled with the belief that the causal activity in question involves a *particle,* without changing the significance of "particle" in a way that renders the debate between Hertz, Thomson, and others effectively meaningless. It is ironic that Thomson's own son is best known for work showing that the electron can be diffracted like a *wave.* But the most damning evidence against the electron involves evidence that cannot be explained simply with the notions of either wave or particle but ends in paradox *whatever sort of motion of matter in space* we use as our model of explanation. Such evidence is a more than adequate justification for calling for a complete reappraisal of the conclusions made by Thomson and others.

After the "discovery" by Thomson of the electron, physicists began to work feverishly to develop a theory of the internal structure of the atom. Over the next fifteen years, in a veritable race for the moon, a series of models of the internal structure were hypothesized, empirical consequences were deduced from the model, and these were confronted with the empirical data. By the dawn of the twentieth century, the first coherent picture of atomic structure seemed to be materializing. The atom was thought of as being composed of a positively charged sphere - accounting for most of the atom's mass - with negatively charged electrons embedded in it. The "plum pudding model" of the atom, as it became known, assumed that matter was more or less evenly distributed within the atom, as was the positive charge. Though it was no longer considered to be the fundamental unit of matter, the atom was still conceived as a densely solid piece of matter with no empty spaces in it. The heterogeneous nature of this internal structure (the presence and configuration of embedded electrons in a positively charged sphere) suggested various ways in which the characteristic emission spectrum of an element might be generated. It was speculated, for example, that if an electron was disturbed by external stimulation, then it might oscillate momentarily before returning to its original position. The rate of oscillation of the electron (and hence the frequency of light it emitted) could depend on factors such as the original position of the electron, and the density of distribution of positive charge in the sphere.

Little further progress was made until it was decided to probe the structure of the atom with positively charged helium atoms, known at the time as "alpha particles". Rutherford had observed that such particles underwent deflections when they passed through thin layers of matter, so perhaps a systematic analysis of such deflections could reveal something of the number of component parts within the atom, how they were distributed, and what properties they had (such as mass and electric charge). This was the beginning of a strategy of particle bombardment that has continued relentlessly to this day.

In a celebrated experiment carried out by Geiger and Marsden under the direction of Rutherford, alpha particles were shot at a thin piece of gold leaf. Given the relative mass of these particles, it was expected that they would pass through the metal with only minor deflections. When the experiment was performed, most of the particles passed through as expected with little aberration. Some particles, however, were deflected at large angles, whilst others rebounded directly at the alpha source, a fact that Rutherford famously declared was "almost as incredible as if you fired a fifteen-inch shell at a piece of tissue paper and it came back and hit you." This led to the conclusion that the atoms of the gold leaf were composed mostly of empty space, with the bulk of their mass concentrated in a very confined region, and it was this bulk that had deflected the hefty alpha particles. The area of confined mass would become known as the nucleus.

The gold leaf experiment seemed to confirm that most of the mass of the atom was concentrated in a small fraction of the area occupied by the atom. Influenced by this evidence and by Planck's theory of radiation, Niels Bohr proposed a theory of atomic structure that would go on to attain the kind of prominence enjoyed by few physical models in the history of science. The nucleus, it was supposed, was composed of matter of unknown structure with positive charge, whilst the electrons were thought to roam in the outer areas of the atom. The belief that electrons existed in these outer reaches was reinforced by the empirical fact that they could be procured easily from the surface of certain metals by low energy radiation (a phenomenon known as the "photo-electric effect"). According to the new planetary model, the concentration of positive charge in the nucleus held the negatively charged electrons in orbit around it, much as the gravitational "attraction" of the sun held the planets in orbit.

The theory had shortcomings, and within a few years had already been modified considerably, but the basic elements of Bohr's model continue to dominate our way of conceiving of the atom. The unit is still visualized in terms of a positively charged nucleus surrounded by electrons arranged in shells. These electrons are considered not to orbit the nucleus in the classical sense, but to exist in "stationary states" that are naturally stable and involve no radiation of light. Atoms absorb and emit radiation only when electrons move between two stationary states. When this happens, the difference in energy of the stationary states corresponds to the energy of the light emitted or absorbed.

The relatively rapid success and widespread acceptance of Bohr's model derived from its ability to provide a theoretical underpinning for the peculiar structure of the Rydberg formula. Balmer had predicted the visible wavelengths emitted by hydrogen using a strikingly simple mathematical expression, later generalised by Rydberg so that it could describe the emission spectra of other elements as well. Bohr's model provided a plausible explanation for the simplicity of the expression: the simple integers that occurred in the formula were understood to represent fundamental energy levels within the atom. This explanatory capacity led physicists to manifest a curious tolerance towards a series of features in the model that grated against the intuition. Nowadays, although it is common to read that the Bohr picture of the atom is "obsolete", or that it has been superseded by the valence shell model of the atom, the fundamental elements of the Bohr model remain at the heart of atomic theory. And the features of the model that grated against the intuition a century ago are grating still.

When it was hypothesized that atoms had a nucleus composed of a number of positively charged particles (the proton) bound together, the question arose as to how these particles could overcome their mutually repulsive electrostatic forces. In the 1930s, the notion of a *nuclear force* was developed to account for the stability of the nucleus. To facilitate such a claim that went against everything we knew about electrostatic charges and their powerful mutually-repulsive influences, this completely novel causal player was postulated. The justification for introducing this new force of nature was the known "fact" of the stability of the nucleus. But the very *existence* of a nucleus of tightly bound positively charged particles was a conjecture from empirical evidence that could have been interpreted in other ways.

The nuclear force needed to be many times greater than the force of electrostatic repulsion if it was to explain the supposed properties of the nucleus. Surely such an enormous force would have empirically measurable consequences that would allow us to develop hypotheses about its nature and its origin? The complete lack of any such consequences, instead, led to the ad-hoc conjecture that the range of this colossal force did not extend beyond the nucleus. The postulation that the force had such an incredibly limited influence was a supposition that was custom-made to fit the empirical data, and was not based on independent empirical findings that would have allowed us to quantify or even confirm the existence of the force.

Paradoxically, the very lack of empirical foundation for the postulation of the nuclear force has led to relatively unbridled speculation about its nature. The force is now believed to be a residue of the "strong interaction," a force that binds quarks together, and the quarks, in their turn, make up the protons and neutrons in the nucleus. The strong interaction is mediated by another particle called the "gluon." The properties of both the quark and the gluon have a very tenuous connection to the empirical evidence. One of the properties of the quark is called "colour confinement," a characteristic that allegedly makes the quark impossible to isolate and observe directly. The trouble is that entities whose properties and states cannot be observed directly can be attributed very unusual characteristics indeed.

As with the other reinterpretations of empirical evidence made in this book, we do not contend that the quark and the gluon have no relation to the empirical evidence.

Indeed, they form a rigorously consistent and genuine bridge between empirical data and a picture of the atom that is extremely likely to be *wrong*. The theoretical bridge has validity insofar as it provides a comprehensive account of how such theoretical entities, *if they existed*, would give rise to the empirical evidence. But the probability that such entities are actually present in the world is exceedingly remote - the spheres of Ptolemy's *Almagest* are much more defensible in comparison. And just like those spheres, the likelihood that gluons and quarks are figments of our imagination is outstandingly high. Any candid appraisal of their status will acknowledge that, although compatible with the empirical data, they are cut dramatically adrift from it. It does not need to be said that the *compatibility* of theory with empirical data is a very different matter to the *plausibility* of theory in confrontation with the empirical data.

The challenge confronting us then is this: can an account of atomic phenomena be coherently developed without invoking ad-hoc features whose very function is to explain away the empirical data – empirical data that makes some aspects of current theory look very far-fetched in the first place? Can we describe the atom in a plausible and relatively simple way without postulating forces or entities that cannot be verified independently? What we are after is nothing more and nothing less than what the scientific spirit, deep down, naturally yearns for: a theory with characteristics that point to maximum epistemic content, a description that has the greatest chance of corresponding to how things really are.

In this chapter we present an account of the structure of atoms without invoking any new entities or forces beyond a very simple unit of matter (called the "genesis-unit") and the fundamental polarity which constitutes it. The polarity within this unit will be used to explain the way in which hydrogen atoms fuse together to form more complex elements, the pattern of chemical bonding in general, and the structure of the periodic table. We will describe in a very natural way how units with this simple internal polarity can exert a *neutral* electrostatic influence on each other when they are separated by a great distance, net electrostatic *attraction* on each other when they are located at closer proximity, and extreme *repulsive* influence when they are at close range.

1.2 Neutral influence, net attraction and extreme repulsion

The claim that the nucleus consists of a number of particles with a net positive charge is a hypothesis that is not in any way forced upon us by the evidence. The fact that many (if not most) physicists consider the hypothesis to be an unavoidable and empirically-compelled conclusion is a measure of the extent to which this view is irrationally entrenched in current theory. Far from being forced by the evidence, the conclusion has fostered the development of an increasingly implausible theoretical structure, an unwieldy framework that should prompt us to go back and question the original conjecture. Indeed, the entire complex story of quarks and gluons has its origin in the issue of how identically-charged nucleons can overcome their net forces of repulsion to cohere together in the nucleus, and at the same time generate the varied empirical data that is produced by the atom.

The proposal here is that the reader resist the pressure to accept the established view, at least for as long as it takes to read these few pages, and consider for a while the very real possibility that the positively charged nucleus, like the luminiferous ether and Ptolemy's spheres, is one of the phantoms of science, an entity with an impressive array of theoretical properties, all of which are meticulously related to the empirical evidence but which possesses no reality whatsoever. Instead, we present a very different hypothesis for the structure of the atom. We hope that the reader not be dissuaded by the simplicity of the model and we ask that an effort be made to read this chapter at least. By that point, the capacity of our model for explaining the structure of the atomic table will hopefully have become apparent. At that stage the reader can tell for himself if the hypothesis rings true or not.

We posit that the hydrogen atom is a simple and indivisible unit that is constituted by an internal polarity. Larger atoms are composed of fusions of these basic units. The polarity consists of a "positive" component - the B pole - which has been generated from the "negative" component – the V pole. We use the letters B and V ("being" and "void") to avoid confusion with the south and north poles of magnets, or the polarity between positive and negative charges. As we shall see, B and V underlie both magnetism and electricity and permit us to give a unified explanation of both. Thus, we envision that the B pole consists of something *substantial* (almost like a concentrated material particle in the old-fashioned sense) whilst the V component consists in some sort of *privation* of substance. Now that is not to say that the V component is simply nothingness, akin to the Newtonian concept of space as an empty void. That kind of nothingness, after all, might well exist in this portion of reality even if a B component had *not* been generated there in the first place. We envision that the V component is a void of a more radical nature – an "area" of reality where a B component has been generated *from* a previous state of nothingness. In this sense, the V pole sounds a little like what is usually referred to as "anti-matter". It is what is left when matter is generated from nothingness. But, unlike a particle of anti-matter, *it cannot exist independently* of the adjacent B pole whose existence is actually giving rise to it in the world in real time. Apart from this passing mention, we ask that all talk of "anti-matter" be left there. This concept has no place in our framework.

The unit that comes into being as a result of the generation of this polarity will be referred to as the "genesis-unit". The reason we use this term is primarily because the unit – as we shall see as we go on - will be invoked to account for the appearance of virtually all other material phenomena. This unit will be the basic locus of all causal activity in the world, whether that influence be gravitational, magnetic, or electrical. All causal activity arises in a unit of this sort and has its effects in one or more other such units. No more primitive source or target of causal activity exists in the world. The unit is not itself composed of smaller components of matter. It is the most basic instance of matter that exists, and, in fact, *constitutes* matter. The hydrogen atom is composed of a single genesis-unit, whilst heavier elements are constituted by multiple genesis-units fused together.

According to this account, there are really *three* different kinds of reality within the genesis-unit. Firstly, there is the positive B component concentrated presumably at one end of the unit. Then there is the negative pole from which the positive pole has been generated. Finally, there is nothingness, an area (or areas) distinct from the positive component that has come to be, but also radically distinct from the "hole" that has been left when the positive component was generated. If we allow that the positive component is concentrated at one end of the unit, whilst the negative component is a little more dispersed throughout the rest of the unit (although predominantly concentrated at the end opposite to the positive pole) then our model acquires surprising explanatory power. The picture we are presenting of the basic hydrogen atom retains its overall simplicity but we are now making the plausible assertion that the generation of the B component does not result in a simple, undifferentiated, homogenous "hole" in reality (the V pole). Rather, the process of generation of the B pole demands that this positive entity be, as it were, dragged up out of a well of nothingness. The deepest part of the well, i.e., the end of the resultant unit of matter that is most distant to the new positive pole, will constitute the most significant portion of the V component in the atom. In the intervening area between the extremity of the V pole and the positive pole, there will be a progressively diluted combination of void and nothingness (where void is conceived of as that which results when the B component is generated from nothingness).

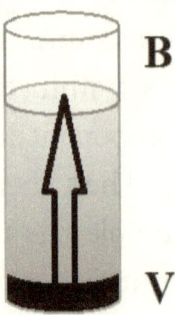

B

V

Figure 1.1 *Schematic representation of the gradient of dispersion of the V component in the genesis-unit. The B pole is drawn up from nothingness, thus creating a well with a negative void at the deepest point, and a progressively weaker concentration of negative void and nothingness in the central portion of the unit.*

If this unit constitutes matter but is not constituted itself by smaller pieces of other material, then what exactly does it consist of? It is proposed that the unit is constituted by the very work impulse that gives rise to the polarity within the unit. The polarity is maintained by the continuous presence of the work impulse at the heart of the unit, and this becomes the fundamental building block of the universe. The empirical advantages of this claim will become more apparent shortly. We can speculate on what the polarity within the unit consists of, using analogical language that might help us in thinking about the dualism that obtains. What is more important, however, is what comes into being as a result of the action of this work impulse. From this moment onwards, a material entity is present in the system that interacts with the other entities that are present in the system. The reader is asked to be patient with the ethereal nature of this description of the fundamental building block of matter. What we are trying to do is make sense of empirical reality by constructing a framework upon simple principles. We need to develop the framework a little more as we go along, and then see if it has the capacity to explain the empirical evidence in a plausible way. The reader will soon have the opportunity to evaluate the explanatory merits of the system.

Figure 1.1 attempts to depict what such a unit might look like. The B pole is drawn up from the well of nothingness and thus is shown at the top of the unit. At the bottom of the well is the densest area of void. In between, there is a gradient consisting of a progressively lesser concentration of void. But the gradient is not composed of a mixture of V component and B component: rather it consists of a combination of void and *nothingness*. *All* of the B component is located at the top of the unit. The V component is mostly concentrated at the bottom, but is partially dispersed through the portion of the rest of the unit that lies beneath the B pole. The *total* magnitudes of the B and V components are, of course, perfectly equal: one was generated from the other and therefore they have a strict relationship between them of equality in magnitude yet contrasting in nature. The insistence that the gradient have these characteristics will permit an explanation of the various phenomena of attraction, repulsion and atomic cohesion that prompt the postulation of multiple forces in the standard model.

No doubt the reader is surprised (and possibly horrified) at the absence of the electron in this model. A little more patience and you will see how this simple picture can account for all the empirical data that prompt the postulation of the electron. The relation of protons and neutrons to the genesis-unit can be described in even simpler terms. But first we must consider how genesis-units interact with each other. Take first of all the case where two hydrogen atoms (or "genesis-units") are located at a considerable distance from each other (see Figure 1.2). In this illustration, no attempt is made to show the perfect gradient with which, presumably, the V component diminishes as we move towards the B pole. It is merely to suit the artistic and (as we shall see soon) the mathematical limitations of this author that the V component is depicted as being dispersed into four progressively more insignificant portions. The important point is that the *total* magnitudes of the B and V components of each unit are perfectly equal. In this case, where there is a significant separation between A1 and A2, the distances between the like poles of the different units (e) can be considered to be identical to the distances between the unlike poles (d). The slight difference in length between d and e will give rise to a differential between the attractive and repulsive impulses that is of negligible import (we must imagine the units to be far more greatly separated in space in proportion to their own size than is shown in the diagram). Therefore each unit will present itself to the other as an electrostatically neutral entity.

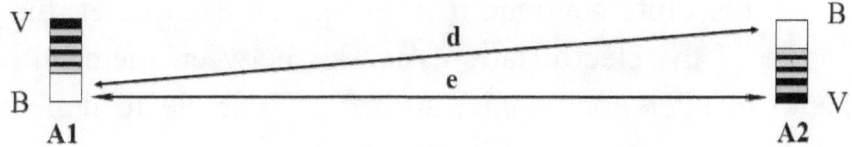

Figure 1.2 Two units at considerable distance exert no net
electrostatic influence on each other

If two genesis-units have their polarities oriented in opposite directions, then the difference in distance between the like poles and the unlike poles will become more significant as they converge on each other. This is depicted in Figure 1.3. As the units continue to converge, a point will eventually be reached where the growing differential between d and e will entail that the magnitude of the attractive influences between the B of A1 and the V of A2 (and vice-versa) will significantly outweigh the repulsive influences between their opposite poles. At this distance, electrostatic bonds between atoms will be possible. Of course, there will be a *range* of distances at which electrostatic attraction will be felt, and the extent of this range can be established by empirical means.

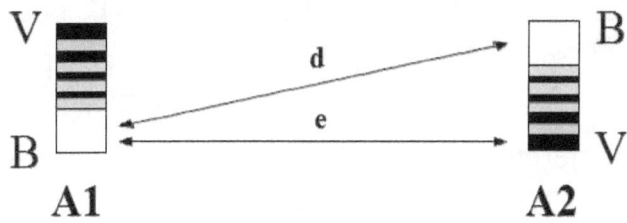

Figure 1.3 Genesis-units exerting net attraction on each other

As the units continue to converge on each other, the nature of the electrostatic dynamics between them will begin to alter once again. An effort is made to depict this new situation in Figure 1.4. As stated earlier, for simplicity the V component is shown as being dispersed throughout the unit in four separate portions, v_1, v_2, v_3 and v_4. Consider first of all the repulsive influence between the B of A1 and the B of A2. Because each B component is located in a geographically concentrated area, we can express the repulsive tendency between the units as: Cb^2/d^2, where C is the constant of proportionality for electrostatic influence, b is a measure of the electrostatic influence of the B component of each genesis-unit, and d is the distance between the B pole of A1 and the B pole of A2.

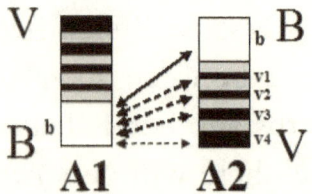

Figure 1.4 *Closer proximity leads to net repulsion*

The attractive tendency between the B pole of A1 and the V pole of A2 will be a bit more complicated to calculate. The V component is dispersed, which means that the differences in the *direction* of attractive influence between the B of A1 and the dispersed components of A2 (i.e., v_1, v_2, v_3 and v_4) will become more significant as the units converge, leading eventually to a drastic reduction in the net attractive influence. Let us consider this in more detail. The total attractive influence between the B of A1 and the V of A2 can be written as a sum of the individual influences in the following terms: $bv_1 + bv_2 + bv_3 + bv_4$, where bv_i is a measure of the magnitude of electrostatic attraction between the B component of A1 and the v_i portion of the V component of A2. Using the standard expression for electrostatic influence, this total is:

$$Cbv_1 \cos\theta_1/e_1^2 + Cbv_2 \cos\theta_2/e_2^2 + Cbv_3 \cos\theta_3/e_3^2 + Cbv_4/e_4^2;$$

where C is the constant of proportionality for electrostatic influence, e_i is the distance between v_i and the B component of A1, and θ_i is the angle between the vector bv_i and the vector bv_4 (for calculation purposes we are taking the direction of the vector bv_4, the most substantial portion of electrostatic attraction between A1 and A2, as the norm for the direction of attractive influence).

Depending solely on the *relative height of the genesis-unit* and the *pattern of dispersal* of the V component throughout the unit, there can be no doubt that convergence of two units can diminish the magnitude of the overall attractive influence with respect to the magnitude of the repulsive influence. The mathematical formulation of the relationship between the height of the unit, the pattern of dispersal, the separation of the units and the consequences for electrostatic influence will be left to others. This author is not capable! However, enough has been said to illustrate the point that a relatively simple model of the atom can furnish an explanation of the empirical fact that atoms attract each other at a certain distance but repel at closer range.

Slightly different models can also offer plausible explanations, one of which we will consider here. Figure 1.5 displays a pair of genesis-units that are more elongated than the previous model. At the poles there will be a net attraction between the pair, but this will be offset by the larger repulsive tendencies towards the centre. As in the previous model, the V component is dispersed throughout the length of the unit, with the greatest portion concentrated at the end. The B component is entirely concentrated at the other end of each unit, but its magnitude will be precisely equal to that of the total V component. This goes without saying, since the V component results from the generation of the B. Say that one quarter of the total V component is concentrated at the end of any given unit. The net attraction at each poles will thus be approximately proportional to $BV/4$. If the unit is sufficiently elongated and the V is evenly dispersed along its length, then we can discount the attraction between the B of A1 and the portion of the V that is not located at the pole of A2. Taking into account both ends of the composite system, the total attraction will be approximately proportional to $2(BV/4) = BV/2$. Since the total B and V components are equal in magnitude, this can be simplified to $V^2/2$.

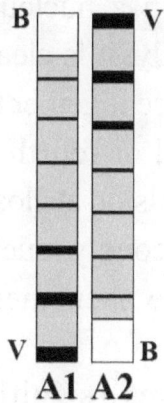

Figure 1.5 *Modified model of the atom demonstrating the net repulsion of two units.*

Now let us consider the net repulsion. It will be approximately proportional to $(3B/4)^2$. This follows from the fact that three quarters of the B component of A1 is dispersed along its length in such a way that it will be more or less directly contiguous to the three quarters of B component of A2. Thus the net attraction between A1 and A2 was proportional to $V^2/2$, whilst the net repulsion was proportional to $9V^2/16$, which, evidently, is greater than $V^2/2$. Of course, we have no empirical justification for asserting this particular level of dispersion of the V component, although it should be possible to derive the actual level of dispersion by empirical means. We only wish to illustrate that it is relatively easy to develop a model of the genesis-unit that can explain why two units can repel each other when they are sufficiently close to each other.

The reader might justifiably ask why our model is anxious to account for *repulsion* between the component parts of complex atoms when we have already rejected the view that atoms contain a nucleus of similarly-charged particles? Because, firstly, it is clear that *pairs* of genesis-units do not normally fuse together to form a complex atom (as we shall see, a third or fourth genesis-unit is usually required). A natural repulsion at close proximity would help to explain that fact. Secondly, such a repulsive tendency aids us in explaining the phenomenon known as "nuclear fusion".

Nuclear fusion requires a significant quantity of energy to happen, but (in the case of elements lighter than iron-56) it leads to the release of an even greater quantity of energy. If atoms were composed of genesis-units with balanced *non-dispersed* polarities then they should fuse together (with the B component of one aligned to the V component of the other) at relatively *low* energies, but this, manifestly, does not happen. The alternative model we are presenting (as in Figure 1.4) allows that atoms will experience a state of mutual repulsion at close range. To overcome this repulsion, the atoms will need to be hurtling towards each other with the kinds of high kinetic energies that are typical of particles in the interior of stars. Once the atoms have collided and are positioned even closer together than the mutually repulsive range depicted in Figure 1.4, then their electrostatic situation might be imagined as illustrated rather imperfectly in Figure 1.6.

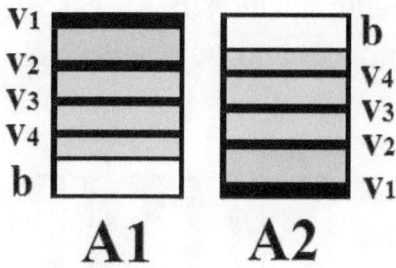

Figure 1.6 The unstable state of two units in close proximity

When two units are aligned in this way, their situation is made unstable by the fact that the v_2, v_3 and v_4 of A1 will repel the v_4, v_3 and v_2 of A2, and this repulsion exceeds the combined attractive influences of the two bv_1 attractive components. The diagram may not demonstrate this point unambiguously, but it is easy to imagine genesis-units of a certain height with a certain distribution of the V component which would entail that, when units are aligned side by side, the repulsive influence of the dispersed V components will exceed the combined attractive influences at either end of the units.

The net repulsive influence entails that the units will not stay aligned like this for very long after their initial collision. Indeed, we find that the isotope of helium with just two "protons" (^{2}He) is extremely unstable and has only been observed on rare occasions for exceedingly fleeting moments of time. On our understanding, ^{2}He is actually composed of two genesis-units forced together. Such a relationship cannot persist for long given their mutual repulsion. How then can multiple genesis-units be fused into an atom of helium, which (in the case of ⁴He) is believed to contain two protons and two neutrons?

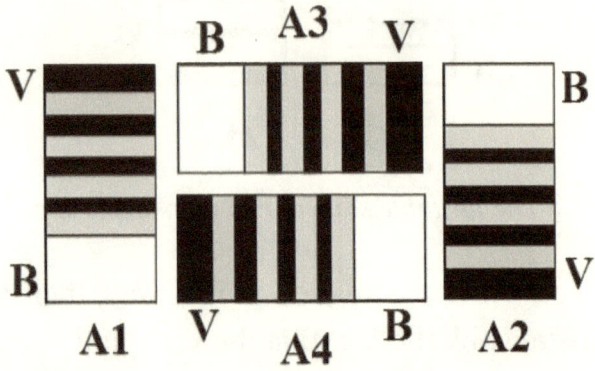

Figure 1.7 The fusion of four genesis-units to form helium

Figure 1.7 shows how such a fusion is possible. We imagine that the high kinetic energies of hydrogen atoms in the interior of stars will create the situation where multiple atoms will be thrown energetically together in every sort of relative alignment possible. Whenever four atoms are thrown together in an alignment similar to the one shown in Figure 1.7, they will immediately bond together in a stable manner. There will be a certain net repulsive impulse between A3 and A4, but this will be more than compensated for by the "capping" influences of A1 and A2. Indeed, the net attractive electrostatic influences of the overall structure entail that, once the initial force of repulsion is overcome, the process of fusion will release electrostatic energy as the poles of the capping genesis-units come into contact with the opposite poles of A3 and A4. This is the fusion energy that is the powerhouse of stars.

It is also clear from this diagram that a complex structure with just three units (i.e., without either A3 or A4) will be possible. And, of course, ^{3}He is a stable isotope of helium that occurs naturally in the universe. But it will be a rare element because, in the high energy situation in the interior of stars, it will have an extremely high probability of picking up a stray hydrogen atom and becoming ^{4}He.

The question that immediately arises is why such a structure that is actually composed of *four* genesis-units should have an atomic number of just 2? During this chapter the claim will be developed that the chemical properties of complex atoms are determined by the *residual* electrostatic properties of the genesis-units composing the atoms. This will be explained in more detail as we go along. Considering Figure 1.7, it can be seen that the electrostatic influences of A3 and A4 are very much directed towards other genesis-units *within* the structure. The B of A3, for example, is balanced on one side by the V of A1 and on another side by the V of A4. The poles of A1 and A2, by contrast, are only counterbalanced on one side only. This means that these units will exert more residual electrostatic influences *outwards* towards other atoms potentially in range. By "residual" we mean the electrostatic influence that is *not* already counterbalanced by an opposing influence within the complex atom.

As we shall see, the very compactness and symmetry of the helium atom means that it exerts relatively low residual impulses in any direction. It shares this characteristic with other members of the group of noble gases. Nevertheless, from outside the helium atom, it is A1 and A2 that "stand out" as far as electrostatic influence is concerned. This means that it behaves in certain experimental situations (such as the famous experiment by Moseley demonstrating a strict relationship between the emitted x-ray spectrum and the atomic number of an element) as if it were an entity that had *two* principal causal players within it. Data of this sort has traditionally prompted the inference to protons. In our model, the two players that dominate the x-ray spectrum of the helium atom are A1 and A2, since the influences of A3 and A4 are mainly directed to counterbalancing opposite influences within the helium atom itself.

On this model, what is normally described as a "proton" is a genesis-unit whose electrostatic influences are not mainly counterbalanced by the opposing electrostatic influences of other genesis-units in the complex structure. "Neutrons", by contrast, are genesis-units whose electrostatic influences are directed to other units in the composite atom. We see then that the electrostatic capacity of the "neutron" is mainly employed with the structural task of holding the atom together, whilst the "proton" has sufficient "residual" electrostatic influence to generate the kind of empirical data recorded in a pioneering way by Moseley. It is the number of "protons" then that will dictate the atomic number of the element.

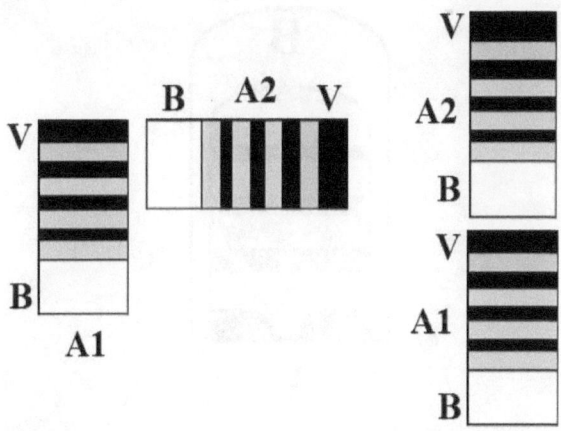

Figure 1.8 *Alternate fusion patterns that are unstable*

Figure 1.8 shows two different ways in which hydrogen atoms could conceivably fuse together. To the left, A2 is shown fused to A1 in a perpendicular alignment, with its B being held to the V of A1 by simple electrostatic attraction. To the right, the same poles fuse together in a vertical alignment. Are such alignments found in nature? If not, this would present a challenge for our model and may indicate that the shape of genesis-units is not accurately represented by simple cylindrical "wells". The non-occurrence of such alignments would suggests two possibilities: firstly, perhaps the real shape of the genesis-unit is such that alignments of this sort would actually involve bringing portions of like poles together (this possibility is explored in Figure 1.9); or, secondly, perhaps the true shape of the unit permits alignments of this sort to happen in a fleeting manner, but there is something about the alignment that makes it very likely (in high energy situations) to fragment into individual units or transmute into an alignment of a different sort.

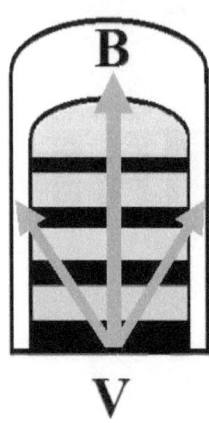

Figure 1.9 *Alternative attempt to depict the genesis-unit.*

Figure 1.9 presents an alternative image of the atom. The B pole is drawn up out of nothingness, creating a cylindrical well, but the B component is not located entirely at the top end of the unit. In this model, the real kernel of the V component is located at the *centre* of the bottom end, and the outermost "skin" of the cylinder which is most distant from this kernel consists in B component. However, if the B component is concentrated *principally* at the top end, then the unit will have a definite, unitary direction of alignment and will give rise to the sort of empirical phenomena that we are anxious to explain, such as "magnetic spin". The fact that most of the B component is concentrated on top, and the fact that the V component is dispersed throughout the rest of the unit, will entail that atoms will exhibit the diverse phenomena of neutral influence, net attraction and extreme repulsion described earlier in this chapter. But, on this model, it would be more difficult for two hydrogen atoms to bond end to end, which seems entirely possible for the original model. Consider the arrangement depicted on the left side of Figure 1.10. The proportion of B component located on the periphery of the V end of A1 counteracts the attractive influence of the B pole of A2 and makes for an unstable bond. However, when four units are thrown into the arrangement depicted to the right, the previous instability prompted by the repulsion of the B pole of A2 and the B component on the periphery of the V pole of A1 is now counterbalanced by the proximity of the V pole of A4. Indeed, the very symmetry of this arrangement ensures that the various electrostatic influences are perfectly counterbalanced, giving rise to a distinctly stable bond.

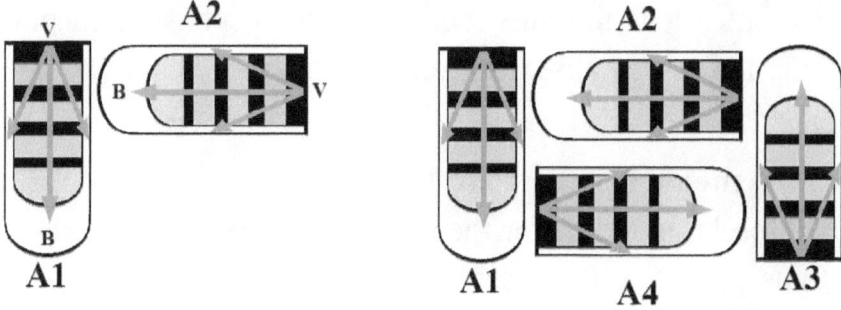

Figure 1.10 *To the left is depicted an unstable arrangement of two genesis-units. The B pole of A2 is repelled to some degree by the portion of B component on the periphery of the V pole of A1. To the right we see a stable configuration of four genesis-units to form helium.*

As this chapter progresses, it will be argued that symmetry of atomic structure - and the consequent electrostatic equilibrium that it guarantees – is the key to understanding chemical bonding and the mysterious orderliness of the periodic table of elements. But we must leave it to those with greater mathematical and geometrical expertise to work out the shape and distribution of the B and V components within the genesis-unit. The criteria for working out that shape and distribution are purely empirical. Among these are the observation that two hydrogen atoms will not fuse together in an atom without the presence of other "capping" units. The initial energies necessary for atomic fusion to occur, and the final energies released when the fusion of helium is achieved, are the key data for understanding the initial electrostatic repulsion between genesis-units and the final state of electrostatic equilibrium, resulting in the emission of energy. This data should yield the distribution of the B and V components around the genesis-unit, since it is this distribution that is responsible for the pattern of energy absorption and emission.

1.3 Elements heavier than helium

Moseley's findings on the x-ray spectra emitted by various atoms constitute the essential starting point for developing a model of the atom. Moseley placed samples of elements in an evacuated glass tube. The samples were subject to causal excitation, resulting in the emission of x-ray radiation. The radiation was diffracted through a crystal, allowing its spectrum to be analysed. It was found that there was a strict mathematical relationship between aspects of the emitted spectrum and the atomic number of the element used in the sample. This establishes beyond doubt that the atomic number of an element is no mere convention. It is not simply a way of categorizing elements in order of their relative masses. The number itself corresponds to a definite physical feature of the atoms of that element, and the consensus in the scientific community has been that the feature in question is the number of protons in the nucleus of the atom. As already stated, our approach interprets the atomic number to correspond to the *number of genesis-units* whose electrostatic influences are not counterbalanced by the electrostatic influences of other units in the complex structure. Using this approach, we intend to account for the structure of the periodic table and explain the bonding characteristics of the various groups of elements.

The *fusion* that takes place when helium is formed is something radical and bears little relation to the kind of attachment that takes place when atoms settle into the states of electrostatic equilibrium that occur during *chemical bonding*. Indeed, as the famous experiment carried out by Geiger and Marsden under the direction of Rutherford showed, matter that appears solid is actually made up mostly of empty space. This is usually interpreted in terms of atoms with a central nucleus and electrons spinning in a void, but it can just as well be understood in terms of a model of matter in which atoms are *smaller and more compact*, but held at a distance from each other by the electrostatic tension that keeps them in equilibrium. Our model of the atom helps to explain why these vast empty spaces exist within portions of matter. Atoms are held at a certain distance from each other because at that point mutual net attraction is the case; closer proximity would result in mutual repulsion. At this point of relative electrostatic equilibrium, the influences of the B and V components of the genesis-unit are reciprocated, even though the atoms that are "attached" to each other will not even be in physical contact.

The *fusion* of genesis-units into heavier atoms differs radically from this sort of attachment and results in a structure that is much more compact.

The standard model asserts that the heavier atoms arise from the process of nucleosynthesis in which elementary particles are fused together to form nuclei. The nuclei later capture electrons in order to create atoms that are electrostatically neutral. The fusion of nuclei and the capture of electrons proceed in stages, with the electrons taking up positions in various shells around the nucleus, beginning with the innermost shell and only proceeding outwards when the inner shells are full. The shells each have a maximum capacity as far as electrons are concerned, and this capacity fulfils certain mathematical regularities depending on the number of the shell. Whenever the mathematical regularities fail to hold, reasons for the anomalies are given, some of which are plausible, some less so.

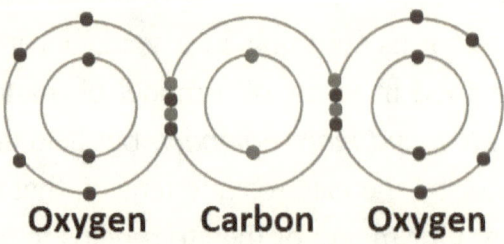

Oxygen   Carbon   Oxygen

Figure 1.11 Carbon dioxide – a classic example of a so-called **"covalent" bond.** *Oxygen, we are told, normally has six electrons in its outer shell and requires two more for "completion". An atom of carbon, instead, is short four electrons in its outer shell. If an atom of carbon donates two of its electrons to two different oxygen atoms, receiving two electrons in return from each atom, then all three atoms will end up with full outer shells.*

Along with this picture of the formation of the elements, there is an intuitively-attractive account of how chemical bonds are forged between atoms of different elements. Elements tend to bond with other atoms in such a way, we are told, that they end up with outer shells that are completely full of electrons. Thus an element that happens to have only one electron in its outer shell (such as sodium) might "donate" this electron to an atom whose outer shell lacks a single electron in order to be full (for example, chlorine). As a result of the donation, both atoms become electrostatically charged in opposite ways and are attracted to each other to complete the bond and form the chemical substance known as sodium chloride, or common salt. Other elements bond by *sharing* the electrons in their outer shells. Figure 1.11 shows a simple depiction of how two oxygen atoms share electrons with a carbon atom so that all three end up with full outer shells.

Using this model of atomic bonding, the structure of the periodic table began to make sense. The periodic table itself had been drawn up in the previous century when chemists like Dmitri Mendeleev and others laid out rows of elements in order of ascending weight. A new row was begun when the chemical characteristics of the previous elements in the table began to repeat. As a result, the columns of the tables ended up containing elements with similar chemical properties. One of the most significant columns was that of the noble gases, a group of colourless, odourless elements that are extremely stable. A simple structuring of the table would place these elements in the right-most column, whereas the highly reactive alkali metals would be placed in the left-most column (containing elements that each had an atomic number one greater than the closest noble gas). In the early decades of the twentieth century, physicists such as G.N. Lewis began to realise that sense could be made of some of the regular features of the periodic table if they were understood in terms of elements with similar numbers of electrons in their outer shells. The noble gases were assumed to have full outer shells, meaning that they had no inherent tendency to seek chemical bonds that would lead to outer shell completion. The alkali metals, by contrast, all had only one electron in their outer shells and were highly reactive. By "losing" that electron, an element such as sodium (with eleven electrons originally and only one in its outer shell) assumed an electron configuration similar to that of the noble gas argon (which has ten electrons, eight of which are in the outer shell).

This impressively simple account of atomic bonding and the structure of the periodic table leaves many questions unanswered. Presentations of the subject make assertions like: "Because the second shell can only hold a total of eight electrons . . ."; but it is far from clear why the shells have certain capacities and why these capacities are determined by the particular mathematical formulae that govern them. There are also grave problems with the postulation of the electron itself, as we will argue later. One thing that *is* clear, however, is that the mathematical formulae and the numerical relationships that are captured by the structure of the periodic table *must* have a concrete grounding in the nature of the various elements and in their bonding characteristics.

Can we develop a more coherent and ordered theoretical grounding for the periodic table than the standard picture based on electron shells? The model presented below is rudimentary and is the fruit of a modest period of reflection on the structure of the table, but it nevertheless demonstrates that reasonable alternatives to the electron picture are possible. The synthesis of hydrogen atoms into more composite structures is described in terms of a few simple patterns of fusion.

Figure 1.12 illustrates two ways that genesis-units will be depicted in this chapter, depending on their overall role within the complex atom. To the left we have an illustration which shows the different polarities of the various genesis-units in the composite structure. On the right is a schematic diagram that will be useful when we come to consider the chemical bonds formed between elements. The grey horizontal lines represent the *binding* units, i.e., those genesis-units whose role it is to hold the structure together and whose electrostatic influences are largely counterbalanced by other influences within the structure. The ovals represent the genesis-units whose electrostatic influences are *not* largely counterbalanced by other forces within the atom. Thus they will have a primary role in determining the chemical properties and emission spectra of the element. It is the number of these kind of units that will correspond to the atomic number of the element. The fact that the oval units are depicted in this diagram larger than the binding units has no correspondence in reality. The two sets of units are identical in nature but happen to play a different role in the cohesion of the atom. But the practice of depicting the outer units as larger will help us later to highlight the structural role they play in chemical reactions with other elements.

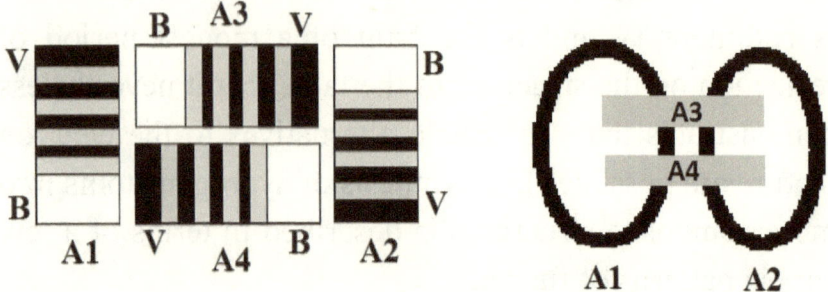

Figure 1.12 The helium atom. *According to our model of atomic fusion, helium is composed of two genesis-units held together by two "binding units" – i.e., genesis-units whose role it is to hold the structure together and whose electrostatic influences are largely counterbalanced by other influences within the structure. On the right is the simpler schematic diagram that we will find useful later when we consider chemical bonding. The grey horizontal lines represent the binding units.*

The helium atom fused from four hydrogen atoms becomes the fundamental building block or corner-stone upon which the stable heavier atoms are generally formed (although the isotopes of some elements may not necessarily be formed around a ^{4}He atom). ^{4}He is an extremely stable entity, a prime case of the kind of equilibrium enjoyed by atoms with an even number of genesis-units. An even number indicates greater reciprocity of electrostatic influences between the genesis-units. According to our model of causal interaction, causal tendencies such as attraction and repulsion always proceed in the direction of restoring equilibrium in the system. A balanced and compact unit such as the helium atom lacks the internal disequilibrium required to attract an entity with the opposite sort of disequilibrium. In the extreme circumstances that prevail in the interior of stars, this atom now interacts with other hydrogen atoms. The fusion of a new hydrogen atom onto the pre-existing structure will be more enduring if it is accomplished using two binding genesis-units to hold it to the stable helium atom (see Figure 1.13).

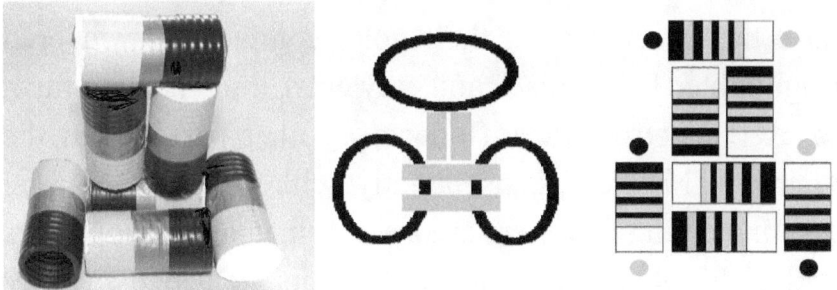

Figure 1.13 Lithium. *On the left is a three dimensional model made from garden hose. The simpler diagram in the centre highlights the genesis-units whose electrostatic influences are not covered (henceforth referred to as "proactive" genesis-units"). We see that a genesis-unit has been fused to the helium atom by means of two binding units. If only one binding unit is used, then the lithium atom will be prone to bind to other atoms, producing heavier elements. But a lithium isotope without this extra binding unit is possible. ^6Li exists naturally and is stable. The image on the right is a more elaborate illustration of lithium showing the proactive units whose poles are "uncovered" and thus have the observable effects that lead us to assign the atomic number "3" to this element.*

There are two stable isotopes of lithium. The most common has two binding units holding the final hydrogen atom onto the atom of helium (^7Li), whilst the other isotope uses only one binding unit in this role (^{6}Li). Presumably the reason why ^{7}Li is ten times more abundant in nature is the fact that ^{6}Li does not have the electrostatic equilibrium that this extra binding unit brings. Thus it has a higher probability of being transformed into one of the heavier elements or fragmenting back into a helium atom and two stray hydrogen atoms. The image to the right of Figure 1.13 shows the lithium atom in more detail. We see how the three genesis-units at the corners of the triangle have the extremities of their poles completely uncovered (highlighted by the black and grey dots), thus enabling them to exert electrostatic influence on other particles in the vicinity. This atom, consequently, will generate empirical data indicating that there are *three* principal causal players within this element, at least as far as its influence on *other* atoms is concerned, prompting scientists to assign it atomic number 3. Henceforth we will refer to the genesis-units that have this role within the atom as being "proactive". The four *binding* units at the centre, by contrast, have all the extremities of their poles covered: their electrostatic influences are counterbalanced by the other units in the structure and they will not have a noticeable influence on the observable behaviour of the atom.

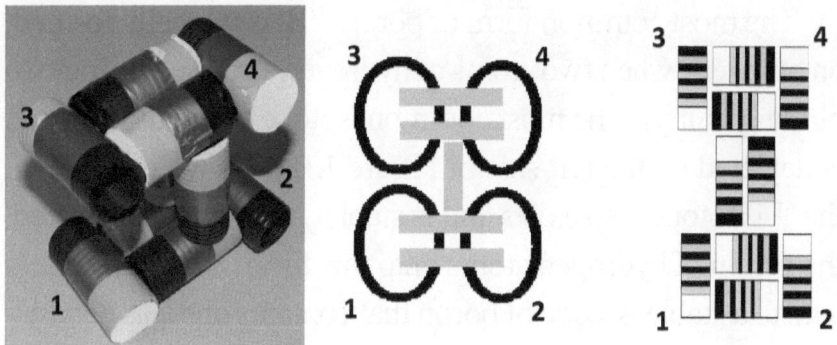

Figure 1.14 Three views of beryllium. *On the left is the 3-D model of ^{10}Be, the most stable isotope of beryllium. Notice the way the two "binding" units are positioned in the interior of the atom. In the centre is a schematic flat diagram depicting the most common form of the element, ^{9}Be, consisting of two helium atoms fused together with a single binding unit. To the right we have the more elaborate flattened diagram of ^{10}Be which shows the polarities of the genesis-units. But the shortcomings of this flat diagram are evident if we compare the positioning of the individual units with the correct positioning on the 3-D model.*

The next element on the atomic table is beryllium. The most common form of this element results when two helium atoms are fused together with a *single* binding unit, as depicted on the left of Figure 1.14. The most stable isotope of beryllium is ^{10}Be. This would consist of two helium atoms fused together with *two* binding units, as depicted on the right of the diagram. It is not surprising to learn that ^{8}Be is an extremely unstable isotope of beryllium with a half-life of only 6.7×10^{-17} seconds. The tendency of the relatively imbalanced ^{8}Be to degenerate into two much more electrostatically balanced ^{4}He atoms is just too powerful to allow the isotope to exist for long.

The most common form of boron, ^{11}B, is typically formed, on our view, when two atoms of hydrogen are fused (using two binding units) to the most common isotope of lithium, ^{7}Li (^{11}B is depicted on the left side of Figure 1.15). As we saw earlier, the ^{6}Li isotope also exists and is stable. As might be expected, the fusion of hydrogen atoms onto this form of lithium should lead to a stable isotope of boron that contains one less genesis-unit than the more common form. And we find that ^{10}B does indeed exist in nature as a stable isotope of boron, but ^{11}B is about four times more abundant in the universe.

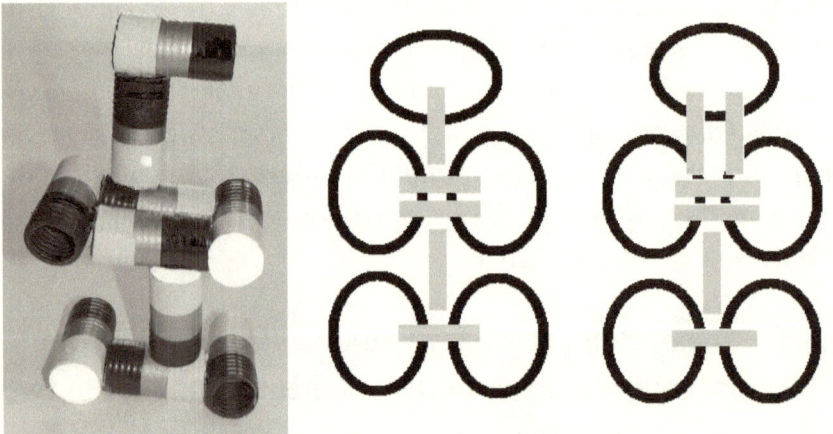

Figure 1.15 Isotopes of boron. *The most common form of boron, ^{11}B, is depicted on the right. The original ^{7}Li atom can be seen at the top of the structure. To the left and centre we have the stable isotope, ^{10}B, formed by fusing genesis-units to the ^{6}Li form of lithium.*

The most common form of boron (^{11}B) has an uneven number of genesis-units. As we shall see as we go along, genesis-units tend to be arranged in pairs in composite atoms because that is how their electrostatic influences prompt them to align themselves. The general tendency is for the units to position themselves so that they mutually balance their reciprocal influences to the greatest possible extent. In the high-energy inferno in which new atoms are formed, elements that have an *uneven* number of genesis-units often show a disposition to "pick up" a *single* genesis-unit and transmute into new elements. The unpaired genesis-unit in the original atom is like a hook for any stray unit that comes available for fusion. This propensity is not present in all atoms that have uneven numbers of genesis-units, because it depends on how those units are arranged and how well the electrostatic influences are already reciprocated. In the case of ^{11}B, there is indeed this tendency to pick up an extra unit, giving rise to the formation of ^{12}C, by far the most common form of carbon (Figure 1.16). Of course, different stable arrangements of units to form carbon are also possible. ^{13}C is a stable isotope, and like ^{9}Be would involve another of the pairs of genesis-units in the structure fusing together with the aid of two binding units, as was the case with ^{4}He.

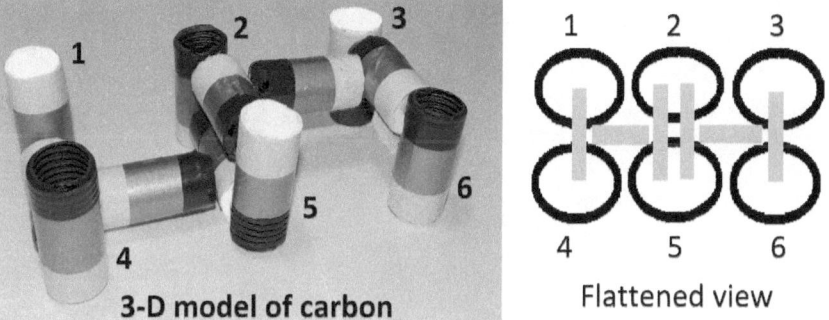

3-D model of carbon Flattened view

Figure 1.16 Carbon. The six "uncovered" genesis-units at the periphery of the atom give the element its atomic number. The most common form of the atom (^{12}C) uses six binding units as well.

Oxygen is an interesting case. It has an atomic number two higher than carbon and an atomic weight that is four atomic mass units greater. Evidently, two proactive units have been fused to carbon using two binding units. But where exactly on carbon have these units been fused? The standard approach to the atom works out where each new electron is placed as the process of nucleosynthesis progresses and the latest electron is captured. We shall compare our view to this standard approach in more detail later when we turn to the real meaning of the Pauli exclusion principle. According to Hund's Rule, orbitals (which can hold a maximum of two electrons with opposite magnetic spin) are each occupied *singly* with electrons of parallel spin before double occupation occurs. If we were to adapt this rule to our model of atom formation, then the genesis-units fused to carbon to form oxygen should be added on in the manner shown in the atom to the right of Figure 1.17.

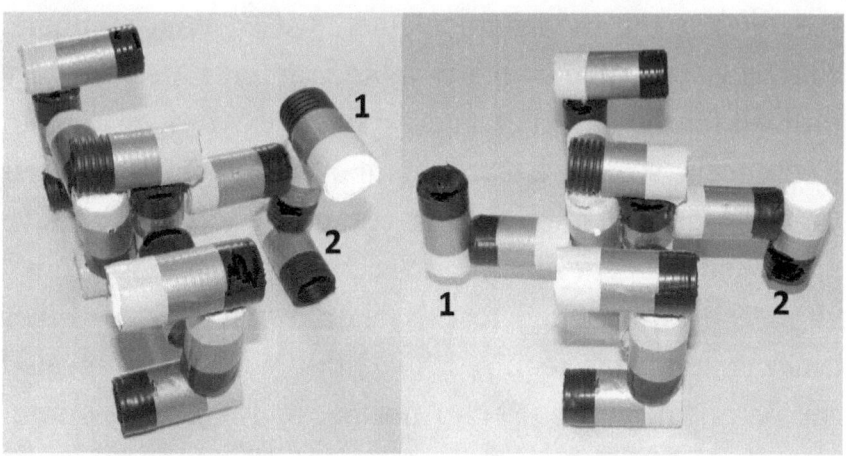

Figure 1.17 Oxygen. *The version of the oxygen atom to the right shows a single proactive unit fused with a single binding unit onto each side of carbon. The chemical bonding behaviour of oxygen, however, leads to a rejection of this model of the element. The correct version to the left shows two proactive units fused using two binding units onto the same side of carbon.*

Here, a binding unit and a proactive unit are fused onto each side of a carbon atom, in a bow to an adapted form of Hund's Rule. We will see later, however, how a structure of oxygen of this sort does not help us to explain the behaviour of oxygen when it comes to chemical bonding. If the structure of oxygen is as shown to the left of Figure 1.17, then we can do a better job in accounting for this behaviour. It may seem unwise at this point to disobey a respected rule of thumb like Hund's Rule solely on the basis of the structural features of a primitive model. Hopefully some of the strengths of our approach will become more evident by the time we come to discuss the exclusion principle.

The structure of lithium is our model for the formation of elements that have uneven atomic numbers, whilst carbon (i.e., ^{12}C) is the model for elements with even numbers. The basic principle is that two binding units are generally utilized for the fusion of a new proactive hydrogen atom to a composite structure that is composed of *pairs* of genesis-units in relative stability. For composite atoms that have a non-paired genesis-unit, by contrast, a new hydrogen atom can be attached to this "hook" without any other binding unit being required. The lithium model is repeated for boron, nitrogen (^{15}N), fluorine, sodium, aluminium, phosphorous, chlorine and potassium. The carbon model is repeated for oxygen, neon, magnesium, silicon, sulphur, argon (^{36}Ar) and calcium.

This approach to fusion helps to explain a pattern that is repeated regularly in the periodic table. As we move progressively through the table from the lighter elements upwards, the mass of the next element in succession rises by one atomic unit, then by three atomic units, then by one atomic unit, then three, etc... Our model of atomic fusion readily suggests various explanations for this 1-3-1-3 pattern. The explanation given above perhaps comes to mind most readily: namely, elements that have an even number of proactive units are relatively stable and will have little residual electrostatic influence that would make them prone to picking up another genesis-unit; thus, two binding units are required for the fusion of a new proactive unit; elements with uneven proactive genesis-units, by contrast, will have a relatively high magnitude of residual electrostatic influence, making it very easy for them to fuse another proactive genesis-unit without the aid of extra binding units. Another explanation also comes to mind: perhaps each element is not typically formed by fusing genesis-units onto the element directly below it in the table; perhaps it is formed by the fusion of *four* units onto the element that came *two places* earlier in the table; thus aluminium (atomic number 13 with atomic weight 27) is formed by the fusion of four units onto sodium (atomic number 11 with atomic weight 23); silicon (atomic number 14 with atomic weight 28) is formed by adding four units onto magnesium (atomic number 12 and weight 24). This latter explanation has a certain consistency in that it accounts for significant stretches of the periodic table in terms of a pattern of fusion in which *pairs* of proactive units are appended to pre-existing elements using *pairs* of binding units. Of course, there are exceptions to this rule, but these can be accounted for without having to stretch credulity too much.

Two elements are worthy of comment. The most common isotope of nitrogen is ^{14}N, whereas our first explanation would predict that the formation of nitrogen would proceed by using two binding units to fuse a proactive unit to carbon (^{12}C), thus leading to the formation of ^{15}N (and the second explanation would predict that ^{15}N would have been formed by fusing four units onto ^{11}B). It is interesting that ^{15}N *is* indeed a stable isotope of nitrogen, a fact that is consistent with our approach to atomic fusion. The abundance of ^{14}N in the universe, however, confirms that a single binding unit is sometimes sufficient to fuse a proactive genesis-unit to a composite atom consisting of stable pairs of genesis-units. Perhaps nitrogen differs simply because of its greater size. Empirical considerations such as size can be expected to exert an influence on the process of atomic fusion. General rules or principles may well hold in limited situations but will be modified or abrogated by changing empirical conditions. ^{14}N may have superior stability to ^{6}Li and ^{10}B because it is larger. It may be less likely to have a new genesis-unit fused to it immediately because the complex internal relationship of the seven genesis-units in the structure may confer on it a greater degree of internal cohesion than is the case with ^{6}Li and ^{10}B.

The other element that requires a comment is argon. According to our model, it ought to be possible to form argon from chlorine without the aid of an extra binding unit (or, alternatively, form argon from sulphur using four units in total). Chlorine has an uneven number of genesis-units and therefore has the "hook" that dispenses with the need for additional binding units. The standard periodic table tells us, however, that argon has an atomic weight of forty atomic mass units, indicating that four binding units were utilised to fuse the hydrogen atom to chlorine, an atom with an *uneven* atomic number! This seems a strong counterexample to our model. But when we examine the matter more closely, we discover that the abundance of ^{40}Ar in the earth's atmosphere is an accident of the peculiar history of development of our planet. It was not formed by atomic fusion at all but has its origin in radioactive potassium - ^{40}K. ^{36}Ar, on the other hand, is by far the most abundant isotope of argon in the sun and the stars, just as our model would predict. According to the simple model of atomic fusion that we are following, argon ought to consist of eighteen proactive units and eighteen binding units, conferring on it an atomic weight of thirty six atomic units. And this is precisely the weight of the most common isotope of argon.

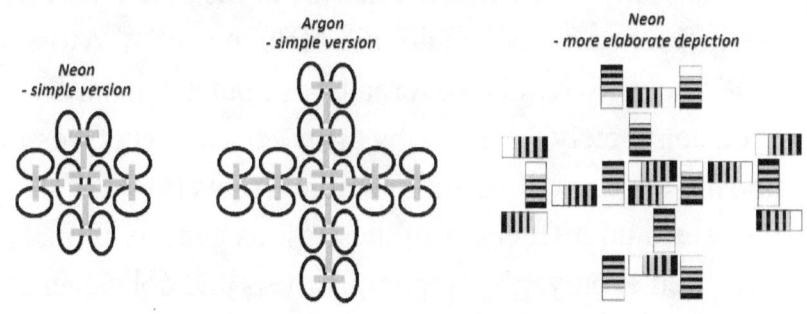

Figure 1.18 The structures of neon and argon. *Both have exceptional stability because of their regularity of structure. The original helium atom is positioned at the centre of the arrangement. Each proactive genesis-unit is one of a pair, guaranteeing that its B-V polarity is well reciprocated. The overall symmetry of the structures ensures that residual attractions between the polarities of different sets of pairs are also reciprocated. Overall, there is relatively little disequilibrium in the composite atoms that would make them attractive to other atoms that have the opposite kind of disequilibria, thus they are almost completely non-reactive.*

Figure 1.18 presents hypothetical structures for neon and argon. Every proactive genesis-unit is paired off, which means that their B-V polarities are reciprocated by the other member of the pair. However, we must assume that the influences of the B and V poles of a particular genesis-unit cannot be completely covered by the opposite poles of its partner, no matter how closely fused that partner might be. Take a pair composed of genesis-units A1 and A2, where A2 is situated to the right of A1. If a further genesis-unit is positioned to the left of A1, then, presumably, it will also be influenced by the polarity of A1, since A1's sphere of influence is precisely that – spherical in nature – and a partner can only be located to one side or the other. We refer to this influence as the "residual" influence of A. Most of A1's B-V polarity will be covered by A2 but it can hardly be covered completely. That is why *symmetrical* structures for composite atoms will be more stable. As well as ensuring that the residual influences of the various genesis-units are reciprocated evenly, giving compactness and cohesiveness to the structure itself, symmetry also makes the composite atom less likely to bond with other atoms. A symmetrical structure will be more balanced as far as the B-V polarities of its constituents are concerned. According to our model of causal interaction, the causal dynamics of any system is fundamentally directed towards achieving equilibrium. The most reactive substances are those in disequilibrium since they naturally give rise to processes that evolve towards equilibrium. Neon and argon are already stable and have little tendency to interact with other objects because they have little need to evolve towards equilibrium.

The structures of potassium and calcium are compatible with our model of atomic fusion. Potassium utilizes two binding units to fuse one proactive unit to ^{36}Ar, the most common isotope of argon. Calcium is formed when a single genesis-unit fuses to one of the "hook" units in potassium without the need for an additional binding unit. Alternatively, calcium may be formed by fusing two proactive genesis-units to ^{36}Ar using two binding units. This would give rise to ^{40}Ca, the most common isotope of calcium.

As the elements get heavier, more binding units are sometimes used for the fusion of new proactive units. Presumably this is a function of the changing empirical conditions as atoms become more massive. But the pattern of fusion that we have outlined is still followed in various stretches of the table, even for heavier elements. Titanium (^{48}Ti) is composed of twenty-two proactive genesis-units fused together with twenty-six binding units. Each proactive genesis-unit is paired off, so two binding units are utilized to fuse a new proactive unit and form the next element on the table, vanadium, which has twenty-three proactive units and twenty-eight binding units. Chromium is then formed by simply adding a proactive unit to the hook unit of vanadium without the need for an extra binding unit. To create manganese from chromium, two binding units are again used to fuse a proactive unit, and then iron is formed simply by adding a genesis-unit to the hook unit of manganese. In turn, cobalt is formed from iron with the aid of two binding units.

It is not too surprising that the pattern of fusion that we have proposed should be generally applicable only to the lighter elements. Heavier elements are much more likely to be formed by the fusion of atoms that are *already composite*, instead of involving simple fusions of individual genesis-units to composite atoms. But even if our model of fusion cannot be extended to all of the heavier elements, our model of atomic stability and cohesiveness should still hold good. Atoms that have even numbers of proactive genesis-units arranged symmetrically ought to be less reactive. Let us see how this can be applied to the next noble gases - krypton, xenon and radon.

Krypton has an atomic number of thirty-six, whereas argon had exactly half that number. Our hypothetical structure for argon (see Figure 1.18) had the pairs of atoms arranged in a more or less two-dimensional cross, with a pair of proactive genesis-units at the centre and two pairs arranged along each of the four arms. Krypton is formed from *two entire crosses* like these fused on top of each other. In that case there would be ample scope for the composite atom to absorb variable numbers of binding units at the various points of contact between the crosses. And indeed there are multiple stable isotopes of krypton, one of which is ^{80}Kr. This, we imagine, could be composed of the two crosses of argon (two atoms of argon composed of 36 genesis-units each) plus a binding unit for each pair of genesis-units in the arms of the cross to fuse the two argon crosses together (eight binding units) to give a total of 80 genesis-units.

The next noble gas is xenon and it has an atomic number of 54, exactly triple that of argon. We can imagine that it might be formed from three of the argon crosses fused together, thus giving it the same symmetry enjoyed by the lighter noble gases. All of this is perfectly compatible with our model of atomic equilibrium. When we come to radon, however, composed of eighty six genesis-units, there is no obvious way of arranging this number into a symmetrical structure. And perhaps there is no need to do so, for when we look at the case of radon we discover that it is not generated by atomic fusion at all, and neither does it enjoy the stability and cohesiveness of the other noble gases. Radon is characterised as a noble gas because it is inert and does not combine readily with other elements. Like the noble gases, it is colourless and odourless. But it has its origin in the radioactive decay of other elements, and it does not have a single stable isotope. Its longest-lived isotope, ^{222}Rn, has a half-life of less than four days. Thus, the lack of a symmetrical structure for radon cannot be taken as a counter-example to our model of the stability of the noble gases, for radon is far from stable. This is undoubtedly due to many factors apart from the asymmetry, not least among them the process by which radon is generated.

1.4 Chemical bonds between atoms of different elements

According to our model, the formation of the elements at the beginning of the periodic table involves the progressive fusion of new genesis-units to composite atoms that were already formed in the same fashion. We cannot rule out that some atoms (such as beryllium, perhaps) may derive from the fusion of two or more ready-made helium atoms, or indeed heavier elements. Once the genesis-units have been fused together, they are more or less in equilibrium, depending in part on whether or not they form a pair within the composite structure. But the equilibrium of a genesis-unit within the structure will also depend on the symmetry of the structure as a whole, as we have discussed earlier. Even if a genesis-unit is part of a pair, all of its B-V polarity will not be reciprocated perfectly by the polarity of its partner. Any causal agent in disequilibrium will exert its influence in all directions. The partner of a given genesis-unit, A1, may well reciprocate most of A1's influence, greatly diminishing it, but the fact that the partner can only take up a position on one side of A1 means that A1 will still exert an effect on the other side if a third genesis-unit comes within range.

The upshot of all of this is that the bonding behaviour of an element on the macro-level will mirror the structure of the atoms of that element at the micro-level. A composite atom with a non-symmetrical structure will be more inclined to bond with other elements in such a way as to produce a compound structure that has symmetry. If the combination of two elements does not produce a certain level of symmetry, then they will not mix at all. Elements that already have symmetrical structures will be less inclined to bond with any other substance, although particular empirical conditions could coerce a symmetrical element into a bond of some sort.

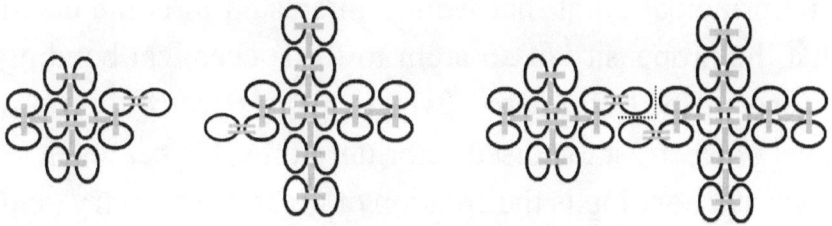

Figure 1.19 Atoms of sodium (left) and chlorine (centre). *Sodium is formed from an extra proactive genesis-unit fused to a neon atom, whilst chlorine resembles an atom of argon with one proactive unit missing. These incomplete symmetries make both elements reactive because of the poorly reciprocated electrostatic influences of the unpaired genesis-unit in each atom. Thus they are prone to bond with other elements. To the right, we have a depiction of the sodium chloride molecule. The bond is held together by the residual electrostatic disequilibria of the "hook" genesis-units of sodium and chlorine.*

The classic case of a chemical bond is that of sodium chloride, or common salt. According to the standard approach, sodium has a single electron in its outer shell, whilst chlorine lacks a single electron for the completion of its outer shell. Our model agrees that sodium has something "extra" whilst chlorine has something "missing", and in both cases that something is the genesis-unit whose absence or presence would confer on the respective elements a much greater symmetry. Figure 1.19 attempts to represent what is involved in that evolution towards symmetry. It is important to emphasize that the depicted symmetrical structures are hypothetical and schematic in nature. The business of discerning the real symmetries that exist in composite atoms like those of sodium or chlorine cannot be accomplished without detailed examination of the empirical evidence.

The principal point that we are interested in is the claim that the propensity of an atom towards chemical bonding is driven by asymmetry (which gives rise to a certain level of electrostatic disequilibrium); thus, the net result of chemical bonding is the creation of greater symmetry (and the better reciprocation of electrostatic influences). In this sense, our model has superior explanatory power to the standard approach. The standard electron shell model fails to give a coherent account of why the shells have various capacities and why atoms tend to bond in such a way that the maximum capacity of the outer shell is achieved.

Simple examples of the covalent bond are water (H_2O) and carbon dioxide. Oxygen is composed of four pairs of proactive genesis-units arranged in a cross-like formation, but with one arm missing. If two hydrogen atoms bond to oxygen then it achieves the kind of symmetry enjoyed by neon.

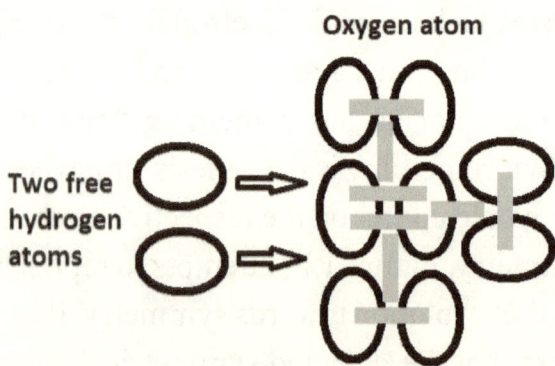

Figure 1.20 H_2O. *Two hydrogen atoms bond to oxygen to form a symmetrical structure similar to neon*

Earlier we had a brief look at the standard picture of the carbon dioxide molecule. The bond was considered to involve the sharing of pairs of electrons by each of two oxygen atoms and a single carbon atom. This filled the outer shell of all three atoms to eight electrons each. Our symmetry model of bonding sees the carbon dioxide molecule in very different terms, but agrees that oxygen is lacking two "somethings" and that carbon can supply that lack. Figure 1.21 shows two possible arrangements of oxygen and carbon atoms in such a way as to achieve greater symmetry.

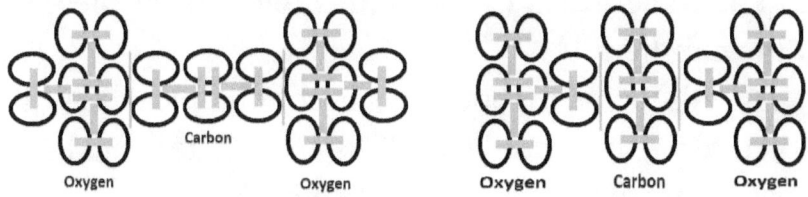

Figure 1.21 Schematic representation of the formation of the carbon dioxide molecule. Oxygen lacks a pair of proactive genesis-units for symmetry. The arrangement to the left shows how a single carbon atom could bond with two oxygen atoms to form a symmetrical arrangement. The arrangement on the right may also be possible.

Another straightforward example of this tendency towards symmetry is the molecule of methane (CH_4). Four hydrogen atoms will confer on carbon the perfect symmetry of neon (see Figure 1.22).

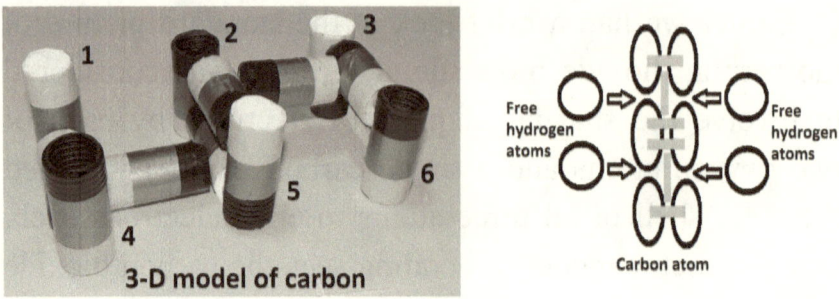

Figure 1.22 Carbon and methane (CH$_4$). *On the left is a 3-D model of carbon, whilst, on the right, four hydrogen atoms bond to carbon to form a symmetrical structure like neon.*

1.5 The structure of the periodic table and the meaning of the numbers

By this point it should be becoming clear how the structure of the periodic table is to be interpreted. The causal dynamics that underlies the process of atomic fusion is driven by the question of equilibrium. The progressive fusion of genesis-units onto composite atoms will not result in a stable and permanent element unless the new atom possesses a degree of internal equilibrium, and this will be largely dependent on the mutual reciprocation of causal influences by the various genesis-units making up the composite atom. A symmetrical arrangement of pairs of genesis-units is the key to the stability of an atom, but elements do not require a perfectly symmetrical structure in order to enjoy a relatively permanent existence. A composite atom has a greater or lesser symmetrical structure, and its relative symmetry at the micro-level will dictate its degree of reactivity with other elements. All stable chemical bonds lead to a composite structure that is more symmetrical than their constituent atoms.

Different elements with similar features (as far as their symmetry or lack of symmetry is concerned) will have similar chemical properties. Each of the alkali metals from lithium onwards has the same structure as one of the noble gases with an extra proactive genesis-unit appended to it. Thus they are highly prone to bond with the halogens, which lack a genesis-unit in order to achieve symmetry.

Given this pattern of atomic bonding, it comes as no surprise to discover that the noble gases take on a paradigmatic quality as far as the formation of molecules is concerned. The noble gases have symmetry at the micro-level. Elements that have one more proactive genesis-unit than one of the noble gases will tend to bond to elements that have one less proactive genesis-unit in order to produce a symmetrical structure. Atoms that have two genesis-units more than a noble gas will bond to elements that have two units less (as is the case with the formation of beryllium oxide, for example).

Many bonding patterns can take place that increase symmetry but do not result in a structure that resembles a noble gas. Dilithium is composed of two atoms of lithium bonded together. Before the bond, the lithium atoms would each have had a genesis-unit that was not part of a pair. Once the bond is formed, these odd genesis-units will be attached to each other to form a non-fused pair, increasing the overall equilibrium of the structure. Bonds of this sort are easily explained using the symmetry model, but require a more complicated treatment using the valence electron approach.

It is important to keep in mind that the elements have no natural tendency to form bonds that increase symmetry *per se*. What drives the causal dynamics of entire systems is the tendency towards *equilibrium* between the various constituents of the system. Composite atoms like sodium or chlorine have internal disequilibria arising from the fact that their component genesis-units have influences that are not being reciprocated. The disequilibrium underneath is the motor that drives the causal dynamics of the system and causes sodium and chlorine to bond together in such a way that the various influences are reciprocated. Reciprocation is more likely when the arrangement of the structure is symmetrical. Thus symmetry is an indirect consequence of the natural evolution of a system, whereas lack of symmetry is an indication that an atom is likely to evolve in a certain direction.

-3	-2	-1	0	+1	+2	+3	+/- 4
				1 H			
			2 He	3 Li	4 Be	5 B	6 C
7 N	8 O	9 F	10 Ne	11 Na	12 Mg	13 Al	14 Si
15 P	16 S	17 Cl	18 Ar	19 K	20 Ca		

Figure 1.23 Periodic table based on the symmetry model of atomic bonding.

Figure 1.23 presents a periodic table for the first twenty elements. This is arranged in a way that highlights the tendency of the elements to increase their equilibrium through improved symmetry. Group 0 is the noble gases, elements that have their proactive genesis-units arranged in perfect symmetry. Groups +1, +2 and +3 have one, two or three elements fused onto a symmetrical arrangement. Groups -1, -2 and -3, by contrast, are lacking one, two or three proactive genesis-units for a symmetrical structure. Group +/-4 has elements that have four genesis-units fused to a symmetrical composite atom, but can achieve symmetry if their bonding behaviour allows them to "gain" four more.

The number eight has a special significance in our attempt to organise the structure of the lower part of the periodic table. The standard model understands this feature in terms of the claim that the second electron shell has a capacity of eight electrons, whilst the s and p subshells of the third electron shell have a combined capacity of eight. According to that model, the elements are built up by adding protons (and neutrons) to the nucleus and electrons to the various shells. The elements up to argon are formed by progressively filling the first and second shells, and then the s and p subshells of the third electron shell.

Our model accounts for these numbers in a simpler and more concrete manner, citing equilibrium (and hence symmetry) as the explanatory key for understanding the structure of composite atoms. Helium was composed of two proactive genesis-units, and this corresponds to the standard model's claim that the first shell has a capacity of two electrons. Around this helium core, four pairs of proactive genesis-units can be arranged symmetrically, corresponding to the standard claim that the second electron shell has a capacity of eight electrons. The cross-like structure of neon can once again become symmetrical if we add a pair of proactive genesis-units onto each of the four arms of the cross, corresponding to the s and p subshells of the valence model. In fact, extending the arms of the cross in a symmetrical banner *always* involves adding eight proactive genesis-units (one pair for each arm of the cross) on to the structure that was already there. Krypton, on our understanding, involves creating a composite atom from two argon crosses fused together. This is a leap of eighteen genesis-units from argon.

The structure of the standard periodic table perfectly mirrors this progressive development. The first row of the table has just two elements (hydrogen and helium) because the first milestone of symmetry is reached when helium is formed from two proactive genesis-units. The second row of the table has eight elements because the second milestone of symmetry is the cross-like structure of neon, formed by adding eight proactive units to helium. The third row again has eight elements because the next symmetrical structure can be realised by adding a pair of units onto each of the four arms of the cross of neon. The fourth row, by contrast, has eighteen elements because the next symmetrical structure involves fusing onto argon an entire cross-like structure composed of eighteen genesis-units.

The electron model is more laborious in the way it accounts for the structure of the table up to this juncture. The first electron shell has a capacity of two, and that explains

64

why helium ends up in the rightmost column. The second shell has a capacity of eight, explaining why the second row has eight elements. The third shell is supposed to have a capacity of eighteen electrons, but its s and p subshells have a joint capacity of eight. The stability of argon (with eight electrons in the third shell), and the fact that the third row of the table has only eight elements, is thus "accounted for", but in a contrived way. Similarly, Krypton is reckoned to have only eight electrons in its outer (fourth) shell, but the third shell is attributed eighteen electrons, and this is used to explain why the fourth row of the table has eighteen elements. Indeed the usual description of the fourth row of the table involves a progressive addition of electrons to the third and fourth shells in a somewhat haphazard fashion.

Our model has no difficulty in explaining the presence of eighteen elements in the fourth row of the table. Eighteen proactive genesis-units must be added to argon to achieve the next symmetrical milestone, krypton. Therefore there will be eighteen elements between argon and krypton. In the course of this progressive development, there will be other lesser milestones as well. The first two genesis-units added to argon will form the core of the new-cross like structure that will be eventually fused to the first cross to produce krypton. This pair will render a particular (incomplete) symmetry to the whole structure and there will be consequences for the emission spectrum produced by the element. The fusion of four more pairs in a cross-like form around this central pair will represent a new watershed in symmetry. These various watersheds in symmetry will correspond to numbers like 2 (the pair of genesis-units forming the new core of a new cross-like structure), 8 (the four pairs of units fused to the core pair), 10 (a cross-like structure composed of a core pair with four pairs fused on to it), and 18 (a cross composed of eight pairs fused to a core pair). All of these numbers appear naturally in the periodic table. There are *two* elements in the first row, and *eight* in the next two rows. The fourth row has *ten* elements extra because it involves a fusion of a core pair plus four other pairs to argon. Rows four and five have *eighteen* elements each because they represent the progressive fusion of two more cross-like structures (composed of eighteen genesis-units each) onto argon.

We can see how quite a few features of the symmetry model have a corresponding feature in the electron model, but the symmetry model, at first glance, does not appear to have all the details found in the electron's model treatment of subshells. The first energy level or shell is said to have a single subshell (designated the *s* subshell). The second shell is attributed two subshells (*s* and *p*) of capacities two and six. The third shell has three subshells (*s, p* and *d)* of capacities, two, six and ten. In short, each shell has the same number of subshells as the number of the shell, and the capacity of each subshell increases in increments of four. Where does the electron model get its notion that the various energy levels or shells are divided into subshells, and on what basis does it claim that the subshells have these various capacities?

Before the notion of electron shells was developed, spectroscopists had identified characteristic series of lines in the spectra of the alkali metals. These sets of lines were given the letters *s* (sharp), *p* (principal), *d* (diffuse) and *f* (fundamental). As the electron model was developed, it was hypothesized that these series of lines were caused by the distinct energy levels of electrons located in subshells within each shell. In the mid nineteen twenties, Friedrich Hund designated the subshells themselves with the letters *s*, *p*, *d* and *f* (letters in alphabetical order starting from *g* would be assigned for the extra subshells contained within larger shells). This seemed to represent a significant step forward in theory because it allowed the assignation of precise energy levels to the various electrons using hard evidence from spectroscopy.

Coupled with this development, the Pauli exclusion principle armed physicists with a mathematical tool with which to work out the detailed electron configurations of the elements. According to the exclusion principle, each electron within the atom can be uniquely identified using four numbers, n, l, m_l and m_s. The first number is the *principal quantum number,* and this designates the shell or energy level of the electron. The second number, l, is the *subshell,* designated with the numbers 0, 1, 2, 3, 4, etc., (which correspond respectively to the letters *s*, *p*, *d*, *f*, *g*, etc.). Each subshell in turn is divided into orbitals, and the number of the orbital is the magnetic quantum number, m_l. Any given subshell will hold $2l + 1$ orbitals (where l is the number of the subshell), whilst the orbital itself can hold a maximum of two electrons. The fourth number, m_s, is the *spin quantum number.* This can only have two values, up or down.

The exclusion principle states that no two electrons can have the same set of quantum numbers. Two electrons occupying the same orbital will have the same principal quantum number, subshell, and magnetic quantum number (the orbital that they occupy). Hence, in order to have a *unique* combination of quantum numbers, they will have to possess opposite spin. This principle was of assistance to physicists in building up the picture of the electron configurations of the elements in the periodic table. As electrons were assigned to each orbital and subshell, the exclusion principle, used in conjunction with the rules stipulated by Hund, provided a tool for discerning when the various energy levels were filled and when to begin assigning electrons to a new subshell. It also provided some sort of story for why the various shells had the particular capacities that they had. For example, the second shell has a principle quantum number of two. This means that it has two subshells (0 and 1). The 0 subshell has one orbital, whilst the 1 subshell has three orbitals. Each orbital can hold two electrons, each with opposite spin. This means that the second shell has a maximum capacity of eight electrons. Of course, this falls far short of an *explanation*: it does not provide a causal story for why a shell has the particular capacity that it has.

The evidence from spectroscopy did not always fit in with the dictates of the exclusion principle, nor with the other mathematical principles that were being applied to the periodic table, so various explanations had to be developed to account for these anomalies. The shielding influences of electrons in the more complicated shells, or the interfering influences from the protons in heavier nuclei, were thought to be responsible for the fact that the empirical data did not always respect the mathematical model. Such rationalizations of uncooperative empirical data continue to be highly plausible. It seems natural that simple mathematical regularities should tend to break down in the case of larger atoms, extra electrons, heavier nuclei, and, in general, a proliferation of "interfering" factors.

It is not our task to investigate the extent to which the electron model "coerces" the empirical data to fit mathematical regularities. However, our symmetry model does demand a re-evaluation of the mathematical framework of the electron approach. This includes the rejection of the stipulation that an electron shell has a maximum capacity of $2n^2$. Such mathematical regularities *do* indeed approximate to the physical reality of atoms, but they have limited application only. The first, second and fourth stages of symmetry correspond respectively to the fusion of two genesis-units, the addition of eight further genesis-units, and the complete fusion of another eighteen genesis-units (i.e. $2n^2$ for $n = 1$, $n=2$ and $n=3$), but this actually omits the authentic third stage of symmetry (the formation of argon) which involved the fusion of eight genesis-units onto the cross-like structure of neon which comprises ten units. We find a confused correspondence to this fact of nature in various features of the standard model's view of the evolution of the periodic table. For example, it assigns a shell configuration of 2, 8, 18 and 8 electrons respectively to krypton. In reality, the milestones of symmetry are achieved when composite atoms gain 2, 8, 8 and 18 proactive genesis-units.

Let us consider these contrasts in a little more detail. The electron model prescribes that shells have a maximum capacity of $2n^2$ electrons. As heavier atoms are progressively formed, the inner shells are supposed to be completely filled first before a new shell is begun, but this principle breaks down on the fourth row of the table with potassium. As the fourth row progresses, electrons are variously added to both the third and fourth shells until we arrive at krypton. At that point the third shell is eventually filled with eighteen electrons and the outer shell is filled with eight.

Consider how this account contrives the hypothesized structure of particular atoms to suit the mathematical regularity, $2n^2$. The third row of the table finishes with argon, which is supposed to have two electrons in the first shell, eight in the second and eight in the third. When the fourth row of the table is begun with potassium, a new electron must be added to the electron configuration of argon. But to which shell should the electron be allocated? According to the $2n^2$ regularity, it should become the ninth electron of the third shell. Indeed, one would expect that the fourth row of the table should involve the sequential filling up of the third shell to eighteen electrons before beginning a new shell. But the first row of the periodic table has the alkali metals, the group with the most distinctive properties of any group in the table. Potassium is highly typical of the alkali metals. Indeed, it was not conclusively distinguished from sodium until the nineteenth century. The three alkali metals above potassium all have a *single* electron in their outer shell, so it seemed essential to assign the extra electron in potassium to a new *fourth* shell, even though this violated the principle that inner shells be filled to capacity before new shells are begun.

Once the mathematical principle of $2n^2$ was violated, the assignation of electrons to the other elements in the fourth row of the table became problematic. It was crucial that the principle be resurrected before arriving at krypton, so decisions had to be made as to how the electrons were to be allocated for the rest of the fourth row. Potassium, as we said, had been given a configuration of 2,8,8,1 in the respective electron shells. Calcium came next, but it couldn't be given a configuration of 2,8,9,1 because it was too stable to have a single electron in its outer shell. So it was assigned 2,8,8,2, while scandium was considered to exhibit 2,8,9,2. The next eight elements were variously assigned electrons to both the third and the fourth shell until arriving at copper, at which point the third shell was finally filled with eighteen electrons.

This is surely a case of a flawed mathematical principle conflicting with sound empirical considerations. The decision to assign a single electron to the outer shell of potassium was a good one, based on the observed chemical properties of potassium and its undoubted relationship to the alkali metals. But the decision to revert to allocating electrons to the third shell (for elements after calcium) was prompted by adherence to an unsound mathematical principle that did not have a basis in material reality.

Consider how the symmetry model naturally accounts for the development of the periodic table as far as its fifth row. For what the electron model refers to as "shells", our approach will refer to as "layers". Whenever a layer of genesis-units is completely filled, symmetry prevails. The first layer, or symmetrical milestone, is the pairing of proactive units found in helium. This corresponds to the first row of the table and to the first electron shell in the valence model. The second layer involves fusing eight proactive units to helium, resulting in the symmetrical cross-like structure of neon. The fact that eight such units must be fused to helium means that there will be eight new elements between helium and neon, accounting for the fact that the second row of the table has eight elements. The same will be true for the third row of the table because the next layer is achieved with the fusion of eight proactive units onto neon to produce argon. The fourth and fifth layers are different to the previous three in that they consist in the fusion of an entire cross-like structure of eighteen proactive units onto the previous symmetrical milestones to produce krypton and xenon respectively. The fourth and fifth rows of the table are thus composed of eighteen elements. All of this provides a natural and unforced account of how the table of elements is structured.

1.6 A new version of the exclusion principle

Earlier we presented diagrams of the configurations of genesis-units fused together to form composite atoms. The main purpose of these diagrams was to present the basic structure of this new approach to the atom. We wanted to show, in particular, the way in which the *binding* genesis-units hold the *proactive* genesis-units in pairs to form the composite entity. When we look at the relationships between the pairs themselves, however, we begin to see the shortcomings of these diagrams. These deficiencies are all the more evident if we compare the diagrams to the 3-D models of the atoms. Take the case of neon, composed of a central core pairing with four other pairs arranged around it. The core pair of genesis-units will have their V-B axes aligned in opposite directions (see Figure 1.24). This is what bestows such equilibrium on the pair. When the next proactive genesis-unit is added on to form lithium, the V-B axis of the new unit will most naturally align itself in the opposite direction to the closest genesis-unit of the core pair (see genesis-unit number 3 in Figure 1.24). The tendency towards equilibrium will dictate the way in which each new genesis-unit aligns itself to the composite structure. When genesis-unit number 7 is fused to the structure (which creates nitrogen out of carbon), the new unit will not attach itself permanently to genesis-units 4 or 6 since this would produce a longer structure with less symmetry and more internal disequilibrium. The various influences emanating from the V-B alignments of the six existing proactive units in the structure will ensure that the most stable new structure consisting of seven genesis-units will be one where the new hydrogen atom is fused to either unit 1 or unit 2. We cannot rule out the possibility that (within stars, for example) hydrogen atoms can be fused briefly to units 4 or 6, but such a structure will be imbalanced and will not endure as a stable element.

Our assertion that the first eighteen elements form somewhat flat, cross-like structures is based on the claim that the V-B directional alignment of the individual genesis-units is a crucially important causal factor in the formation of composite atoms. A new proactive hydrogen atom will be fused with an opposite V-B alignment to that of its closest proactive unit in the composite entity. Sometimes this will involve being fused *across* the ends of a pairing (as we saw with lithium). The overall result is a progressive weaving of genesis-units to form definite cross-like structures. One of the upshots of this pattern of evolution is that each member of a pair of proactive genesis-units will have an opposite V-B alignment. This corresponds to the standard claim that the pair of electrons in each orbital have opposite magnetic spin.

It is the tendency towards electrostatic equilibrium that underpins the validity of the Pauli exclusion principle. Each new genesis-unit appended to a structure will uniquely alter the equilibrium of the structure, for better or worse, and this will have undoubted empirical consequences that will be manifested in spectroscopy. The first three layers of symmetry involve fusing genesis-units in pairs around a central helium core to produce a rather two dimensional structure. The fourth and fifth stages involve forming successive layers on top of this structure. Whether a composite atom has one, two, three, four or five layers will have an influence on how it bonds with other atoms in the absorption and emission of light. The role any individual genesis-unit performs in the interplay of influences within the atom will depend on its precise position in the structure and the orientation of its V-B polarity relative to the rest of the structure. The position and orientation of a unit can be identified using a unique combination of parameters, just as the Pauli exclusion principle states that electrons can be uniquely identified by the four quantum numbers that we discussed in the last section. From our point of view – even though we reject electrons – it makes perfect sense to accept this principle as mirroring the fact that each genesis-unit in a composite atom will have a unique role in the way that it contributes to the overall symmetry of the structure.

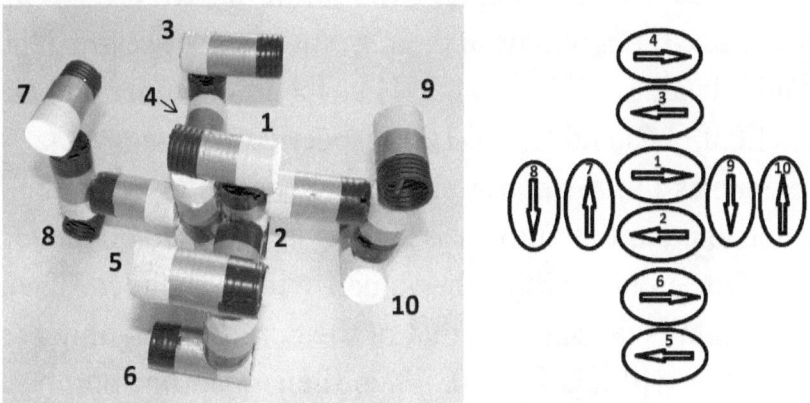

Figure 1.24 Neon. *3-D model to the left whilst the diagram on the right shows the alignments of the V-B components in the proactive genesis-units.*

If we take the elements up to argon, there are four structural features of the position of the genesis-unit within the composite atom. These features are the *layer* in which the genesis-unit is to be found (corresponding to the principal quantum number of the standard approach), the location of the genesis-unit on a particular arm of the cross (this corresponds to the notion of the subshell), the orientation of that "subshell" with respect to the helium atom at the core of the structure, and the *alignment of the*

V-B axis of the genesis-unit relative to the genesis-unit that it is paired with (corresponding to magnetic spin).

Let us see the numbers associated with the proactive genesis-units in the neon atom. Genesis-units 1 and 2 are located in the inner core of the atom. Their layer (corresponding to the principal quantum number) is 1. They are not located on the arm of a cross, so the second quantum number will be 0. Similarly, their orientation relative to themselves can be assigned the value 0. They have an opposite V-B alignment so their "spins" can be assigned the standard values +1/2 and -1/2. Thus their quantum numbers are (1,0,0,+1/2) and (1,0,0,-1/2) respectively.

Genesis-units 3 and 4 are on the second layer so their principal quantum number is 2. They are located on the top arm of the cross to which the value 1 will be assigned. Their orientation is the same as that of the core genesis-units, so this value will again be 0. Thus their quantum numbers are (2,1,0,-1/2) and (2,1,0,+1/2) respectively. The arm on which genesis-units 5 and 6 are located will be assigned the value 2. Their numbers will therefore be (2,2,0,-1/2) and (2,2,0,+1/2).

Units 7, 8, 9 and 10 are also on the second layer, and the arms on which they are located will be designated with the values 3 and 4 respectively. All of these units are oriented orthogonally to the core units in the atom, so this value will be assigned the value 1. Thus the quantum numbers for units 7, 8, 9 and 10 will be (2,3,1,+1/2), (2,3,1,-1/2), (2,4,1,-1/2) and (2,4,1,+1/2).

The number of possible layers in any element from hydrogen to argon is 4. From that point onwards, a second cross begins to form and the new composite atoms now have a double layer. Perhaps it would be simplest to assign the value 5 to the inner pair of proactive genesis-units in the second layer of potassium (the element which follows argon in the periodic table). From argon to krypton, using this convention, four new "layers" are added to the composite atom, and four again between krypton and xenon, and so on. We will require more than twenty layers if we are to assign quantum numbers to all the elements in the table.

This is in contrast to the Bohr model of the atom in two contrasting senses. Firstly, the standard method of working out the electron configuration of atoms will typically have *lower* principal quantum numbers than our model. For example, the standard electron configuration for lead (number 82 in the table) has only 6 principal energy levels, whereas our model has fifteen layers for that metal (three superimposed crosses with 4 layers each, and a fourth superimposed cross of 3 layers). Secondly, the Bohr model allows that an electron can be energised so that it jumps through a potentially infinite number of energy levels. This highlights one of the major ways in which our approach to the atom diverges from the Bohr model: it is not the case that "genesis-units" are roughly analogous to "electrons" in everything but name. When we are discussing the *electronic configuration* of an element, there is a sense in which the layers of our model are analogous to the energy levels of the Bohr model. But when we come to the absorption and emission of light, we shall see how this entire discourse of electrons jumping through energy levels can be made redundant by our much simpler account of optical phenomena. The notion that electrons could be raised through a potentially infinite series of energy levels was already a stark warning that there was something seriously implausible about the Bohr model of the atom. In our model of light, as we shall see further on, elements remain as they are with the same number of layers, regardless of how much light they have absorbed.

As we have outlined earlier in this chapter, elements are formed by a weaving process in which cross-like structures develop outwards from a helium core. Each cross will have four arms. Thus the second quantum number will have four possible values between 1 and 4.

The third quantum number refers to the orientation of the plane of the V-B axis of the unit to the plane of the V-B axis of the core pair of units. Here, we are not interested in the actual *direction* that a particular unit is pointing in (that direction is represented by the fourth quantum number - magnetic "spin"), but only the *plane* in which the axis lies. In three dimensional space, only three planes are possible relative to the plane of the core units. Thus the third quantum numbers has value from 0 to 2, where 0 represents the plane of the core units.

In the case of neon, the proactive genesis-units all lie in just two of the three possible planes. In other elements, such as carbon, all the units lie in the same plane. Figure 1.25 shows the stable isotope of boron, ^{10}B (the most common form of boron is ^{11}B). Genesis-unit 5 lies in a different plane to any of the proactive units in neon and would be assigned the value 2. In order to make carbon from this isotope of boron, two genesis-units would have to be fused to the ends of genesis-unit 5, both of which would lie in the same plane as the core units and consequently have a third quantum number with value 0.

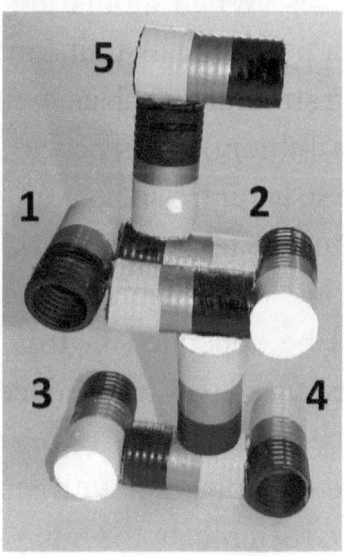

Figure 1.25 Boron revisited. *The third quantum number of the core units is assigned the value 0. Genesis-unit 5 lies in a different plane and the value 2 is assigned to its third quantum number.*

That takes us to the fourth quantum number. Whenever we speak of the property of magnetic spin, we are speaking of the *relative* orientation of something with respect to the alignment of something else. Within each pair of proactive genesis-units, only two possible V-B directions are possible, and they will always line up opposite to each other.

Sometimes the Pauli exclusion principle is presented as if it had some role in explaining why the electron shells have the capacities that they do, but it is not at all clear how a mathematical formalism that *describes* empirical data can hope to *explain* that same data. Nature does not adhere to mathematical regularities for the sake of regularities. Nevertheless, it must be said that the exclusion principle does point to a fact of nature that has genuine causal consequences. When atoms are being formed, it often happens that new genesis-units are fused coercively onto a pre-existing structure (such as the genesis-unit fused to helium to form lithium). This process does not aid the equilibrium of the structure, but the *next* genesis-unit added to lithium will naturally take up a position that will restore equilibrium. During this ongoing process of evolution, genesis-units will be appended in this two-fold pattern. An odd unit will be fused that will reduce equilibrium, and then the pair will be completed to restore equilibrium. Each new pair will take up a position with respect to the whole in such a way as to reciprocate the many causal influences within the atom in as complete a way as possible. This is what the exclusion principle is pointing to: the process by which this tapestry of equilibrium is woven, with each new unit contributing uniquely to the overall state of the atom.

1.7 Which approach has the greater explanatory power – the standard framework or the genesis-unit model?

Earlier, the reader was requested to follow our line of reasoning with patience. The genesis-unit model still has much need of development. We have not seen how it deals with the empirical evidence that is usually interpreted as indicating the existence of the "electron"; nor have the new understandings of electricity, magnetism and the transmission of light been yet presented. The project has only just begun, but at this point it is still possible to evaluate its relative capacity to explain the structure of atoms, the arrangement of the atomic table and the patterns of chemical bonding exhibited by the elements. Whether an explanation is considered a *good* one depends on many factors such as its simplicity and coherency, its ability to predict unexpected phenomena, and a host of other characteristics of good explanation, the relative merits of which have been debated since Robert Boyle's list of such criteria appeared in the seventeenth century. Here we will dwell for a moment on the simplicity and coherency of the genesis-unit model.

The starting point of the model is the postulation of a simple polarity within the hydrogen atom. On the basis of this polarity, we can describe how composite atoms are formed, and why they exhibit the properties that determine their position in the periodic table. The progression from simple hydrogen atoms to more complex structures is driven by a *single* dynamical influence: namely, the propensity of genesis-units to reciprocate the electrostatic influences of other units. Consider how this contrasts with the complexity of the standard model. Here, electrons are postulated to be in orbit around a nucleus composed of particles that somehow adhere together despite the enormous forces of mutual repulsion that ought to prevail between them. To explain the unlikely cohesion of the nucleus, a new force is postulated, carried by hypothetical particles with some properties that, in principle, cannot be observed. The very survival of the atom, on the standard view, is so improbable that we would be forgiven for wondering if this theoretical framework makes any significant contact with the physical world at all. Can it really be the case that such a host of improbable particles and forces are conspiring together to form a stable and enduring entity, the ground of real things?

By contrast, the genesis-unit model postulates a single dynamical influence. And this influence has enormous empirical support, once we perform a simple re-examination of electrical and magnetic phenomena, apart altogether from the structure of the periodic table and patterns of chemical bonding. It will become clearer as we go along how properties such as magnetic spin cohere perfectly with the postulate that the primitive V-B polarity is at the root of the atom, dictating how each genesis-unit aligns itself with others. Composite atoms are formed by genesis-units progressively being fused to a pre-existing structure. The way that each new genesis-unit aligns itself in the composite atom is determined by the role it plays in reciprocating the electrostatic influences of the units already present. In comparison, the valence electron approach to atomic structure has to do a lot of huffing and puffing to account for the arrangement of the periodic table. Under the standard model, a myriad of rules need to be invoked to explain why atoms have the electron configuration that is claimed for them. And the problem with some of the rules is that they lack genuine explanatory capacity. We assign unique quantum numbers to each electron in the atom, and this practice helps us to discern the characteristics of any given electron. But *why* each electron in the "cloud" must have a unique set of quantum numbers is not explained. It almost seems, sometimes, that we are being asked to believe that electrons conform themselves to numerical regularities just for the love of numbers. Once we appreciate that each genesis-unit plays a unique *structural* role in the electrostatic equilibrium of an atom, then we can see why it might possess a series of unique numbers that determine its electrostatic position with respect to the whole. And the numbers testify that there is nothing indeterminate or cloudlike about this position.

The genesis-unit account of chemical bonding is also simpler and more coherent than the standard approach. The usual explanation of the formation of methane is a classic example of the current theory in action (see Figure 1.26). Each methane molecule is composed of one carbon atom and four hydrogen atoms. The single electron allegedly possessed by each hydrogen atom is supposedly found on the first shell, which can only hold a maximum of two electrons (according to the expression describing how shells are filled, 2^n, where n is the number of the shell in question). The carbon atom is said to have four electrons on its third and outer shell, a shell that can hold up to eight ($2^3=8$). Each hydrogen atom attains a "share" in an electron from the carbon atom, thus attaining the extra electron needed to fill its shell. The hydrogen atoms also donate their electrons to the joint "pool" and in this way the carbon atom gains the four electrons required to fill its outer shell. The result is a stable molecule composed of atoms whose outer electron shells are full.

Ionic bonding occurs when an atom that lacks electrons in its outer shell receives an electron (or electrons) from an atom (or atoms) with electrons "to spare". The electrons in the outer shell of an atom are known as the "valence electrons," and atoms are considered to have a tendency to form bonds that bestow on them a full or closed outer shell. This is a general overview of how such molecules are formed, even if it does not do justice to the many developments that have taken place in the field throughout the twentieth century and the detailed mathematical analysis provided by the contemporary theory of molecular bonding. However, there can be little doubt that contemporary theory is driven by the qualitative picture of the planetary model and the belief that atoms have a tendency towards closed outer shells.

The shortcoming of this approach is that it gives us no clear idea *why* the atoms strive for full outer shells. Neither does it tell us how this striving actually plays itself out in physical terms. What is it about the hydrogen atom that impels it to "share" an electron with carbon? Is this tendency towards attaining a full outer shell emanating from some unknown causal dynamics in the nucleus? A further difficulty with the whole notion of sharing is the problem of understanding how an entity (the electron) that allegedly has the potential to undergo wild fluctuations in its position can be *shared* in any meaningful sense with another atom whose electrons exist in similar far-flung states.

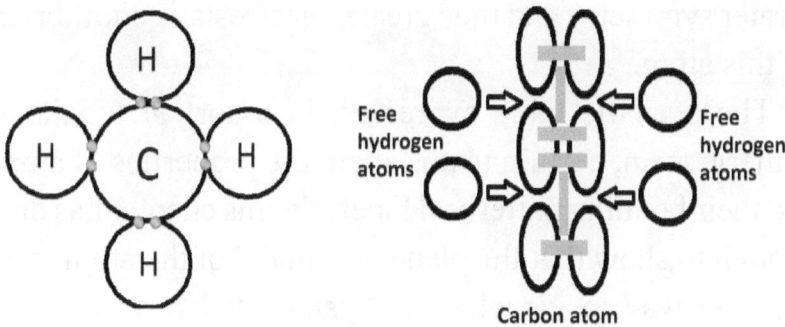

Figure 1.26 Methane. *To the left we have the standard picture of the methane molecule. To the right is the genesis-unit model of the bond.*

The genesis-unit model is able to describe chemical bonding in very simple and coherent terms. Atoms have greater or lesser electrostatic equilibrium depending on how their genesis-units are arranged. The pattern of the units in certain atoms will prompt them to bond with other atoms in such a way as to attain greater electrostatic equilibrium. Atoms with symmetrical configurations will tend to have the electrostatic influences of their constituent genesis-units reciprocated by other units within the structure. Elements with a protruding unit on one side will tend to bond with elements that have a corresponding "hole" on one side. There is no mystery as to why these bonds occur because they are driven by the pure electrostatic influences of the individual genesis-units within the composite atoms. We do not need to wonder why four hydrogen atoms would have a tendency to "share" their electrostatic capacities with a single atom of carbon. If we examine the structure of carbon as depicted to the right of Figure 1.26, it becomes apparent how the presence of four new genesis-units would confer greater symmetry and thus greater electrostatic equilibrium on this atom.

The periodic table expresses, in a sort of tabular or numeric form, certain truths about the properties of atoms and their bonding patterns. Hopefully this chapter has done enough to show that the planetary model of the atom is not the only way to visualize a physical structure of matter that could give rise to these periodic properties. Despite the prominent visual role played by the planetary picture in the electron shell account of atomic bonding, a distilled version of the theory can be produced in which the notion of electron energy levels disappears altogether. Little attention has been given to the line of thought, developed in various quarters during the course of the twentieth century, stating that the relationships expressed by the periodic table are purely numerical, and the term "electron" could be replaced by another term without any negative repercussions for our understanding of how atoms bond together. Our real knowledge in this field is largely numerical in nature, and these numbers do not point unambiguously to a unique physical realisation such as the planetary model. Indeed, before the electron was hypothesized at all, it was already understood that the bonding tendencies of the various elements depended on the group they occupied in the periodic table. Instead of saying, as we do now, that "carbon has four valence electrons that it can share in covalent bonds with other atoms", we can simply substitute for the word "electron" any other property or structure within the atom that we hypothesize to be responsible for bonding of this sort. In our model, that equivalent structure is the proactive genesis-unit.

It is also highly plausible to claim that the patterns of fusion exhibited in the formation of atoms should be mirrored by the chemical bonding patterns between atoms of different elements. The electrostatic "hooks" or lack of them that facilitate or hamper atomic fusion should naturally give rise to the same bonding pattern at the level of the chemical behaviour of the elements themselves.

Moreover, our attempt to characterize each new element largely in terms of its number of proactive genesis-units fits in with the work of Moseley and others. Even though lithium, for example, has an atomic weight of *seven* atomic mass units, just *three* of these units exert significant electrostatic influences beyond the interior of the atom itself. Carbon might weigh *twelve* atomic mass units, but only *six* of its genesis-units are proactive. Therefore, it comes as no surprise that the diffraction of x-rays by a sample of lithium should show a mathematical regularity in relation to the number three and the extent of diffraction by carbon should be in relation to the number six. It is the polarity in the proactive genesis-units that is responsible for the diffraction, after all. The binding units have their electrostatic capacities reciprocated from within the atom, and thus cannot diffract the x-rays, just as our model would expect.

Chapter Two

TOWARDS A NEW MODEL OF LIGHT: AN EXPLANATION OF ATOMIC SPECTRA AND THE PHYSICAL BASIS OF THE RYDBERG FORMULA

"The aim of science ['natural philosophy'] is not simply to accept the statements of others, but to investigate the causes that are at work in nature."

Albertus Magnus

Overview of this chapter and its principal claims

1. The chapter begins with some foundational material of a conceptual sort. We argue that objects have an individuality, separability and ontological grounding that is not reducible to their spatial aspect. The autonomy and separability of objects is not *founded* upon the ontological fact of having a particular location in space. *Rather their representation as having a particular location in space is founded upon ontological facts of a non-spatial sort.*

2. Armed with this critical evaluation of the spatial aspect of objects, we turn to the standard picture of the emission and absorption of

light, which involves electrons jumping from one energy level to another within the atom. If a negatively charged electron moves closer to the positively charged nucleus, then it loses energy and a photon is emitted. Conversely, if an electron absorbs a photon, then it gains the energy associated with standing further away from the nucleus. The polarity between the positively charged nucleus and the negatively charged electron is the "spring" that stores or releases the "energy" we call light. As is well known, this comfortable picture does not stand up too well to closer scrutiny. Why is the electron restricted to fixed shells or energy levels within the atom? Why does it not emit a continuous spectrum of light as it falls from one energy level to another? Are we really to believe that a particle can disappear from one shell and reappear instantaneously at the other without traversing the area in between?

3. A new model of light is presented without the "damned quantum jumping", whilst maintaining the fundamental insight of the older model. Electrostatic equilibrium/disequilibrium is still presented as being the spring that holds and transmits this causal influence in the world, but we no longer speak of the *internal* excitation of units of matter. Now the emission and absorption of light involves *atoms* forming or dissolving relationships of heightened electrostatic tension with each other. The principal distinction in this chapter is that between *protomagnetism* (the fundamental polarity *within* the atom that comes into being when the atom is constituted by the work impulse separating B from V) and *electrostatic tension* (the attractive/repulsive secondary dynamics that comes into play between the B and V poles of *different* atoms).

4. An account is given of how and why atoms in these states of heightened tension tend to form groups that are *orthogonally* aligned to each other. This results in mini-systems composed of 2^2, 3^2, 4^2, etc., atoms, accounting for the peculiar form of the Rydberg formula.

5. A statistical argument is presented to defend the view that a pair of atoms can be the target of *multiple impulses* of light simultaneously, thus giving rise to mini-systems aligned in this way.

2.1 The fundamental locus of causal activity in the universe

This chapter presents a new model of the nature and transmission of light. Again, the reader is asked to be patient for the first part of the presentation. The new model is not described in any of the preliminary sections of what follows because some groundwork must be done first. It is necessary to say a few words about the nature and significance of the genesis-unit and the meaning of the term "energy". It will also be indispensable to devote some time to the understanding of space that underlies our approach. If the reader is anxious to see how the model works, then he could go directly to section 2.8, keeping in mind that he will have missed some of the conceptual content that is foundational for our framework. Then, if he wants to examine the philosophical assumptions underneath the scheme, he can return to this point.

It will be assumed that the basic locus of causal activity in the universe is the simple and stable "genesis-unit" that we have introduced in the first chapter of this book. At the heart of the unit, there is a polarity that is maintained by the continuous action of a work impulse, separating the two poles from each other. For convenience, using the first letters of the words "being" and "void", the poles have been labelled B and V. This polarity at the heart of the atom can be referred to as the "protomagnetic" polarity. All phenomena – electromagnetic, electrostatic and gravitational - can be understood as originating from this fundamental polarity.

It might be helpful to take a step backwards and see where all of this is going. Bohr's model of the atom was really a model aimed at showing how the atom *absorbed and emitted* the discrete frequencies of electromagnetic energy that it did. The atom was conceived of as a sort of energy-absorbing-and-emitting spring, and it was the *opposite polarities* of the electron in the outer reaches of the atom and the proton at the centre that stored the energy in this spring. When the electron absorbed a photon of light, it gained the energy necessary to pull itself away from the electrostatic attraction of the nucleus. The spring within the atom was thus extended and the atom gained energy. Emission involved the tension on the spring being lessened, the electron drawing closer to the nucleus, and light being emitted.

Bohr's model gave us a way of understanding how the spring absorbed and emitted energy (by the electron jumping from one energy level to another). Fundamental to this picture was the belief that the electron and the nucleus possessed electrostatic charges opposite in kind. It was this opposite polarity, and the drawing closer or moving away of opposite charges, that accounted for the atom's capacity to *store* energy within it. Our picture of the atom turns this view on its head. It is not the *relative configuration* of opposite charges that gives the atom the ability to store energy. Rather, the prior existence of action (the work impulse that constitutes the atom) *gives rise to the polarity* within the atom. There is no need to dwell too much on this contrast for the moment until other aspects of our model are developed and clarified. But we shall see later how this feature gives our model a superior philosophical basis and greater explanatory power than the traditional approach.

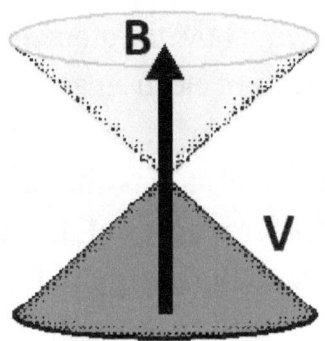

Figure 2.1 Representation of the basic unit of matter - the fundamental locus of causal activity in the universe. In this schematic diagram, a work impulse separates B from V, the lower cone representing the void from which the "matter" of the upper cone is drawn. The unit is constituted by the work impulse that draws B from V. The polarity between B and V is the protomagnetic polarity that underlies all electromagnetic and electrostatic phenomena. The vertical arrow represents the magnitude and direction of the B component of the atom, the fundamental work impulse that constitutes the unit. Diagrams such as this one are attempts to visualize how the unit is constituted. As we go along, our way of visualising the atom can be improved by attention to the relevant empirical data.

The B and the V that we claim are generated by the constitutive action of the atom are *not* opposite electric charges, but neither can they be identified in a simplistic way with opposite magnetic poles. Are we, then, postulating the existence of a new kind of force or mechanism in nature that has no direct empirical justification? Absolutely not! The physical phenomena that we usually interpret to be caused by either electricity or magnetism can all be understood as various manifestations of this more fundamental influence in different situations. Our framework will have greater simplicity and explanatory capacity than the standard view that proposes separate treatments of electrical and magnetic phenomena and falters in its effort to offer a coherent account of the relationship between them.

No doubt, many readers are already wondering how we can claim that such a basic unit is the basis for all empirical phenomena when "everyone knows" that there exists in nature an entire zoo of more fundamental particles. Whilst such a reaction is understandable, it is also naïve. It fails to appreciate that the postulations by the status quo of these particles, even of the electron itself, do not follow *necessarily* from the empirical evidence. We aim to show that the evidence can be interpreted not just differently, but *better*. Everything we experience can be generated by the causal activity of a single fundamental unit of matter.

So what then is the matter that this unit is made of, on our framework? Matter is not ultimately an inert *substance* of some sort. Nor is matter composed of other more basic pieces of matter, as the infinite-sounding regress of modern particle physics would sometimes lead us to believe. And it is not made up of multiple different "elementary" particles with strange properties that stand in complicated relationships to each other. Matter, rather, is composed of these simple genesis-units that have a very straightforward inner constitution based on a dual principle that is held in existence by a work impulse. With these genesis-units and their internal polarity we intend to explain *all* of the phenomena that lead us to hypothesize mistakenly the presence in nature of the entire zoo of subatomic particles (protons and neutrons *do* exist, as we shall see, and are simply different manifestations of the genesis-unit itself). Causal events that are usually thought of as involving *sub*atomic particles will be re-interpreted as causal impulses arising in genesis-units and giving rise to causal effects in other genesis-units. We will give an account of magnetism and electrostatic charge, as well as a new model of the transmission of light, all in terms of the properties and configurations of genesis-units. To do that, of course, we will have to develop our model considerably.

2.2 Primary and secondary causal dynamics in the atom

The unit that comes into existence by the action of the work impulse separating B from V has both primary and secondary causal dynamics associated with it. The *primary* causal dynamics of the atom is the work impulse that holds B and V in separation from each other. This is what the standard approach to physics tends to call the "rest energy" of a piece of matter. Once these poles have been generated, they give rise to a *secondary* form of causal dynamics which takes on a veritable life of its own. This life involves all of the electromagnetic and electrostatic phenomena associated with a piece of matter. Gravity, too, will be understood as a residual effect arising from the same dynamics.

How and why does this secondary causal dynamics arise? We can use many valid analogies from everyday experience to illustrate the phenomenon. A refrigeration unit expends energy in separating heat from cold. The energy used to create this separation is analogous to the primary work impulse at the heart of the atom that separates B from V. Once the heat and cold have been separated from each other, they can become sources of secondary causal dynamics. Imagine a fridge with an open door placed in a room that has a uniform temperature. The refrigeration unit cools the inside of the fridge, but creates greater heat in the room. This "polarity" in temperature causes air to move in the room. There is a tendency for the "poles" of the process (the cold interior of the fridge and the warmer exterior) to interact with each other so that equilibrium is restored. In the same way, there is a tendency for the B pole of a genesis-unit to act on the V pole of its neighbour so that equilibrium is restored. But now imagine an enormous room with thousands of fridges with open doors. The primary causal dynamics at work within each fridge is the refrigeration unit separating hot from cold. The secondary dynamics will no longer involve simple tendencies towards equilibrium of the cold interiors and warmer exteriors of each *individual* fridge. The *entire system* will now be in interaction with itself on this secondary level, a complex interrelationship of the hot and cold "poles" of all the units that are present in the system.

During this process of early evolution of the system, the genesis-units are formed by the action of the primary work impulses that generate each atom. The work impulses create a polarity within the atom and these polarities give rise to secondary causal dynamics between the various atoms of the system. In our refrigerator analogy, the secondary dynamics resulted in the "poles" (hot exteriors and cold interiors) of the fridges entering into causal interaction with each other. This led to the continual movement of air between the poles. In the case of atoms, such continuous causal activity is manifestly *not* the case, not least because the primary work impulse within the atom is somehow *static* in nature. It does not continually expend energy, in contrast to the persistent activity of the compression unit in the fridge. So just what does this secondary causal dynamics of the genesis-units consist of?

The best way to illustrate this secondary process is perhaps with a consideration of the way molecules cohere together in liquids. Each molecule in the interior of a quantity of liquid is pulled or pushed equally in every direction by the molecules that surround it. The molecules on the surface do not have molecules surrounding them on all sides with the result that they cohere more strongly with the molecules that are just below them. This creates a "surface tension" that behaves like a type of skin on the top of a liquid, allowing certain objects to float on the surface even though they are denser than the liquid itself.

We can quantify surface tension by assigning it a magnitude in terms of energy per unit area. This is despite the fact that in normal circumstances - when the liquid is in an undisturbed state - the tension that prevails on the surface does not involve the individual molecules of liquid expending any energy at all! Energy is only expended when the liquid is disturbed and the molecules re-establish equilibrium between themselves, leading once again to tension on the surface. But in the undisturbed state, when the liquid has "settled down", no further energetic action is required to maintain the surface tension, nor, indeed, to maintain the coherence of the molecules in the interior of the liquid (here, for convenience, we are imagining an ideal liquid that is not subject to gravitational or other influences). The coherence of the molecules in a liquid at rest is the product of a secondary dynamics that works itself out *during the process in which the liquid comes to rest*. The dynamics gives rise to an equilibrium in the liquid that maintains itself without any further expenditure of energy until such time as the liquid is disturbed.

In an analogous way, the stability of groups of genesis-units is the product of a process of the "settling down" of the system in which the individual genesis-units attain equilibrium with each other. We shall consider in detail shortly how this equilibrium is achieved. Once the balance is attained, no energy is expended to maintain that balance. But if the equilibrium is disturbed, an impulse with causal potential evolves through the system, and this is what we call the transmission of light.

There is a tendency on the part of the B pole of any given unit to collapse into the V pole of other units adjacent to it. Why does a general collapse not take place? In the refrigerator example, a general collapse *does* take place – hot and cold currents of air from the various poles move towards each other, tending to destroy the polarity. But atoms have a perfectly-proportioned resistance that helps to maintain the polarity within them. The resistance derives from the fact that each genesis-unit possesses both principles in equal measure. The B pole of an *adjacent* unit might well have a tendency to collapse into the V pole of a particular unit, but the B pole of that unit will resist and repel that tendency to a perfect degree.

2.3 The capacity of genesis-units to absorb or emit "energy"

We wish to consider the situation that arises when genesis-units become "energised". This will lead us to formulate an explanation as to how elements have certain emission spectra, permitting us to account for the Rydberg formula. But first of all, a word about what it means to absorb or emit "energy". The fundamental locus of causal activity is the genesis-unit, and it is in terms of the basic characteristics of this unit that we intend to define the meaning of the terms "matter" and "energy". Energy has always been a difficult concept to define in precise terms. There is a tendency to think of it as being something ethereal in nature, a capacity to do work that can take on many forms while being utterly formless in itself. In our scheme, energy is to be understood as an abstract term that has *no genuine existence at all* in the world. It is a useful conceptual tool for describing and quantifying the various magnitudes of work done by genesis-units in various situations, but it has no independent presence in reality. The fact that genesis-units can do varying magnitudes of work does not imply that these units have a capacity to expend varying degrees of energy. The dynamics of these units, as we shall see, is very simple and unitary. It is the shifting relations between *combinations* of units that give rise to the broad spectrum of causal magnitudes in the world.

This approach involves a re-thinking of what is meant by "energy". We wish to resist the temptation to think of energy as a multiply-divisible "fuel" that empowers the work impulse in the genesis-unit. Rather, we wish to consider the *work impulse itself* as an indivisible event, something that creates the B out of the V, thereby constituting the fundamental unit of matter. This action, as it turns out, will become the "powerhouse" that ultimately drives all causal processes in the universe. But it is only in abstract terms that we can speak of the work impulse as having "energy". The genesis-unit itself is indivisible in character, and this (as we shall see) is the origin of the quantum nature of micro-physical events. In the early twentieth century, the discovery of the apparent quantum nature of energy prompted great surprise in physics. Despite the enormous theoretical development of quantum theory over the past one hundred years, it is far from clear why causal events at the micro-level should have this quantum nature. In our model, the origin of the quantum nature of phenomena lies in a simple characteristic of the fundamental constituent of matter. But we shall argue later on that there *is* a sense in which the work capacity of a genesis-unit can be arbitrarily partitioned. Certain types of causal capacity (such as electrostatic attraction) can be *distributed* by a single unit over multiple other units, leading to a partitioning of that capacity without infringing in any way on the integrity and indivisibility of the work-impulse of the genesis-unit. For this reason, it will often be convenient to continue using the term "energy", but – and hopefully this will become clearer as we go along – the notion of energy as an independent constituent of the world will be rejected.

As we saw in Chapter One, atoms will begin to exert a net attraction on each other once they have attained a certain proximity. But if they converge on each other too much, the manner in which the V component is distributed along their length will prompt them to repel each other. The fact that atoms attract each other at a certain distance and repel at a closer distance is one of the relatively few things we can be sure about in atomic theory. This means that there is this "middle distance" at which atoms attain a state of equilibrium with each other. The figure that appeared in Chapter One is reproduced here as Figure 2.2 in order to illustrate this point. When two atoms have settled into such a state of electrostatic equilibrium with each other, no "energy" is expended by the bond. In other words, no work is done by either atom to maintain the bond. The surface tension of a liquid is our standard analogy for visualizing this characteristic of the atomic world. Such tension arises from the dynamics between the individual molecules of liquid and constitutes a resistance to disturbance of the equilibrium that has been established between the molecules. As such, the tension can be abstractly described as possessing a magnitude of "energy", but in reality no energy is expended at all by the system while it remains in its stable state.

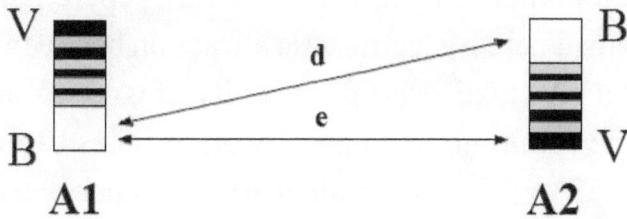

Figure 2.2 *Atoms A1 and A2 have converged to a point where there is a net attraction between them. The fact that d is longer than e entails that the net repulsion between the like poles will be less than the net attraction between the unlike poles*

One of the fundamental characteristics of physical reality is its tendency towards equilibrium. Any disturbance of equilibrium will give rise to an evolution of the system until this imbalance is "resolved". Under Maxwell's theory of electromagnetism, such tendencies towards equilibrium gave rise to stresses and strains in the universal medium that permeated all of space. Causal influences were propagated through the medium by virtue of the effects generated by these stresses and strains. In contrast to Maxwell's view, our view is that the system we call physical reality has a *natural* tendency towards equilibrium. Electrostatic imbalances prompt the system to evolve in such a way as to tend towards greater equilibrium, but these imbalances are *not* transmitted by the tensions or strains in the medium. Rather, they can be directly embodied in *concrete relations* between the only causal players in the world – genesis-units.

Let us see how this tendency towards equilibrium plays itself out in situations where the electrostatic bond is broken between a pair of atoms. No work was being done by the atoms to maintain themselves in the original bond, but once the bond is broken, then there are repercussions for the system. There is electrostatic attraction between A1 and A2, and it will take a certain quantity of work to fragment them. The minimum amount of work it takes to fragment A1 and A2 will be equivalent to the "energy" of their electrostatic connection. Once fragmented, the system as a whole enters into a higher state of disequilibrium (because the bond constitutes a state of relative equilibrium), but this change does not happen in an instantaneous way. Beginning at the point where the atoms were originally bonded, there is an evolution through the system of an influence of a very particular sort. This does not involve waves, fields or particles. To understand the nature of this evolving influence, we need to discuss the nature of the "space" that surrounds the genesis-units and through which the influence is transmitted. The next few sections take up this task. The intention is to genuinely deepen our understanding of the nature of this evolving influence we call light, including the reason that its velocity appears to be a universal constant. For the moment, the reader is asked to hold on to this picture of the transmission of light as originating in a bond being broken between two atoms in equilibrium, giving rise to an evolution of the system that is directed to maintain or restoring equilibrium.

2.4 What implications does spatial separation really have for causal interaction?

Despite the widespread supposition that the absolutist conception of space has been "superseded" by the relativistic conception, we continue today to adhere to the broad outlines of the classical mechanical framework. In our attempts to visualize what is going on inside the atom, in our efforts to construct a model of particle interaction that will explain the genesis of the empirical data that is emanating from quantum physics, we basically work out of the same model that first became fully concrete in seventeenth-century science. Even if spatial position and velocity are no longer thought of in absolute terms but in relative terms, we still think of space as a container of material objects whose *existence* (if not all of its properties) is independent of the objects that it contains.

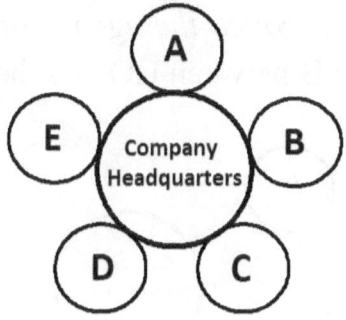

Figure 2.3 *The notion of a causal configuration*

A simple analogy can be used to illustrate an alternative to the absolutist conception of space that has come to dominate our construal of the "being" of physical objects. Here we introduce the notion of a "causal configuration". A causal configuration can be thought of, in abstract terms, as a space in which the "position" of objects is determined by their causal relations with each other. Take the example of a company that has various branches in different locations and a single headquarters. Let's say that the various branches have no contact or influence on each other, but receive all of their operating instructions from headquarters. In this case, the causal configuration of the company could naturally be represented as in Figure 2.3. Spatial contiguity in the diagram represents *connectivity* in causal terms. Headquarters exerts causal influence on each of the branches, so it is natural to place it at the centre. It would make no sense to place, say, B, between HQ and C, because that would seem to indicate that the causal influence between HQ and C was somehow influenced or mediated by B. If, however, B has a satellite branch that is controlled by HQ *through the agency of B*, then it would be proper to place B between HQ and the satellite branch.

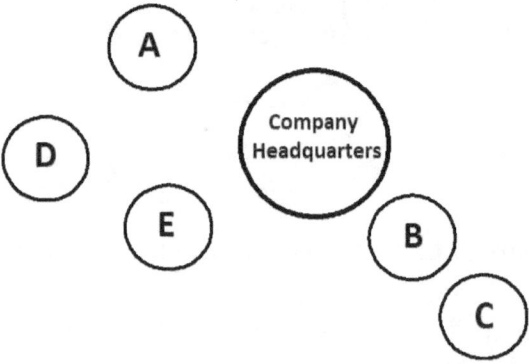

Figure 2.4 The geographical configuration of the company

Now imagine that company branches B and C are in fact located in the same town, on the same street. Despite this proximity, they have no direct contact with or influence on each other, as far as the causal dynamics of the company is concerned. In this case, it would be potentially misleading to represent B and C adjacent to each other in our company causal configuration space, since this spatial representation is not interested in expressing geographical issues. The geographical proximity of B and C could be described as a feature of their *geographical* configuration. Figure 2.4 shows the geographical configuration of company headquarters and the various branches. We see that HQ is not at the centre and that some branches are located between it and other branches. It seems clear that both of these configurations contributes in its own way to our understanding of the state of affairs in question - one configuration provides a mapping of geographical information and the other provides a mapping of causal dynamics. It is also clear that neither view separately, nor both views combined, manages to give a particularly comprehensive understanding, or full knowledge, of the complete physical reality of the company and its various branches.

The geographical configuration of the company is analogous to the configuration of objects in container space in the classical mechanical framework, whilst the causal configuration of the company is analogous to the true network of causal relationships that obtain between objects. Now consider an observer who only has access to the *geographical* configuration of the company but not to the causal configuration. He observes that when certain phenomena of type O appear at HQ, other phenomena of type E appear at C, despite no change of any sort being noticed at B, which stands between HQ and C. After repeated observations of this sort, he makes an inductive inference that HQ is having a causal influence on C, and that this causal influence operates "at a distance" (by "acting at a distance" we intend to express the notion that HQ acts on C without any observable disturbance on B, which stands between them. Genuine action at a distance, in normal physical terms, would involve a cause having an effect at a distance without any physical disturbance of the space in between). If our observer is a gifted mathematician, then, after an extensive collection of empirical data, he may be able to formulate this inductive generalisation into a law in which one of the terms - call it G - represents the force "acting at a distance" between HQ and C.

Such an inductive law will manage to "save the phenomena" in that it will correctly predict/describe relations between phenomena at HQ and phenomena at C. However, not only does it fail to shed light on the authentic causal dynamics in operation in this situation, but it gives the completely false impression that the real causal mechanism behind the phenomena is a force acting at a distance (a force that can *penetrate through the reality of B* without disturbing it). Those of us who have access to the map of the genuine causal configuration of the company know that no such force exists, or is necessary, to explain the causal dynamics in operation.

This simple analogy is intended to illustrate the point that the causal configuration of things need *not* coincide with the spatial (geographical) configuration, as is tacitly assumed by the container approach to space. In other words, if two spatially separated objects - a causal source and a causal target – are having a causal relationship, then the container approach takes it for granted that instances of matter in motion (particles or waves) are traversing the causal "chasm" between them. The analogy encourages us to consider that the causal configuration of a system may actually be *distinct* from its spatial configuration. It exposes the naivety embodied in the container view's presumption that the spatial aspect of an object reveals the causal characteristics of that object in a comprehensive way. Most importantly, from our point of view, it supports the view that causal interactions between two objects that are spatially separated are possible without any intervening intermediaries, such as waves or particles. The best way to describe such causal relationships between spatially dispersed objects is with the terminology of causal "processes" that will be used extensively in this book.

2.5 Spatial location is derivative of prior causal relationships

(Let us recall that if the reader is finding this conceptual groundwork too tedious, then he can skip to section 2.8 to see the new model of light in action). According to the container model, objects have positions in space, and these positions help to define the causal relationships that will unfold in the system. But what if spatial extension and location are *derivative* of the causal relationships that already obtain in the system? In this section we will present the broad outlines of an imaginary world in which this is how things are. The perceptual apparatus of the beings that inhabit the world will be described in simple terms. We will see how these beings develop a spatial representation of objects that have deeper causal relationships that go far beyond their spatial aspect. It is hoped that the common-sense character of this imaginary world might throw some light on our world and the significance of the spatiality that dominates our perspective on reality.

To begin, there are two basic elements to the beings' viewpoint on their world: the kind of experience they have of the *objects* that exist in the analogical world; and how they construe the *causal relations* between these objects. In the imaginary world, as it turns out, one basic law of nature governs many of the causal interactions that occur, and this is that *objects tend towards equilibrium in their properties.* So, for example, if object A is hotter than object B, then A and B will interact with each other in such a way that their temperatures tend towards equilibrium. Typically, when two objects interact with each other, *effects* are produced. Effects are changes that occur in objects as a result of an interaction. For an effect to be produced in an object, work has to be done on the object, and work requires an interval of time (the principle that work requires time is a venerable one in physics). The perception of an effect is thus a sure indication that a causal process has *already* taken place, culminating in the effect. If an effect impinges on the sense organ of an observer, then these effects can be classified as *manifestations* of the causal interaction that has taken place.

The beings that inhabit the imaginary world have only one sense organ. It is not important to know how this sense organ works, nor how the beings transform sensory evidence into "theory" or "understanding." What is more relevant for our analogy is to know the kind of thing the sense organ is stimulated by. It turns out that the sense organ is passive and will only be stimulated if it is impinged on by the sort of effects of interactions that we have just spoken about. There are various types of effect-producing interactions between objects, but one type is so prevalent that it dominates the beings' experience and is the basis for their particular viewpoint on the world. This type of interaction takes place when a body exerts a certain type of causal influence – let's call it "causal-interaction-L" - on another body. This second object need not be substantially or noticeably changed by causal influence L, but the effect that is produced by the interaction impinges on the beings' sense-organ and gives the beings an "experience" of the object. In reality, the beings are not experiencing the object directly, but are experiencing the effects of an interaction that the object is involved in.

Two related points arise from the above description of the imaginary world that are significant. Firstly, any effect that is registered by the beings' sense organ is a culmination of a *prior causal process* that led to the effect. Thus, the sense organ detects momentary *events* and not *the natural causal processes themselves.* Secondly, it can easily happen that the beings misinterpret these effects as being properties of the objects themselves, instead of being momentary *manifestations* of processes that the objects are involved in.

Given that the beings' sense organ can only detect effects of prior interactions between objects, we can set down a kind of first principle for understanding the beings' perspective on their world as follows: *experience requires effect-producing interaction.*

Given the law of nature outlined earlier (the tendency towards equilibrium) and this epistemological principle, what kind of *knowledge of objects* are the beings capable of attaining? Take an object, A, in causal-interaction-L with another object, B. Typically, when any two objects are in this type of interaction-L, they will produce multiple effects E_1, E_2, E_3, etc. These effects can be registered by the beings' sense organ, giving rise to a "representation" of B. If object B has sufficient stability, then the multiple effects produced by the incidence of energy from A will tend to converge on each other and give the beings the impression that they are experiencing a macro-object. In reality the beings are not experiencing B directly at all but the *effects* produced by B when it is in causal-interaction-L with A.

If what the beings have experienced is the effect of an *interaction,* then this cannot be a direct experience of the fundamental properties of the objects involved in the interaction. The situation could be likened to a company whose office communicates with the outside world only by fax and never allows anyone to enter the office and see its interior at first hand. An outsider's view of the office will be completely in terms of faxes, and these may or may not reveal something accurate and profound about the inner workings of the office.

This takes us to a point that indicates how the beings can be confronted with empirical evidence that could easily lead them to develop paradoxical descriptions of the world they live in. The objects in the imaginary world *have a reality that goes beyond what the beings can know of them by just taking their effects into account.* This is a plausible claim for the imaginary world, and indeed, for any world. The effect-based perspective is really a view of how things *impinge* on the beings. It cannot be a view of how those things are in themselves apart from their impinging. To assert that there is *more* to reality than can be accessed from the effect-based perspective seems reasonable. Otherwise we would be asserting that things have no reality apart from what can be experienced of them. The approach we are taking thus avoids the horns of a dilemma. We are not claiming that the beings, with their limited sensory capacity, have the ability to know reality fully (naive knowledge claim). Neither are we reducing reality to what can be known with such an imperfect sensory capacity (limited reality claim). Instead, we are emphasizing the partial nature of the beings' knowledge, along with the overwhelming likelihood that reality transcends what the beings can know of it (limited knowledge/transcendent reality claim).

If the beings' notions of the properties of objects are mediated completely by effects, then these properties could be described as *effect-properties* or *manifested-properties* (as opposed to the much more significant effect-*generating* properties of the realities in question). Such manifested-properties form the basis of the beings' *spatial* viewpoint on the world, which does not, and *cannot*, penetrate to the full story of the causal dynamics of the world. Having said that, the effect-based spatial viewpoint that the beings enjoy is *not* an illusory view of the world, nor a subjective representation of reality. It is an accurate and reliable description of the world from the point of view of the *momentary effects produced by interactions* between entities. That being said, any account of reality that has to rely totally on effects does have a marked *incompleteness* and will face grave difficulties in accessing the effect-generating properties of a given system.

In the analogy, the causal interactions between objects are not directly apprehensible, but their *collateral effects* do manage to impinge on the sense organ of the beings. On the basis of these collateral effects, the spatial viewpoint on the world is constructed. In our own world, the container approach tends to view the separability of physical states of affairs (i.e., distinct locations of objects in space) as the *ontological basis for the individuation of physical bodies* and the attribution of unique properties to them. But what if the spatial aspect of objects in our world is an edifice constructed on collateral effects, as was the case in the analogical world? If this is so, then it would be a grave mistake to assume that the spatial aspect of objects (their position and velocity in space, their distances from other objects, etc.) provides us with comprehensive and unambiguous data for attributing causal properties to them.

The analogy helps to highlight the *naivety* with which we tend to treat the spatial properties of objects. There is an endemic inclination to look on spatial properties as if they were *inherent* properties of the objects in question, and a general lack of attention paid to the obscure ontological significance of the spatial elements that dominate our viewpoint on the world. What if the identity and individuality of objects is *not* grounded in their spatial separability from each other? What if their identity is grounded in something different, and it is on the *basis of the prior existence of this individuality* that bodies, when they begin to interact with each other, take on a spatial aspect?

If spatial separability is not the basis for ontological individuation of objects, then what is? This is not a question that needs to be immediately addressed in order to be able to assume that *spatiality arises out of a prior ontological grounding and not vice-versa.* Eminent thinkers in the past have had good reasons for not granting objective ontological significance to spatial terms, and we intend to follow suit. If, as a result, we manage to develop a coherent realist interpretation of atomic phenomena, then we will consider that to constitute acceptable philosophical justification for our assumption.

2.6 The nature of the spatial perspective is related to the nature of the causal relation that the spatial perspective is built upon

In the naively-realist ontology of the container view, the structure of our spatial perspective reflects the structure of the world. Objects just *are* located in space; their extension in space *is* the true mode of their substantial being (even if the magnitude of that extension will depend on the reference frame of the observer who measures it); the void that we perceive between objects *is* a genuine void, with an independent existence and mode of being all of its own. In a challenge to this fairly naive view, we wish to affirm that our spatial perspective on things *does* contain elements that genuinely reflect aspects of material reality, but our perspective is gravely dependent on the way that objects causally impinge on our senses.

Far from being a reliable representation of the physical world (as is too-readily assumed in the container conception), the spatial perspective suffers from a number of major deficiencies. Firstly it is based on the *effects* of a causal relation between the world and our sensory apparatus. Thus it has no access - and no sophistication of instrumentality can give it access - to the real nature (or effect-generating properties) of objects themselves. Secondly, the structure of the perspective, as just stated, is fatally dependent on the *nature of the particular causal relation that underpins the perspective*. To see how this is so, consider again for a moment the illustration used earlier of a company headquarters and its various branches. Say that - from the point of view of the command structure of the company - there are very close causal links between the headquarters and each branch. When an order is issued from HQ to a particular branch, it is carried out with immediate effect. In other respects, the causal interactions between the divisions of the company are slower and less clearly defined. When goods are being moved between the various branches, for example, it can take considerable time.

Now imagine a being whose perceptual apparatus only gives him access to the effects produced in the branches by orders issued from HQ. He has no cognitive access to the geographical configuration of the company, nor to the ontological reality that constitutes the various parts of the company, nor to any other effect that might issue from the various causal relations that take place between the divisions of the company. All he can perceive are the effects that are generated by the command structure of the company. On the basis of these effects, the being's cognitive apparatus represents the company as being a very tightly connected conglomeration of objects, with HQ at the centre, and each of the branches arranged around it, in physical contact with HQ, but separated from each other.

If, instead, the being's perceptual apparatus had given him access only to the effects of movement of goods between the divisions of the company, then his cognitive apparatus would have represented the company in a different way. All of the various divisions would have had connections between them, and HQ would no longer have been represented as being at the centre.

The point of this analogy is to show that *the way that a reality is represented by the cognitive apparatus of a being will depend on the nature of the particular causal influences that the being's perceptual organs are stimulated by, and potential causal influences of this sort are multiple.* In our own world, it is the effects of what we call "electromagnetic radiation" that stimulate our perceptual organs in an overriding manner and dominate the way that we represent the world. *Other* causal influences are also operative between objects, and if our sense organs were stimulated by the effects of one of these other types of causal influence, then our picture of the world would be very different to what it is. Considerations such as these should prompt us to be less naive about the ontological significance of our spatial perspective on material reality.

We have called the causal relation that makes perception possible in the imaginary world "causal-interaction-L". Objects in the imaginary world have an identity and separability whose ontological foundation is not directly accessible to the beings' sensory apparatus. Remember, their direct experience is of the *effects* of interactions only. Object A enters into causal-interaction-L with objects B and C. The effects of these interactions impinge on the sensory apparatus of the beings and they generate a perspective in which A, B and C are separated in space. The beings' perspective on the individuality and separability of objects is attained through the lens of causal-interaction-L, but this is a very limited viewpoint on the actual *causal configuration* of the source objects themselves. The full causal configuration of the system is dependent on *all* of the causal influences present in the system. Causal-interaction-L gives a perspective on the system only as it appears from the point of view of the effects generated by that particular causal relation. Once the beings have attained this particular perspective on the configuration of ontological groundings in the system, then, by observing the movement of these groundings, they can make inferences about other causal relations such as gravity and magnetism. But the way in which objects appear in the perspective is dependent on the effects of causal-interaction-L, which becomes the lens by which all other phenomena are understood.

2.7 The partiality of the spatial viewpoint

We are claiming that objects have an individuality, separability and ontological grounding that is not reducible to their spatial aspect. The autonomy, separability and individuality of objects is not *founded* upon the ontological fact of having a particular location in space. *Rather their representation as having a particular location in space is founded upon ontological facts of a non-spatial sort.*

Objects have individuality and separability, and this identity is the basis of the possibility of causal interaction between them (a simple object does not causally interact with itself, but the parts of a composite object may interact with each other). If the world were not fragmented into separate individual objects then there could be no interaction at all. Given that a system is in a state of substantial disunity (in other words, the system is populated by various objects that have causal autonomy from each other), various types of causal interaction are possible between the various objects in the system. Think of an object that is causally active in four different ways: it emits light; it emits sounds; it emits particles; it pushes against other objects. Think of four types of being, each of which has a sense organ that is sensitive to the effects of only one of these four modes of causal interaction. Each being can potentially develop a representation of his world that is radically different to that developed by the others. The being that is only sensitive to sound could conceivably build up a three dimensional picture of reality based on the sounds that impinge on him. The resultant representation might be a world without colours, whose objects are represented as being spatially extended in proportion to the loudness of the impulse received by the sense organ. Such a picture of the world does not get at the substance of the world, but is a sort of skin built up of manifestations of objects from the point of view of a particular causal relation. But if this picture is the only viewpoint that the being has, then he might easily begin to think that the world of his representation *is the world as it really is*. But consider how different his representation of the world will be from each representation of the other three beings!

No matter how powerful a sense organ might appear to be, no matter how apparently regal the picture that it gives us of the world, *that picture cannot be more than a shadow of how things really are.* Consider a system that consists of a number of objects in relation to each other. Say that the particular system we are considering includes an observer. The observational apparatus of the observer (including measuring instruments) is itself an ontological grounding that is evolving with the rest of the system. It is a fact of the empirical conditions of observation for complex systems that a non-godlike observer will *not* have direct access to the overall causal configuration of the system, and, generally, will have a poor understanding of the causal dynamics of the system (which would require a long-term perspective on the evolution of the system). Given these limitations, it is a feature of the way that human observers try to comprehend their environment that their understanding is built upon *momentary* events, such as the collision of two ontological groundings with each other. To understand the complex dynamics that *led* to the collision would necessitate a long-term perspective and a grasp of factors that are not immediately accessible to the observer. But the *collision itself* is a public event that can be experienced by other observers and can become the basis of a shared "observer-neutral" perspective on the world.

If it is reasonably accurate to state that the human observational apparatus is geared towards registering *momentary* events in the causal evolution of the system in which one finds oneself, then it may be useful to make a distinction between the "snapshot viewpoint" and the "God's-eye viewpoint" on a system. The God's-eye viewpoint would be one in which the entire causal dynamics of the system is comprehended. Such a viewpoint would have access to the dynamics of the system (such as the tendency towards equilibrium of properties) from the moment the system came come into being until it had fully evolved as a result of the causal dynamics at work in the system. The snapshot perspective, instead, suffers from the limitation that its vantage point is from *within the system itself.* Furthermore, it is based on the effects that the causal dynamics of the system is having *in this moment* on the ontological grounding of the observational apparatus of the observer. In such a momentary perspective, *the bare ontological groundings of the objects in the system stand out in featureless spatial relations with each other.* Let us see in more detail how the spatial configuration stands out in relief once the system is apprehended in the snapshot perspective.

The diverse ontological groundings stand in complex natural relations to each other, and these relations are dependent on the overall natural dynamics of the system. We have not tried to illustrate the nature of these relationships, but we will dwell for a moment now on a (purely imaginary) way in which a complex natural relationship between objects could, under certain conditions, take on a spatial aspect. Take the case of three magnetic objects, A, B and C, that are all of identical size. Say that all three objects adhere together as a result of their magnetic attraction. The only thing that prevents direct contact between the objects is a thin film of plastic that separates each object but does not block the magnetic interaction. Say that A is the most strongly magnetized object of the three, with C being the least magnetized. Imagine an observer whose sensory apparatus is sensitive only to stimulation by a peculiar kind of microscopic particle. When such particles are emitted by a source and strike a magnetic object (such as A, B or C), they can then be reflected into the sensory apparatus of the observer. When they stimulate this sensory apparatus, they provoke within the observer a representation of the object they have just struck, and the nature of that representation will be related to the magnetic moment of the object from which they have just come. A particle that, for instance, reflects off A before entering the sensory apparatus of the observer, will provoke a representation of a *larger* object than will be the case for B or C, since A has a larger magnetic moment. If the particle is reflected off a non-magnetic body, however, such as the plastic film that is located between the objects, then the particle will provoke a representation of a void, and the magnitude of the void will be related to the degree to which the object resists being magnetized. The observer's impression of A, B and C will thus be of three different sized objects spread evenly out in space, with empty spaces between them, A being the largest in size, followed by B and then C.

Evidently this is an artificial illustration of our point, and it only shows how an arrangement of objects that is already spatial (A, B and C arranged contiguously) can lead to a *more extended* spatial impression of them on the part of an observer. However the main feature of this analogy that we wish to highlight is that a *magnetic* property (pertaining to the particular causal dynamics of a system) can conceivably be registered as a *spatial* property in the observer's spatial configuration of things. Objects with *greater* magnetic moment are registered as objects that are *more substantial* in size, whilst greater resistance to magnetism is registered as a greater extension of void. In the same way, we claim, causal relations that are *not in themselves spatial* can appear as spatial in our viewpoint.

This talk of "appearance" and "impressions" might lead one to think that spatiality is to be considered a purely *phenomenal* aspect of the observer's perspective on the causal configuration of things. Whilst not denying that there have to be significant phenomenal features present in any observer's representation of his environment, we do not wish to give the impression that spatiality is to be considered a subjective feature of cognition. The spatial relations between objects, instead, are to be considered as registering an objective feature of the relations between objects. In the analogy just presented, spatial extension accurately represented magnetic moment insofar as the *dimensions* of the spatial extension of the represented objects was proportional to the *magnitude* of their magnitude moment. In this sense, the spatial representation was based on objective characteristics of reality. This takes us to a central feature of spatial extension. Spatial relations can be described as "featureless" relations between objects. A purely spatial relation between objects does not specify any relationship or connection between objects apart from the spatial distance that separates them. But this means that an observer's conception of space *does* have one feature, and that is extension. If our approach to understanding spatiality is correct, then spatial *voids* between objects reflect the *magnitude of a natural disunity* between objects that is *not* simply spatial in nature (as we shall see later, the magnitude of spatial extension represented by an observer will also depend on the observer's relative *motion* in the system). The spatial *extension* of a substantial object, by the same token, reflects the magnitude of the number of basic atomic units (in proportion to their density) that are *united together* to compose this macro-object.

To make this last point in a simple way, consider a system composed of three objects, A, B and C, and an observer. Say that at time t the observer performs an observation and registers A as being twice as far away from C as from B. According to our way of describing systems, we would say that there is some ontological disunity in the system that, at time t, gives rise to a configuration of the system in which the ontological groundings of A, B and C stand in certain relations to each other. The nature of these relationships at time t is such that there is twice as much of a certain disunity between A and C as is the case between A and B. Based on the effects generated by the particular causal relation which makes perception possible, the observer registers these disunities as spatial gaps. This representation is objectively accurate insofar as the disunities *do* reflect the state of a certain sort of ontological fragmentation of the substances of A, B and C, and these disunities do have relative *magnitudes* that are proportional to the spatial extension of the gaps. The spatial aspect of objects, thus, can be said to arise when objects are registered by an observer as bare ontological groundings in relation to each other, and these relations are experienced only from the point of view of their relative *extent.*

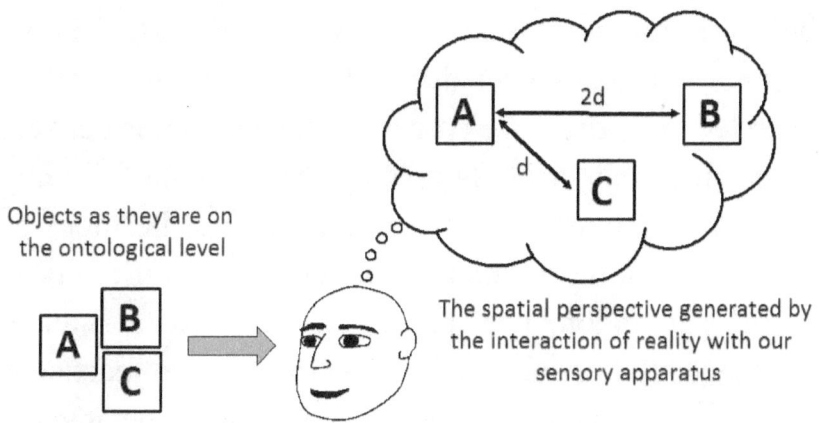

Objects as they are on the ontological level

The spatial perspective generated by the interaction of reality with our sensory apparatus

Figure 2.5 *The generation of the spatial perspective*

Figure 2.5 represents the generation of the spatial perspective in simple terms. At the ontological level, objects A, B and C stand in complex relations to each other. On the basis of these relationships, certain causal interactions take place between the objects. From the point of view of one type of dominant causal relationship, A and C are twice as much ontologically fragmented as are A and B. This happens to be the causal relationship that gives rise to the particular interaction that generates the effects that impinge on our sensory apparatus. These effects in turn generate the spatial perspective. According to the spatial perspective, A and C are represented as having twice the spatial separation as A and B. Thus the *extent* of the relative ontological gaps between the objects (as experienced by an observer with a particular motion relative to the system) is faithfully represented by the spatial viewpoint. We have absolutely no reason for characterising that viewpoint in idealistic or anti-realist terms. There is nothing subjective about this representation of the world. Nevertheless we must not fail to acknowledge that the spatial viewpoint is generated by a *particular* type of causal interaction between objects. It simply does not give us access to the full range of causal relationships between objects. Furthermore we must not *reify* the relations that form the framework of sensory perception. The spatial gaps between A, B and C may indeed tell us something about the *extent* of particular ontological fragmentations between the objects, but it would be a cardinal error to assume that space itself existed in an autonomous way.

We are so accustomed to reifying space that we do not realize the flagrant nature of this error. Imagine a creature whose only form of sensory perception is aural. He hears three notes, A, B and C, and his perceptual apparatus registers the fact that the difference in pitch between

A and C is twice that between A and B. As a result, a spatial perspective is generated in which objects A and C are represented at twice the distance as A and B. The creature cannot be faulted for the limitations of his sensory apparatus, yet at some point in the history of his species we would hope that he would come to the awareness that the contents of the world transcend the contents of his spatial perspective. But if he does come to this awareness, it is also important that he not dismiss the spatial perspective as if it had no genuine connection with the properties of real things. The extent of the spatial separations in his standpoint *does* reflect objective facts about the world (the extent of the differences in pitch between the notes), but the creature will construct a defective physics if he strives to base his understanding of the world on the notion of objects having position in autonomous container space.

The reader will object that *our* situation is different. The space that we experience *must* be real! We can move in the void and measure its extent. This naïve response, however, is not a valid defence of the reification of space. Movement and measurement are simply causal interactions that are possible when there are ontological disunities between objects. They do not entail that the space that appears in our representation of these disunities has an autonomous existence. Differences in the pitches of notes can also be measured, but this does not mean that the corresponding spatial gaps represented by the being had independent reality. The spatial gaps that appear in the spatial perspective *are* symptomatic of genuine ontological fragmentations in reality. In these ontological gaps, causal interactions naturally occur. When we move in "space" we are simply undergoing a series of complex causal interactions with the objects that surround us. Our movement occurs in the ontological gap, but we have no reason to believe that the real nature of the gap is spatial, no more than the spatial gaps perceived by the being in our analogy had autonomous existence.

The space between objects is a feature of our way of representing material reality. We make the elementary error of believing that the space is real, and then we ask ourselves how a causal influence can traverse this space. How can a magnet affect a piece of ferromagnetic material located at a distance, we ask? This prompts us to introduce notions like waves, lines of force and fields. Classical mechanics, in fact, was dominated by the question of how bodies could influence each other by contact action. Action at a distance was considered inadmissible, therefore a causal intermediary was required to carry the influence across empty space. Lines of force and fields represented a solution to this quandary in situations where material intermediaries like particles seemed to be lacking. Beginning in the seventeenth century a grand project was initiated (which continues today in a more sophisticated form) that seeks to reduce all natural phenomena to a description in terms of matter moving in space. In other parts of this book, we have entitled this programme the "matter-in-motion project". But if the spatial gap between objects is merely a feature of our way of representing an ontological separation of *another sort*, then this project is misdirected from the start. Concepts invoked to explain how ontological gaps are bridged - such as fields, lines of force and space-time - become redundant.

2.8 The transmission of light through the causal configuration

Armed with these considerations of the partiality of the spatial perspective, we return to the issue of the emission and absorption of light by the atoms in a gas. The B pole of each unit has a tendency to collapse into the V poles of other units, but this tendency is resisted by the B poles of each of those units. This means that the atoms converge on each other to a certain extent and then form an electrostatic bond at that distance. Now consider a system of gas that is composed of atoms in these stable states. An external causal influence is then exerted on the system, such as the application of heat. The exact mechanics of this influence is not important for the moment, but let us examine what happens when such an external influence causes a rupture in the electrostatic bond between two atoms in the gas.

Before the rupture, this electrostatic bond contributed to the relative equilibrium of the system at a particular moment in time. The reader is asked to imagine the system from the perspective of the totality of the electrostatic relationships between the genesis-units that compose the system. These invisible bonds, in our framework, require no fields or medium to transmit their influence. Recall our discussion of spatial separation in the preceding sections: our picture of the spatial separation between objects is a feature of the way that our sensory apparatus generates our perspective on the world. If two objects appear to us as being spatially separated, then we can take it that there *is* a division of some sort between these objects, and the extent to which they appear separated *is* a relative measure of the extent of this division. But it would be naïve of us to think that the appearance of *spatial void* that is generated by our sensory apparatus is actually a feature of the world. Consequently the fields and lines of force that we think are necessary to propagate an electrostatic or magnetic influence through this spatial gap are superfluous, at least as far as this sort of spatial separation is concerned.

On our view, then, the spatial gap that we represent as prevailing between objects is symptomatic of a different sort of "gap" between objects, and we have no reason whatsoever for thinking that lines of force or fields (which are all envisaged as permeating *space*) are necessary for the propagation of influence across this ontological gap. What we are left with is simply the electrostatic tensions between objects. The totality of the electrostatic tensions between all objects constitutes the electrostatic equilibrium of the system. For our framework, this property - the electrostatic equilibrium of the system - takes the place of all talk of luminiferous ether and the fields of force in Maxwell's system.

Let us return to the rupture that occurs when the electrostatic bond between two atoms is broken. This rupture is relevant for the electrostatic equilibrium of the system, but it does not impinge on the totality of the system in an instantaneous way. Rather, it evolves through the system in a particular direction, beginning from the point at which the rupture occurred. We can envisage this evolution like a concentrated ripple progressing through the system in a single direction. We have no further reason for believing that it requires a medium for its evolution, since we are now aware that the old gap that we once thought to be filled with the old ether is a feature of our mode of representation, and this mode of representation doesn't tell the full story of how things are in themselves.

Terms like "rippling through the system" should not be interpreted to signify some sort of wave motion through the system. This evolution is to be thought of purely in terms of causal influences exerted by concrete objects (genesis-units) on concrete objects, and nothing more.

Let us step back a moment and see where this is going. We have a system composed of genesis-units and only genesis-units. These units are in causal interaction with each other. We (as a consequence of our mode of sensory perception) represent these units as inhabiting *space*, but the impressions of spatial separation and spatial extension are merely generated by the *non-spatial properties* of causal interactions between genesis-units. If a God's-eye viewpoint on the system were possible, then we would see the genesis-units standing in the totality of their true causal relationships with each other, and this viewpoint would not simply be in spatial terms. Now, since it is authentic understanding of the system that we seek (and not just empirical descriptions that save the phenomena), we must ask ourselves a basic question: how should we describe the evolution of a causal influence through the system: in comfortable spatial terms that utilize fictitious notions such as fields or waves? Or in less comfortable terms that seek to comprehend what is happening from a God's-eye viewpoint? The fact that it is impossible for us to attain a God's-eye viewpoint does not mean that we cannot purify our representation as much as possible from perspectival features.

In what follows, we attempt to describe the transmission of light from this broader standpoint. To reiterate: we can only understand light from the point of view of the evolution of the system as a whole. If we seek to understand it from the *relative* standpoint of *particular* observers whose representation is couched in *spatial* terms, then our description will be, to the same measure, incomplete. The electrostatic bond between two genesis-units is broken and this has repercussions for the wider system. What repercussions does it have? What dictates which genesis-units will be affected by those repercussions? The causal influence that arises from the breakage of the bond is not simply an influence that originates in the breakage of a *particular* state of electrostatic equilibrium between two genesis-units. From a God's-eye viewpoint, the electrostatic equilibrium of the *system as a whole* is something that is in a constant state of evolution, and is the sum total of all the particular bonds, attractions and repulsions between individual genesis-units. When any bond is broken, the entire system evolves towards a resolution, together with *all of the other* electrostatic dynamics that are playing out in the system in that moment.

Take the breakage of the bond between atoms A1 and A2 at time t_1. A causal influence emanates through the system from this point at that moment. At time t_2, the causal influence will have evolved to a part of the system immediately contiguous to the location of the original bond. Here we are using spatial terms, because it is impossible for human beings to describe the world in any other terms. But we must reiterate that "contiguity" is not intended to refer to a *spatial* location beside the *spatial* location of the original bond; rather it refers to a point in the *causal configuration* of things. On the ontological level, things have an inherent separability. Certain causal interactions between objects make us generate a perspective where this separability is represented as spatial separation. Even if the spatial separation is a feature of our way of representing things, those things *still* have some ontological properties that separate them from each other. The evolution of the causal impulse that emanates from the broken electrostatic bond must be thought of as moving through *this* ontological substrate, even if we have difficulty comprehending it in non-spatial terms.

Let us use the term c_1 to designate the location in the causal configuration of the original bond. At time t_2, the influence has emanated to point c_2. If there are other genesis-units (call them Ti, Tj, . .) located at this point, and if certain other conditions are fulfilled, then these genesis-units can be the target of the causal impulse. We would normally call this event the "absorption of light". In reality what is happening is that the electrostatic disequilibrium introduced into the system by the breakage of the bond between A1 and A2 is "resolved" by the change that is now wrought in units Ti, Tj, . . . But what are the conditions that must be fulfilled for this causal process to come to a conclusion, and how do these genesis-units "absorb" the electrostatic disequilibrium caused by the breakage of the original bond?

2.9 The Rydberg formula and the Bohr atom

Earlier, we considered how Bohr's original planetary model of the atom raised questions because it did not explain the stability of matter. Orbiting electrons ought to irradiate energy and spiral into the nucleus. Bohr's tactic of introducing the notion of stationary states (electrons that occupied orbitals without orbiting) would have seemed an unacceptable ad-hoc strategy to save the model if these states did not have the added appeal of providing an explanation for the origin of the emission spectra of elements. The notion of quantized states was able to account for the well-defined lines that characterize atomic spectra. If a sample of hydrogen, for example, is placed in a spectrometer and exposed to white light (light of many different wavelengths), it will absorb most of the various wavelengths and emit light in narrow, well-defined bands of wavelengths, or "lines". The claim that the electron inside the hydrogen atom could only exist in certain quantized orbits was an attempt to explain this fact. According to this idea, an electron in a given orbit would absorb the light that was incident on the atom; the energy gained by absorbing this energy would cause the particle to "jump" to one of the higher permitted states; after some time had elapsed, the electron would fall down to a lower state, emitting radiation of the wavelength that corresponded to the energy difference between the two stationary states. The stationary states were of fixed quantized energy levels, so the difference in energy between these orbitals was also fixed and corresponded to one of the wavelengths of light that appeared in the characteristic atomic emission spectrum.

The fact that Bohr's model was able to account for these spectral lines *quantitatively* was an enormous feather in its cap. Balmer had already discovered that the visible emission spectrum of hydrogen could be described by a stunningly simple formula consisting of a constant and a series of integers.

$$\lambda = B \ m^2/(m^2-2^2)$$

where λ is one of the wavelengths of light in the hydrogen spectrum, B is the constant discovered by Balmer, and m is any integer greater than 2. By inserting different values for m in the formula, the visible spectrum for hydrogen is produced. A few years later, Rydberg generalised this formula so that it could also predict the non-visible wavelengths in the hydrogen spectrum, as well as the spectra of other hydrogen-like elements. The generalised Rydberg formula for hydrogen has the form:

$$1/\lambda = R \ /(n_1^2-n_2^2)$$

where λ is the wavelength of light emitted/absorbed, R is the Rydberg constant, n_1 and n_2 are integers, and n_1 is less than n_2.

This was given a theoretical underpinning by the Bohr model. When a hydrogen atom was stimulated by light of the correct frequency, the electron in the atom (according to the model) absorbed the photon of light and "jumped" to a higher energy level. This transition was typically short lived, and the electron would later drop to a lower level, emitting a photon of light. The different values for n in the formula corresponded to the different energy levels (or "shells") occupied by the electron, and the Rydberg formula gave an expression for the energy of the photon of light emitted by the atom when the electron made this transition between levels. This simple account of the origin of spectral lines was sufficient to make many physicists overcome their scruples about the implausible nature of stationary states and the quantum jumping of electrons between orbits.

2.10 The physical basis of the Rydberg formula

Can the genesis-unit model account for the Rydberg expression in a less ad-hoc manner? Let us return to our picture of the transmission of light. The evolution of the impulse that we call "light" was emitted when genesis-units A1 and A2 had their electrostatic bond fragmented at time t_1. The fundamental nature of this impulse is its potential to establish electrostatic bonds between the atoms that it encounters as it evolves through the system. At some future time, t_i, the impulse has evolved to point c_2 where genesis-units T1 and T2 are located. As we shall see shortly, it is not just likely but *probable* within any body of gaseous material that many similar impulses will arrive at c_2 in the same instant. Now consider what happens if three such impulses arrive simultaneously at that point. These impulses were generated by the breakup of the electrostatic bonds between genesis-units and they have the tendency to *reinstate a relative level of electrostatic equilibrium in the system by bringing target genesis-units into electrostatic bonds.* In a sense, it looks like we are proposing a sort of "law of conservation of electrostatic equilibrium" for the system, akin to the law of conservation of energy, but that is not quite accurate. In the transmission of the impulses we call "light", there is indeed a tendency for a broken state of electrostatic equilibrium to evolve until it is reincarnated elsewhere in the system. But the overall tendency of the system is towards *greater* electrostatic equilibrium and the total quantity of equilibrium is certainly not conserved.

So this is our approach to understanding light: atoms in a state of electrostatic equilibrium are (for some reason – and the reasons can be various) pulled apart; the pulling apart and the consequent disequilibrium introduced into the system causes a reactive "pulling together" in the system, a causal evolution that has a particular direction and will pull atoms into alignment with each other, *if* suitable target atoms are positioned correctly. How do the three impulses we were discussing earlier "reincarnate" at c_2 the electrostatic equilibrium that was broken when the three pairs of genesis-units broke up and the impulses were originally emitted? Clearly, T1 is already in an electrostatic bond with T2 (consider again Figure 2.2 which shows two genesis-units in a state of equilibrium) without the assistance of any external impulses towards equilibrium. What happens, then, is that the T1-T2 bond is the target of the three incoming impulses and, as a result, T1 and T2 enter into bonds with other genesis-units in the vicinity. In fact, it would be better to state that it is the *electrostatic relationship* between T1 and T2 that is really affected by the incoming impulses of light. Once these impulses are "absorbed", T1 and T2 attain a greater electrostatic capacity – they are able to draw other atoms to themselves and hold them in equilibrium. In this case, three impulses of light are incident on T1 and T2: therefore they will tend to draw three atoms each to themselves, as depicted in Figure 2.6.

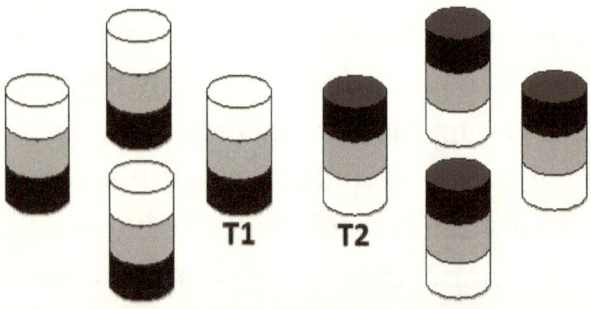

Figure 2.6. Intromission of light gives rise to a group consisting of two pairs of four units each. The electrostatic bond between T1 and T2 is natural and derives from the opposite alignment of their polarities. When three impulses of light are incident on the system, T1 and T2 each gain triple their normal electrostatic capacity. As a result, three atoms each are drawn to T1 and T2.

It is important to emphasize here that we do not accept that the "absorption" of light involves the *internal* excitation of the atom. We reject out of hand the view that genesis-units can hold differing amounts of energy by virtue of changes in their internal constitution. In our framework, each genesis-unit has the same natural capacity to do the same magnitude of work. Sometimes, as a result of the relationship between a genesis-unit and others in the system, the genesis-unit can acquire a heightened capacity to do work. But this is a function of the arrangement of the system, not a change in the internal structure of the genesis-unit.

Let us reflect for a moment on how an impulse of light moves through the system. Consider the atoms, T1 and T2, which are the target of an impulse of light. A system of atoms disassociates in the source material, and the electrostatic tension holding the system together is released. This tension doesn't "traverse space" between the source material and T1-T2 in the target material. The source material and the target material, along with the rest of reality, evolve together on that ontological level which is not apprehensible to sense organs as limited as ours. After a certain time has elapsed, the evolution of the entire system results in the formation of a mini-system of atoms held to T1-T2 by that same magnitude of tension. It doesn't have to have this result, mind you. If the magnitude of the tension is not exactly right, then the evolution of the system could result in the "reflection", "refraction" or "diffraction" of the causal process towards a different final result. But we are considering the case where T1-T2 is indeed thrown by this evolving causal process into a state of electrostatic alignment with a system of atoms. The tension that was released in the source material becomes centred on T1-T2 so that T1-T2 becomes the locus of the new subsystem that is to be formed.

When suitable light impulses evolve to the point where the T1-T2 pair is located, T1 and T2 each gains the capacity to hold other atoms in electrostatic equilibrium with themselves. Looking again at Figure 2.6, T1 can be thought of as the anchor that holds in place T2 and the other three atoms that are gathered around T2. Likewise, T2 is exerting its augmented electrostatic capacity in holding T1 and its square formation of atoms in place. This will be an important point later when we examine the meaning of the Rydberg expression: *the electrostatic influence of T1 must be calculated as being distributed over the four atoms that includes T2, and vice-versa.*

If the groups that attach to T1-T2 are always composed of atoms arranged in *squares* (i.e., groups of 4, 9, 16, 25, 36, etc., atoms), and if we consider the impulse of light that is absorbed/emitted for each *atom* in the group, then we can see how the Rydberg expression arises. Take the case where T1-T2 are each holding four atoms in tension, and then absorb light so that they transition to two pairs of nine atoms each (see Figure 2.7). The Rydberg expression is as follows:

$$1/\lambda = E \,/(n_1^2 - n_2^2)$$

(here we have replaced the Rydberg constant, "R", with "E" because we have designated E as the magnitude of the electrostatic bond between any two atoms). As we understand the expression, $n_1^2 - n_2^2$ expresses the transition from/to a square group of one size to/from a square group of a larger size, where n_1^2 represents the size of smaller group of atoms, and $n_2{}^2$ represents the larger group (in our example $n_1{}^2 = 2^2$ and $n_2{}^2 = 3^2$).

Some of these wavelengths, λ, will satisfy the expression $1/\lambda = E(1/2^2 - 1/3^2)$. According to our way of understanding the physical basis of the formula, however, T1-T2 cannot absorb just *one* of these impulses. It must absorb four simultaneously (or a single impulse of $4E(1/2^2 - 1/3^2)$). And in addition it must absorb five *other* impulses of a different magnitude. Let us see how this works in more detail.

Originally T1-T2 held four atoms each, so their natural electrostatic component, E, was distributed over each group of four, with each atom receiving E/4 from either T1 or T2. But E/4 is not sufficient to enable T1-T2 to hold these groups of four atoms in electrostatic equilibrium. It is the increased electrostatic capacity that T1-T2 has gained with respect to the system by the absorption of "light" from without that enables the pair to hold the groups of four to themselves. Thus the absorbed light contributes E–E/4 towards the holding of each of these six atoms in tension with T1-T2.

Now consider what happens when T1-T2 is the target of more light, causing them to each hold *nine* atoms in equilibrium (remember T1 should be thought of as holding the group of nine atoms that includes T2, and vice-versa). So the natural electrostatic component of T1 is now distributed (with a magnitude of E/9) over each atom in the group that includes T2, and the natural electrostatic component of T2 is distributed over each atom in the group that includes T1. This component is in addition to the total absorbed tension that the group of atoms has received from without and which gives T1-T2 the capacity to exist in this state of heightened tension with the groups of nine. The total tension in the system of twin nines is the sum of the light *already* held by the excited pair when it held the groups of four, PLUS the *new* light that has been absorbed to form the groups of nine. Thus the total absorbed tension amounts to $E-E/9$ for each atom in the two groups.

So what are the individual impulses needed to transform T1 from holding a group of four to holding a group of nine? The *final* quantity of absorbed light for each atom (once the group of nine has been formed) is represented by $(E-E/9)$, whilst $(E-E/4)$ represents the original light that had been absorbed by the subsystem of four for each atom in the group. Thus, for each of the *four original atoms* in the group, T1 needs an extra impulse of $(E-E/9) - (E-E/4)$.

$$(E-E/9) - (E-E/4) = E(1/2^2 - 1/3^2)$$

Four of these impulses will have to be absorbed, one impulse for each atom in the original group of four (or, alternatively, one large impulse four times the magnitude that is then distributed over the group of four).

In addition, T1-T2 must absorb extra light in order to pull the pair of *five new atoms* that get pulled into the group. Once the group of nine is formed, each one of the group is held to T1 by a tension of E, of which E/9 is contributed by T1 itself. Thus five impulses of magnitude $(E-E/9)$ must be absorbed to form the new group of nine (or a single impulse five times that magnitude). $(E-E/9)$ can be rewritten as the expression $E(1/1^2-1/3^2)$, which of course satisfies the form of the Rydberg formula.

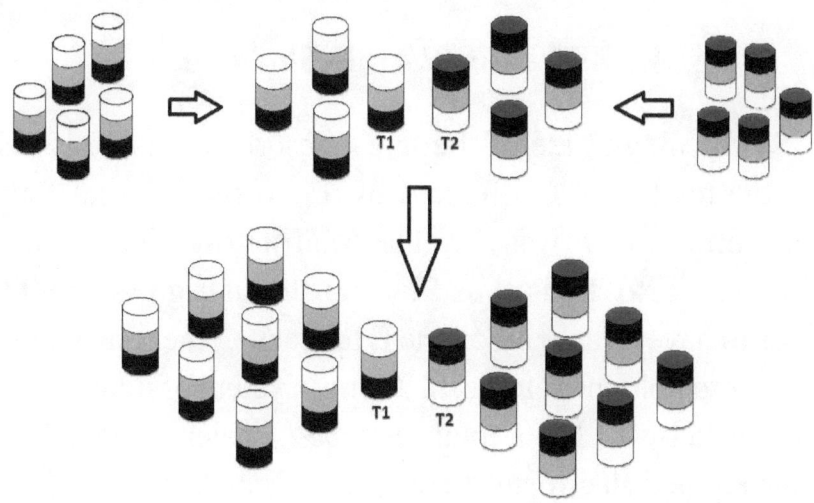

Figure 2.7 A pair of four atoms absorbs light and transitions to a pair of nines

Let us double check to see if these figures tally. When T1 was holding its group of four atoms, it contributed a total of E to the tension of the mini-system (E/4 distributed over each of the four atoms). That means the original *absorbed* tension must have been 3E, making a total tension of 4E being directed from T1 to the rest of the mini-system. This makes sense: if T1 is to hold these atoms in equilibrium with itself then it needs a capacity of E for every atom it holds in tension, and that makes 4E in total.

Later the mini-system absorbs the nine impulses detailed above and it ends up consisting of nine atoms in tension. The nine absorbed impulses are as follows:

$$4[(E{-}E/9) - (E{-}E/4)] + 5(E{-}E/9).$$

This works out as:

$$4E{-}4E/9{-}4E{+}4E/4{+}5E{-}5E/9 = 5E$$

So the magnitude of the latest absorbed tension is 5E and the total tension directed by T1 to the original four atom mini-system was 4E. The total tension holding the mini-system to T1 is thus 9E, and this is right since T1 needs to have a capacity of one E for each of the nine atoms in order to hold them to itself. A mini-system of nine atoms being held by T1 has a total *absorbed* tension of 8E allied to the *natural* electrostatic component E of T1 itself. All of this applies also to T2's relationship with the nine atoms that it is holding in equilibrium.

Thus we have an explanation for the form of the Rydberg expression: the absorption/emission of an impulse of light occurs when an electrostatic bond between atoms is formed/broken; as a result of the absorption/emission of the impulse, groups of atoms either associate in larger square formations, or disassociate into smaller square formations; when such a group of size n_1^2 transitions to a larger group of size n_2^2 (or vice-versa), the magnitude of the impulse absorbed/emitted satisfies the Rydberg expression $(n_1^2{-}n_2^2)$.

The question that must be answered, however, is why the magnitudes of *all* the light impulses on the hydrogen emission spectrum can be described by this formula. In other words, why is it that the groups of atoms that form around T1 and T2 always have these *square* formations?

2.11 Square formations of atoms possess equilibrium

To see why atoms in square formations possess equilibrium, let us return for a moment to the discussion of the point of proximity at which two genesis-units enter into electrostatic equilibrium. In Figure 2.8, A1 and A2 have converged within a range where they exert net attraction on each other. The atoms are aligned in opposite directions. This means that the discrepancy in the lengths of *d* and *e* will ensure that the mutual attraction between the unlike poles is greater than the repulsion between the likes. At some *greater* distance ($e + x$), A1 and A2 would have *first* begun to exert attraction on each other. At this point (if they were free to do so), the units would have rearranged themselves so that their opposite poles were aligned. As a result of the mutual attraction, the genesis-units would have converged to the distance, *e*, but no further, since any greater proximity will lead to net repulsion, for reasons we discussed earlier. So the scenario represented in Figure 2.8 represents the situation that prevails at the *end* of the process in which A1 and A2 exert attraction on each other, converge, and then remain static at the point where further proximity will lead to net repulsion.

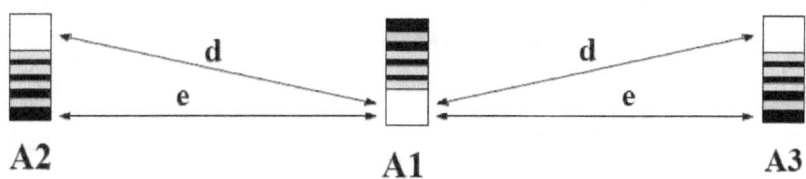

A2　　　　　　　　　　　**A1**　　　　　　　　　　　**A3**

Figure 2.8 A1 has an enormous polarity at its heart and it thus exerts electrostatic influence in all directions.

Now genesis-unit A3 enters into the scene. A1 will *still* exert electrostatic influence in this direction. In theory, its interaction with A2 should lessen its capacity to affect other atoms, but that capacity is certainly not exhausted by the bond with A2. We must keep in mind that the magnitude of the polarity at the heart of the genesis-unit (the "rest energy" of the atom) is many times greater than the influence felt at distance *e* (i.e., at the point when atoms typically settle into electrostatic equilibrium). Thus A1 has this enormous polarity within it and it is natural that it should exert an electrostatic influence in *all* directions.

The impulse that we call "light", however, has an electrostatic capacity of a greatly inferior sort and with a *single* associated direction. This follows from the fact that it arises from the fragmentation of a simple bond between two genesis-units. This bond has a magnitude equal to the electrostatic attraction that is felt between the two units at this distance, and it has a *single direction* corresponding to the exact alignment of A1 and A2 with each other. As this impulse evolves through the system, it has the potential to create bonds of the *same magnitude* between genesis-units that have the *same alignment* as the source genesis-units (from this we already begin to see the real significance of the "polarisation" of light). In our example, T1 and T2 have the same alignment as the source atoms that originally emitted the impulse of light.

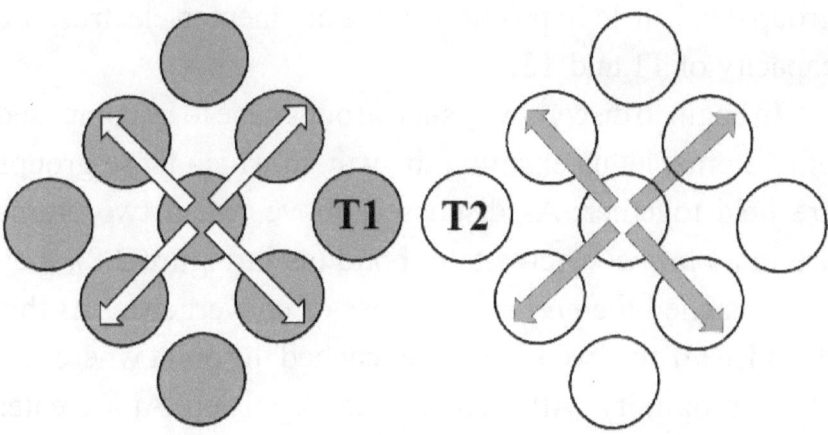

Figure 2.9 Group of atoms (viewed from above) held together by the intromission of light. *T1 and T2 are in a natural electrostatic bond with opposite poles aligned. The other atoms are drawn together by the electrostatic impulses that converge on the T1-T2 bond. T1 can now be thought of as having an augmented electrostatic capacity that permits it to hold in place T2 PLUS the other eight atoms to the right of T2. Similarly, T2 can be thought of as having the capacity to hold T1 and its nine atoms in equilibrium.*

Figure 2.9 shows a group of atoms (viewed from above) that have been brought together as a result of the intromission of light. T1 and T2 were already in a natural electrostatic bond before absorbing eight impulses each (or single influences of magnitude 8E). This light had evolved through the system until it encountered the electrostatic bond between T1 and T2. That union between T1 and T2 is capable of "absorbing" these impulses of magnitude 8E and the upshot is that the T1-T2 relationship behaves as if it had *nine times* the natural electrostatic capacity of a bond between a pair of genesis-units. As a consequence, eight other atoms are drawn to either side of the pair to form a group that is held in position by the augmented electrostatic capacity of T1 and T2.

To begin to discern why such groups have to be composed of n^2 atoms, let us examine the way in which these groups are held together. As discussed above, when two atoms are in a *natural* electrostatic bond (as in Figure 2.8), they have reached the *end* of a process of convergence. At this distance (e), the two units have reached the point where any closer proximity will cause mutual repulsion. At a greater distance ($e+x$), net attraction commenced, causing the units to draw together to distance e.

In Figure 2.9, the atoms that are held by the amplified electrostatic capacity of T1-T2 must be located at a minimum distance of $e+x$ from each other. If they were any closer together, there would be net repulsion because their like poles are aligned in the same direction. But at the distance $e+x$ the discrepancy in the distances between the like poles and the unlike poles is so relatively small that there is equal repulsion and attraction between the units. The lack of mutual repulsion permits the heightened electrostatic capacity of T1-T2 to hold this group in place.

Nevertheless, though there is no repulsion in the sense that the atoms are not actively pushed away from each other, still there is a *resistance to further convergence*. The net attraction - deriving from the incident impulses of light that are drawing the atoms together - coupled with the resistance to convergence, confers stability on the system, but, evidently, a square formation possesses an equilibrium that a non-square group lacks. If non-square groups possessed equilibrium then we would expect them to occur in nature all the time, which would mean that the emission spectrum of hydrogen should contain wavelengths that *do not fit* the Rydberg expression. But the Rydberg expression describes *all* the wavelengths in the hydrogen spectrum. Therefore, we assume that non-square groups do not occur. This is a case where an empirical formula can guide our investigation of what is happening at the causal level.

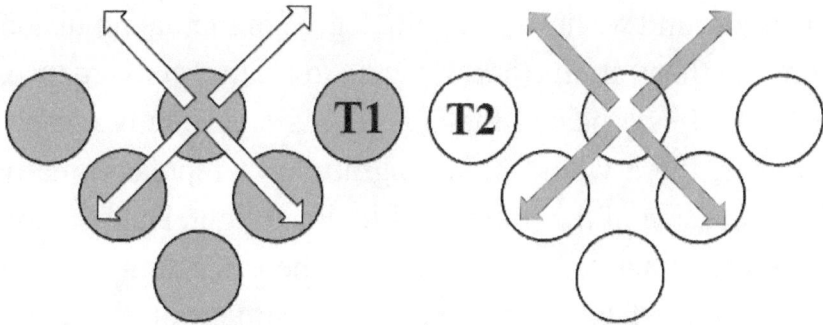

Figure 2.10 Unstable group of atoms. *Again, T1 and T2 are in a natural electrostatic bond with opposite poles aligned. This bond absorbs sufficient impulses of light to draw five atoms towards each side. The form of the Rydberg expression tells us that such a formation will lack equilibrium and will not hold together.*

The search for a complete answer to this question may help us to fill in some details of the structure of the genesis-unit and the way that its polarity is distributed. Already, however, we can imagine how a square group might have greater stability. Consider Figure 2.10. There is an electrostatic tension emanating from the bond between T1 and T2 which draws these two atoms together. The atoms themselves have their own attractive and repulsive influences on each other. As they are drawn together, they settle into mutual equilibrium as these various influences play themselves out.

Now the bond between T1 and T2 is the target of impulses of light, causing five units to be drawn to either side of the bond. These atoms are pulled together, but only as far as the point where mutual repulsion begins. So we have an attractive impulse emanating from T1 and T2 into each side of the dichotomy, just sufficient to hold five atoms to each side, and we have a certain degree of mutual repulsion between those atoms (or resistance to further convergence, at least). If we placed the five atoms sufficiently closely together, there would be a magnitude of repulsion many times the size of the impulse of incident light. But they are drawn together only to the point where the total mutual repulsion perfectly balances the magnitude of attraction provided by the five impulses of light.

As Figure 2.10 tries to express, there is a lack of balance in this formation that is not the case in the square formation of Figure 2.9. In square formations, each atom will be held in place by these reciprocal attractive and repulsive tendencies on two, three or even four sides. This is not true for all of the atoms in Figure 2.10. A full explanation of this phenomenon may be possible by formulating a more detailed account of the way the polarity is distributed in the genesis-unit. For the moment, we will take as empirically justified the assertion that the transmission of light involves the association and disassociation of square groups of atoms.

Even if we are not currently in a position to furnish a complete answer to the origin of the square formations, this approach to light still has theoretical advantages over the Bohr model. That model invoked a causal mechanism (discontinuous quantum jumping between energy levels) that contradicts universal empirical evidence. It attempted to account for the form of the Rydberg expression by claiming that matter can behave in a singular way that is at odds with all experience. With our model, by contrast, the mathematical expression can be satisfied by the simple transition of a larger group of atoms to a group of a smaller size. No suspension of continuity is required, no need for matter to disappear at one point and reappear instantaneously in a different place. Furthermore (and we will discuss this in more depth later), this account of the relationship between light and matter has a unity and simplicity that is lacking in the standard approach. Under our model, it is the polarity between the V and the B that constitutes matter. This fundamental relationship at the heart of the atom leads other atoms to enter into secondary relationships with each other, in which they are drawn to the opposite polarities of other atoms. These bonds, when broken, give rise to the evolution through the system of a causal impulse we call "light". Thus the polarity that constitutes matter is also the root cause of the evolving impulse towards electrostatic equilibrium that is light. There is no great mystery as to how light and matter interact since the same dynamic is foundational for both of them.

In what we have been doing so far, we have tried to present a model of the atom that is as simple as possible, yet has sufficient complexity to account for the wide diversity of atomic phenomena. As we go along, the model will be confronted with the different empirical manifestations of the atom. This will prompt modifications of the model so that it can account better for the experimental data. It might be thought that this strategy involves changing the model in an ad-hoc way so that it fits the evidence, but in fact our methodology is an attempt to develop a model that is less likely to require drastic modifications of an arbitrary sort as new evidence is presented. The key to this approach is to keep the picture of the atom as simple as possible, with the minimum of content that is not prompted by the data. Such a strategy aims to avoid the necessity of the sort of arbitrary additions that were made to atomic theory during the course of its early development. In that case, the *starting point* was a full-blown planetary model of the atom, which went way beyond the available evidence. When difficulties presented themselves, the broad outlines of the model were retained (massive central nucleus, multiple orbitals) but ad-hoc suppositions were made (stationary states, quantum jumping) that effectively contradicted those very outlines.

In our case, we proceed with the simplest possible picture that will account for the data that we wish to explain. The first phenomenon for explanation was the fact that atoms attract each other at a certain distance but repel upon closer convergence. This led us to postulate a picture of the genesis-unit in which the B component was concentrated at one end, whilst the V component was more dispersed. The second phenomenon was the transmission of light and the form of the Rydberg expression, leading to the claim that light involves the formation of square groups of atoms in electrostatic tension. This fact demands that we develop our picture of the atom to account better for these square formations. And we will return to this issue when we consider the movements of charged particles in magnetic fields.

2.12 A look at the way in which groups of atoms store and release "light"

Earlier we stated that when an atomic pair, T1-T2, is the target of impulses of light, it becomes capable of drawing other atoms into electrostatic equilibrium with itself. As the discussion above maintains, it is not the case that the protomagnetic components of T1 or T2 are internally changed or increased. That component remains as it was, generated by the work impulse at the heart of the atom, indivisible. It is the tension that has been absorbed from the external source that makes T1-T2 behave *as if* they had an increased protomagnetic component. But if the protomagnetic components of T1 and T2 do not in fact change as a result of the absorbed impulse, then where does this extra energy "reside" in the system once it has been absorbed? The answer is a familiar enough one and resonates with our understanding of the way that many different kinds of physical systems store "energy". The atoms on the surface of a liquid exist in a state of heightened tension that requires no increase in the internal energy of the individual atoms. This tension, which has the potential to do work, is not stored in *individual* atoms, but in the way that the *overall system* is configured. If I push the north poles of two magnets towards each other and then hold them in position, this mini-system stores energy even though the magnetic moment of the individual magnets is not changed in any way. Systems can increase their potential to do work simply by the re-arrangement of their constituent components. It is not at all necessary that these components themselves undergo an internal change.

The principal turn in this new approach to the emission and absorption of light is the claim that light transmission does not involve *internal* energisation/de-energisation of atoms. It involves, rather, a fundamental propensity at *the level of the system* to evolve towards greater electrostatic equilibrium. In the Bohr conception of the atom, the movement of the electron farther and nearer with respect to the nucleus was the "spring" that stored and released energy; in our model the system evolves in such a way as to give rise to the association/disassociation of atoms in smaller and larger groups, giving the atomic pair at the locus of the mini-system a greater or lesser work potential. The model that we are advocating is able to give a very natural and intuitively-plausible account of the "quantum jumps" in energy, a feature that is left as a brute fact of nature in the standard model. According to our view, the tension that is imparted in the transmission of light has a disposition to form groups that have a square symmetrical form. The quantum nature is a simple feature of basic arithmetic involving whole numbers, because the arithmetic describes the behaviour of *individual* atoms. "Energy" gains its meaning from the behaviour and indivisible properties of individual atoms. Hence there is no partitioning of energy and no expectation of an infinite continuity of that sort in the first place. But what emerges is a unity and plausibility that is greatly lacking in the account presented by quantum theory.

When "light" is transmitted through space, no intermediary such as a particle or a wave is involved in the transmission. Recall once again that we do not accept that the spatial perspective reveals everything about the causal structure of the world. Bodies that appear spatially separated have relations with each other that are not manifested from the spatial viewpoint. Say that the atoms A1 and A2 are in a state of tension because they have their like poles aligned with each other. This tension is at odds with the natural equilibrium of the system and it spontaneously disassociates after a time, leading to the evolution of the tension in the system until it has its effect at a particular atomic pair, T1-T2. This atomic pair is thrown into a state of heightened tension with the system. As a result it is "ejected" from its position of rest in the system, pulling other atoms with it that serve to incarnate the heightened tension in the relationships that now prevail between the atoms of the mini-system. The transmission of the tension from A1-A2 to atomic pair T1-T2 requires no intermediary because these atoms are *already* accessible to each other on a particular causal level that is not evident from the kinds of effects that generate our spatial perspective on things. This is not to claim that an electrostatic change at A can have an *immediate* effect at T. The system is in a state of constant evolution during the course of which the causal influences of all the causal players in the system play themselves out. A change at A cannot make itself felt at T until all of the various intermediate causal interconnections in the system have evolved to the point where A's influence makes itself felt at T. But when the influence is eventually felt at T, it is as if A and T were spatially adjacent (according to our way of visualizing things) and A transmits its tension physically to T.

Think of a catapult that is loaded with a stone inside it, then it is released. The tension of the elastic makes the stone fly. The stone strikes a target whose molecules are *already* in a state of tension with each other. The stone presses into the target, increasing the tension momentarily, even though the "inherent" properties of the individual molecules of the target (mass, charge, natural elasticity) remain the same. The tension is incarnated in the relative positions of the molecules with respect to each other. It has no separate existence apart from the configuration of molecules. It does not subsist as invisible "energy" in the system, but it does confer on the system the causal potential to work "energetically" on other systems or subsystems. Similarly, when the electrostatic tension that is released at A eventually evolves to T, it is as if the atoms at T exert augmented electrostatic influence on other atoms in the vicinity. But all of this is achieved without any intermediary of any sort. An upshot of this approach to light is the claim that if it were empirically possible to create an artificial system in the laboratory where a single atom was completely causally insulated from all other atoms, then it should prove impossible to get this atom to absorb "light". This follows from the claim that absorption of light always involves heightened relationships within *groups* of atoms.

Another analogy might assist in visualizing the dynamics of this model of light. Normally when we press on a balloon and then release the pressure, it returns to its normal size and shape. The compression of the balloon results in the air surrounding the balloon losing a little pressure, whilst the reflation of the balloon results in the surrounding air regaining pressure. In short, the pressure of the balloon is reciprocated by the pressure of the surrounding air. Now we are asked to use a little imagination! Consider a system consisting of a number of balloons inflated to exactly the same degree in a closed container in such a way that each of them exerts equal pressure on the other balloons that are in contact with it. The system differs from an everyday system in that there is no air present in the system at all (outside of the balloons), but neither is there a vacuum. Each balloon is in perfect contact with the other balloons surrounding it. No gaps are present. If one balloon is compressed by the introduction of an external object, this will result in pressure being exerted on the balloons surrounding it.

To make our analogy simpler (and here we have to imagine a strange type of balloon that exerts pressure in one direction only), say that the balloons are actually arranged in pairs and each one exerts a pressure on its partner. Balloon A1 exerts pressure E on balloon B1, and B1 reciprocates that pressure perfectly. Now imagine a situation in which A1 is subjected to a causal influence from outside the subsystem in which it belongs. This influence pushes A1 towards balloons C1, D1 and E1 in such a way that A1 now exerts pressure E on these three balloons as well as on B1. In turn, B1, C1, D1 and E1 each exert a reciprocal pressure E on A1. In the relationship that now prevails between A1 and the other balloons, extra "energy" is being stored in the subsystem. The other balloons are simply exerting their natural internal pressure towards A1, but the heightened state of excitation of A1 means that it now has enough pressure to reciprocate the influence of multiple balloons (whereas previously it only reciprocated the pressure of B1).

This situation is analogous to the absorption of light by a system. When a pair of atoms in an electrostatic bond receives an impulse of light, they are not *internally* excited. Rather, they enter into a relationship of greater electrostatic tension with other atoms in the system. And this, in fact, is what light is – a tension between atoms that is released when light is emitted and reincarnates itself when light is absorbed. Light has no independent existence apart from these relationships of electrostatic tension between atoms. Between the moment of emission and the moment of absorption, we normally think of light as travelling through the intervening spatial distance between source and target. A better way of thinking of this "transmission", as we shall see, is in terms of the evolution of the entire system. This approach will naturally predict a constant rate of evolution of the influence throughout the system.

Once the atomic pair is pushed into this state of greater tension with other atoms in the system, they become capable of "holding" more atoms in a new state of equilibrium than was previously the case. A stable atom is normally held in equilibrium by other stable atoms in its subsystem, each atom distributing the same resistance to causal disturbance equally to every other atom in the group. There is perfect reciprocation between all of the atoms in the group because each makes the same contribution to the state of balanced tension that has been established. But an atomic pair that has "absorbed" light is no longer in a state of equal reciprocation with the other atoms in the system. The increased electrostatic tension between the atoms in the pair means that they behave like *a larger system of atoms*, and the pair can only maintain this tension if it "pulls" other atoms out of the wider system to reciprocate the kind of electrostatic influence that it is exerting.

This is our model for the absorption of light by a system. When light is absorbed, an atomic pair in the system acquires a state of heightened tension with the other atoms in the system. The heightened tension of the atomic pair throws it into disequilibrium with the rest of the system. It now behaves like a mini-system and draws other atoms in the system to itself in a temporary, perhaps fleeting, relationship of reciprocal tension. But if this new relationship is to be stable then the various players in the relationship must make the same combined contribution to the balanced tension. It is clear that the atomic pair that is in a state of increased tension cannot enter into a balanced relationship with *individual* atoms. Such individuals cannot reciprocate the influence of the excited pair. As we have discussed, an atomic pair in a state of excitation enters into a relationship of tense equilibrium only with *groups* that have particular characteristics, and these characteristics are the basis of the integers in the Rydberg/Balmer formulae.

2.13 Can an atom be the target of multiple impulses simultaneously?

How credible is this account of the emission and absorption of light? The most obvious objection concerns the claim that the absorption of light will often involve an atomic pair receiving multiple impulses simultaneously. Take the case of two atoms of hydrogen, T1 and T2, that are already in a state of "excitation", each maintaining a group of four atoms in tension. In order for this group of atoms to transition to a situation where T1 and T2 become the locus of two pairs of nine atoms, *eighteen* impulses of light will have to be absorbed simultaneously. How likely is it that an atomic pair, T1-T2, might be exposed to eighteen impulses of light of just these magnitudes (or a smaller number of larger impulses which precisely add up to the same magnitude)? The claim that the absorption/transmission of light involves atoms receiving/emitting multiple impulses simultaneously could be evaluated empirically without the construction of any new experimental apparatus. A statistical analysis of the number and frequency of impulses of light absorbed or emitted by a material should go some way towards verifying the plausibility of this new theory of radiation.

For the moment, the following can be said in support of the claim. When light is transmitted from a source, according to this new approach, it will involve the disassociation of a particular type of tension in groups of atoms. Thus light transmission typically involves the emission of *multiple* impulses simultaneously. This emission process will perfectly mirror the absorption

158

process just described. If T1 is in tension with a group of nine atoms and later disassociates to a group of four, then five impulses of magnitude $E(1/1^2-1/3^2)$ will typically be emitted, as well as four impulses of magnitude $E(1/2^2-1/3^2)$. Not only will these impulses be emitted simultaneously, but their trajectory will also be similar, if not almost identical. We must assume that the trajectory of emitted light depends fundamentally on the original configuration of the T1-T2 pair in the group of atoms that emits the light. When this tension disassociates, leading to the emission of one or multiple impulses of light, the trajectory of the light is likely to depend on the orientation of the B and V components of the original pair that is the locus of the group. Thus the trajectories of multiple impulses emitted from a T1-T2 bond are likely to be the same.

The transmission of light between atoms of different elements follows the pattern described above. Hydrogen-like atoms (we will consider the nature of such atoms later) display absorption and emission spectra that satisfy a modified version of the Rydberg formula as follows:

$$1/\lambda = EZ^2(1/n_1^2-1/n_2^2)$$

where Z is the atomic number of the element in question. An example of a hydrogen-like atom is ionised helium, He^+. According to our model, if a mini-system consisting of nine He^+ atoms disassociates and makes a transition to an assembly of four atoms, then nine impulses will be transmitted and their magnitude will be four times greater than the impulses released by a similar transition in the case of hydrogen (the atomic number of helium is 2, so $Z^2=4$). That means that the transition from a mini-system of nine to a mini-system of four in the case of a sample of He$^+$ will give rise to the emission of five impulses of magnitude $4E(1/1^2-1/3^2)$ and four impulses of magnitude $4E(1/2^2-1/3^2)$. If these nine impulses are directed towards a sample of hydrogen gas, then it is easy to see the types of mini-system that they are likely to form. If the sample of hydrogen gas already has a mini-system consisting of an atom in tension holding four other atoms, then that mini-system requires four impulses of magnitude $E(1/2^2-1/3^2)$, in addition to five impulses of magnitude $E(1/1^2-1/3^2)$, in order to form a mini-system of nine atoms around the excited atom. The disassociation of a single helium mini-system can provide more than enough tension to "power" multiple transitions of this sort in hydrogen! There were four impulses of magnitude $4E(1/2^2-1/3^2)$. Any *one* of these is sufficient to augment the tension in the group of four original atoms in a mini-system of hydrogen so that they can now participate in a mini-system of nine atoms. Any *one* of the five impulses of $4E(1/1^2-1/3^2)$ just requires a single simultaneous impulse of $E(1/1^2-1/3^2)$ in order to pull five hydrogen atoms out of their stable position in the system to make up the remainder of the new nine-atom mini-system being formed. Such an impulse of magnitude $E(1/1^2-1/3^2)$ is a typical impulse emitted by the disassociation of a nine-atom mini-system of hydrogen. We can expect that there will be many such impulses available in a sample of hydrogen gas as mini-systems form and disassociate continuously.

Let us summarize this example of the emission of light by a sample of helium and absorption by a sample of hydrogen.

A mini-system consisting of two groups of nine He$^+$ atoms (around an excited pair, A1-A2) releases tension and regroups as an assembly of two groups of four atoms (around A1-A2)

The tension in all nine atoms with respect to A1 is consequently reduced. This results in nine impulses being imparted in the direction of a sample of hydrogen gas: five impulses of magnitude $4E(1/1^2 - 1/3^2)$ and four of magnitude $4E(1/2^2 - 1/3^2)$.

One of the impulses of magnitude $4E(1/2^2 - 1/3^2)$ impinges on a mini-system (in the hydrogen sample) consisting of an atomic pair in tension, T1-T2, holding four other atoms.

Such an impulse (of magnitude $4E(1/2^2 - 1/3^2)$) is exactly the right magnitude required to raise the four atoms to the sort of tense relationship with T1 that would be typical of a mini-system of nine atoms around T1.

In order to form the nine-atom mini-system, T1 will need to pull an additional five atoms out of their stable relationship with the rest of the system. This will require five impulses of magnitude $E(1^2 - 1/3^2)$

But the disassociation of the mini-system of helium also released five impulses of magnitude $4E(1/1^2 - 1/3^2)$. If just one of these simultaneously arrives at T1 (and there is surely a high probability of this), then we require just a single concurrent impulse of $E(1/1^2 - 1/3^2)$ in order to pull the five hydrogen atoms out of their stable position in the system to make up the remainder of the new nine-atom mini-stem being formed. Such impulses of magnitude $E(1/1^2 - 1/3^2)$ are a typical emission of hydrogen gas and should be freely available in the system.

It may be challenging to establish empirically whether or not multiple impulses of light tend to be emitted concurrently from a particular mini-system of atoms, but it is relatively easy to perform a statistical examination of the likelihood of a single atomic pair being the recipient of multiple impulses simultaneously. The next section will contain a rudimentary discussion of this sort.

2.14 Statistical likelihood of an atomic pair receiving multiple impulses simultaneously

Imagine a lamp that emits 20W of white light. Say that this light is focussed into a beam that is one metre-squared in cross-section. The energy of a single impulse of light, E, is hv, where h is Planck's constant and v is the "frequency" of the light. Light travels at velocity, c. Thus it is thought of as having a wavelength, λ, which is equivalent to c/v. Therefore,

$$E = hc/\lambda$$

In white light the average wavelength of light is about 565 nm. Thus the average energy of a single impulse of light in this beam will be

$$E = (6.626 \times 10^{-34})(2.998 \times 10^{8})/(565 \times 10^{-9}) \text{ or } 3.516 \times 10^{-19} \text{ Joules.}$$

Our lamp is 20W, which means that it emits 20 Joules of white light per second. In any given second, the lamp will emit an average of $20/(3.516 \times 10^{-19})$ impulses of light, which is equivalent to about 6×10^{19} impulses per second. The light is focussed into a beam that is one metre-squared in cross-section. The photon flux density (photons per unit area per unit time) of the beam will thus be roughly 6×10^{19} photons per metre squared per second.

This beam impinges on a sample of hydrogen gas. We know how many impulses there are in a one metre-squared cross section of the beam per second, so if we calculate the proportion of atoms of hydrogen there are in a one metre-squared cross section of the gas, then we can estimate the likelihood that multiple impulses might impinge on a pair of hydrogen atoms in any given second. To perform this calculation we will need to know the density of the gas and have an idea of the size of a pair of hydrogen atoms in an electrostatic bond. A traditional rough estimate of the size of the hydrogen atom is of the order $6 \times 10^{-30} \text{m}^{-3}$. As the earlier discussion revealed, we do not accept the view that the atom is composed of a positively charged nucleus surrounded by electrons. If our approach to understanding the atom is correct, then the actual size of a hydrogen atom is probably closer to that usually ascribed to a proton. But our model of light claims that light is not absorbed by the hydrogen atom but is incident on the electrostatic bond between a *pair* of atoms that may or may not already be the locus of a mini-system of atoms being held in tension after the intromission of light. The bond between the atomic pair will occupy an area that is certainly much larger than the area occupied by the two atoms themselves since the stable electrostatic bond involves the atoms being positioned at considerable distance from each other (relatively speaking) at the point where mutual repulsion is about to begin. Therefore we can conservatively assume that T1-T2 can receive an impulse of light if the impulse arrives in the area surrounding the atom whose volume is that traditionally thought of as being internal to the atom itself, namely, $6 \times 10^{-30} m^{-3}$.

Say that the density of the sample of hydrogen gas is 10 grams per m^{-3}. 10 grams of hydrogen contain about 6×10^{24} atoms (a mole of any element contains Avogadro's number of 6.022×10^{23} atoms, and a mole of hydrogen weighs 1 gram). We are allowing that each atomic pair occupies an "active" space of $6 \times 10^{-30} m^{-3}$ (in other words, a space in which it can be impinged on by the intromission of light). Thus the 10 grams of atomic pairs of hydrogen occupy the following proportion of the volume of the 1 cubic metre sample of gas: $(6 \times 10^{-30})(3 \times 10^{24}$ atomic pairs$)/1 = 18 \times 10^{-6} m^{-3}$ of matter for every cubic metre in the sample. Please note that, for convenience, we are treating the gas as if it were made up of atomic *pairs*. In reality, there will be many *groups* of atoms in electrostatic equilibrium throughout the sample. These will actually constitute a larger "target" for the incoming impulses of light. We assume that light incident on any part of a mini-system of atoms in electrostatic equilibrium will be absorbed by the atomic pair that functions as the locus of the mini-system, if the alignment of the incident light is the same as that of the T1-T2 pair.

So the 10 grams of atomic pairs of hydrogen occupy 18 x 10^{-6} m^{-3} of matter for every cubic metre in the gas. The hydrogen atoms will also occupy the same proportion of the one square metre cross section of the beam emanating from the 10W lamp. We recall that the lamp emitted roughly 6 x 10^{19} impulses of light per square metre per second. That means that in a given second the chances of each hydrogen pair in the sample being influenced by one of these impulses is:

$$(18 \times 10^{-6})(6 \times 10^{19}) = 108 \times 10^{13}$$

or 1080 trillion impulses per second for every hydrogen atom in the sample.

If each atomic pair is impinged on by 1080 trillion impulses a second, how many of these impulses are likely to be simultaneous? A proper answer to this can only be given by a careful empirical study of the way light is typically emitted by a source. For such a colossal number of causal events, a purely random pattern cannot be ruled out, and this would mitigate against the possibility of a significant number of simultaneous "impacts". But it is easy to imagine how the nature of causal dynamics in the source could lead to regular correlations of emissions. The source can be thought of as being composed of millions of tiny individual causal sources whose activity is far from being independent. In the case of an electric lamp, there is a *unitary causal process* (the electric current) in operation in the lamp that is energising the tiny emitters. The emitters have similar characteristics to each other and will tend to produce their emissions in extremely regular patterns. If evidence can be found for simultaneous correlation between even as little as one hundred impulses in every trillion, then this figure of 1080 trillion impulses would become very significant. It would mean that each atomic pair in the sample of gas would be receiving as many as 100,000 simultaneous "hits" in a given instant, and all from a modest 20W lamp whose beam is spread out over a square metre. This seems more than sufficient to make our model of the emission and absorption of light workable.

In short, the assertion that light absorption consists in atomic pairs or groups of atoms receiving multiple impulses simultaneously is not incompatible with the empirical evidence. As far as this, and other theoretical challenges to our model are concerned, only a proper confrontation of the theory with the empirical data can hope to settle the questions in a proper manner.

2.15 The absorption and emission of light by composite atoms

The foregoing has considered the transmission of light by groups of hydrogen atoms. The heavier elements will generate light in the very same fashion, although the alignments of such atoms in groups will have a more complex form. Within composite atoms, the B-V polarities of the various genesis-units are already being held in check in various ways by the reciprocating influences of other units in the atom. We cannot expect the Rydberg formula to hold straightforwardly in these circumstances, unless in particular cases where a certain (odd) genesis-unit on the periphery of an atom has little of its electrostatic influence reciprocated because of the general equilibrium of the rest of the structure. A multiply-ionised atom could also manifest a spectrum similar to that predicted by the Rydberg expression. As we shall see later, ionisation involves the covering of the electrostatic influences of particular genesis-units within an atom. This means that other genesis-units in the structure will have the liberty to participate in the formation/disassociation of mini-systems in electrostatic tension.

The genesis-unit is the basic unit of causal influence in the world. Any given genesis-unit within a composite atom will be in a particular state of electrostatic balance depending on its role in the equilibrium of the overall structure. It stands to reason that this imbalance will affect the way the atom as a whole bonds to other atoms of the same element. The frequency of the light absorbed or emitted in the alignment or disassociation of atoms of the element will consequently be determined by the exact nature of the equilibrium of the composite atom, and by a particular genesis-unit's "residue" of electrostatic tension that is not being reciprocated by the rest of the structure.

Before absorbing light, a lithium atom will first have to form a natural pair with a second lithium atom. Say that the bond between the pair occurs between the odd genesis-unit in each atom. This genesis-unit is more electrostatically imbalanced than the other two proactive units in lithium. This will affect the way in which the bond between the pair becomes the locus of a mini-system that absorbs an impulse of light. In other situations, one of the *other* genesis-units in the same lithium atom could be the unit around which the bond is formed. Given that the state of electrostatic equilibrium of both of these situations is different, the frequencies of the light absorbed and emitted will also be different. This will give rise to distinct spectral series. The more genesis-units a composite atom has, the more complex its pattern of spectral lines.

As has been the case for well over a century now, it is proper that evidence from spectroscopy should be used to form hypotheses about the structure of the atom. From the above analysis, we can imagine how the distinctive *s*, *p*, *d* and *f* series of lines observed by spectroscopists could be produced. The way that atoms align themselves is regulated by their tendency to form bonds that increase equilibrium. The exact role of a particular genesis-unit in the equilibrium of its parent atom will determine the magnitude of the impulse that we call light which will be required to cause this atom to align itself with others. Thus, atoms will absorb and emit distinctive series of frequencies of light and these frequencies will be related to the configuration of genesis-units in the composite atom. Under some conditions, it will be a particular genesis-unit in the atom that will be the locus of the absorption or emission of the "light" that holds the parent atom in alignment with its mini group. At other times, a different genesis-unit in a different position will absorb or emit the holding impulse. The position of the unit and its place within the composite structure will determine precisely the frequency of the light.

Larger atoms with more complex structures will have a larger range of distinctive series. Thus the *s*, *p*, *d* and *f* series are progressively associated with heavier atoms. But it would require the competence of a spectroscopist to identify the correspondences between these particular series and the hypothetical structure of the atom according to our model. It may well emerge that the standard understanding of subshells has involved forcing the spectroscopic evidence to conform to the mathematical requirement that each shell *n* should have *n* subshells with $2l + 1$ orbitals (where *l* is the number of the subshell). If the symmetry model of atomic fusion discussed in the first chapter manages to account for the spectroscopic evidence in a more natural and simple way, then this would constitute significant empirical support for our approach.

The above broad outline of the way in which elements generate their spectral series differs markedly from the standard framework of understanding. In our account of the structure of composite atoms, it may have seemed at times that electrons and genesis-units were merely different names for roses. Units were arranged in layers around the core like electron shells around the nucleus. Oppositely-positioned units in the same layer corresponded to pairs of electrons in the orbitals of the same subshell. But when we come to the absorption and transmission of light, any superficial similarities between the two models quickly evaporate. In the electron model, a particular frequency of light can be emitted when an electron in, say, the p subshell of the second shell, drops to the s subshell of the first shell. According to our approach, by contrast, a genesis-unit in a certain position of the second layer will have a particular capacity to become the locus of mini-systems of atoms (once, of course, that unit has become the basis of an electrostatic bond with at least one other atom of the same element), whilst a unit in the first layer will have a completely different capacity to form mini-systems. Let us consider this in a little more detail for an atom of lithium.

Lithium is formed by fusing genesis-unit 3 onto the pair formed by units 1 and 2 (Figure 2.11). Units 1 and 2 would originally have had a high state of equilibrium as a result of the fact that they largely reciprocated each other's electrostatic influence. Once unit 3 is fused onto the pair, the equilibrium of 1 and 2 will be disturbed to some degree, but unit 3 itself will be in a much greater state of electrostatic *dis*equilibrium with respect to the composite structure as a whole. Let us use the expression "residual electrostatic disequilibrium" for the magnitude of a genesis-unit's electrostatic disequilibrium that is not reciprocated by the rest of the structure.

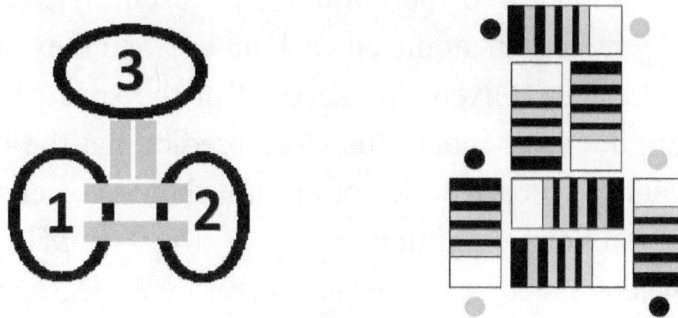

Figure 2.11 A lithium atom

Units 1 and 2 will have very similar residues, leading to a splitting of spectral lines – in fact paired genesis-units in general in composite atoms will give rise to split spectral lines. We will designate these lines with the letter *s*, whilst unit 3 has a residue that we designate with the letter *p*. In order for unit 3 to become the locus of a mini-system, it must first form a pair with the unit 3 of another lithium atom. The bond between the pair will now absorb impulses of a magnitude that stand in a particular relation to *p*. Similarly, pairs formed around units 1 or 2 will absorb impulses that stand in a very precise relation to *s*. Our account of the Rydberg formula for hydrogen required that, in order to hold a mini-system of four atoms in tension with itself, each atom in a hydrogen pair must absorb four impulses of $E(1/1^2 - 1/2^2)$ simultaneously (otherwise, T1 could receive a single impulse of $4E(1/1^2 - 1/2^2)$ or two impulses of $2E(1/1^2 - 1/2^2)$. In either case, T1 then distributes the tension over four atoms). In this situation, the hydrogen atoms at the locus of the mini-system do not form part of a composite atom, so their "residual electrostatic disequilibrium" is E, which corresponds to the natural electrostatic attraction exerted by hydrogen atoms on each other when they settle into equilibrium. Given this account of the spectrum of hydrogen, does our model therefore predict that the lines in the s and p spectral series of lithium should correspond to transitions of magnitudes $s(1/n_1^2 - 1/n_2^2)$ or $p(1/n_1^2 - 1/n_2^2)$, where s and p represent the residual electrostatic disequilibria that generate these various subsystems in tension? Emphatically, *yes*, but it does seem likely that the situation in composite atoms will be complicated by the residual electrostatic disequilibria of other genesis-units within each atom. It might also be possible that a mini-system of atoms with unit 3 as its locus could disassociate to a smaller subsystem that has units 1 or 2 as its locus. In this case a more complicated version of the $(1/n_1^2 - 1/n_2^2)$ expression would be required that utilizes both s and p in its formulation.

It must be left to spectroscopy to settle this question. The challenge here is to work out the residual electrostatic influence of unit 3, once all of the complex dynamics within the atom have been taken into account. This requires identifying a unique number (for the residual influence of unit 3) that stands in relation to various frequencies in the spectrum of lithium, much as Balmer identified a number that stood in relation to all the lines in the visible spectrum of hydrogen (of course, other unique numbers will have to be identified for the residual influences of units 1 and 2). The situation is more complex than hydrogen because hydrogen is composed of a single genesis-unit, with the consequence that all hydrogen atoms have the same residual influence, and the number discovered by Balmer is a simple expression of it.

Let us return to the example of a lithium pair absorbing an impulse (or multiple impulses simultaneously) and becoming the locus of a subsystem. All atoms of lithium will have their own unit 3 with the exact same residual influence. For one of these pairs of atoms to become the locus of a subsystem, it will have to absorb a magnitude of light that magnifies its residual influence by an integral amount such that it becomes capable of reciprocating the electrostatic influence of an orthogonal arrangement of n x n lithium atoms. Say that this subsystem later disassociates to a subsystem of $(n-1)^2$ atoms. There will be a resultant "emission of light" of the order of $p(1/n^2 - 1/(n-1)^2)$. It is possible that this quantity p (corresponding to the residual electrostatic influence of unit 3 in the lithium atom) has already been identified by spectroscopists but has been understood in different terms. If a significant number of the spectral lines in lithium could be related via the $(1/n_1^2 - 1/n_2^2)$ expression to two unique numbers (we can expect that units 1 and 2 will have identical residual influences) corresponding to the influences of the three units that give the atomic number to the element, this would constitute significant empirical support for our model of the transmission of light and of the structure of the atom.

Chapter Three

RESOLVING THE PARADOX OF WAVE-PARTICLE DUALITY

"We choose to examine a phenomenon which is impossible, absolutely impossible, to explain in any classical way, and which has in it the heart of quantum mechanics. In reality, it contains the only mystery".
Richard Feynman in ***Lectures in Physics***

Overview of this chapter and its principal claims

1. "Wave-particle duality" – the notion that light can exhibit mutually exclusive properties in different experimental circumstances – has fostered a marked scepticism with regard to the possibility of developing a coherent model of the transmission of light. Any new model that seeks to avoid this dualistic approach must explain why light manifests characteristics typical of waves (such as frequency, wavelength and interference) in some circumstances, and characteristics proper to particles in others (such as localised position).

2. This chapter will show how the notions of the "frequency" and "wavelength" of light are based on genuine physical attributes of *causal processes* of a certain sort, but they have been mistakenly interpreted as characteristics of physical waves. Instead they are related (respectively) to the relative time-interval required by a causal process to give rise to an effect in target matter, and the

extent that a process will reverberate in the system during that same time interval. This new approach follows from the insight that light is *not movement at all* in the conventional sense of the word. Rather, it involves the *reverberation of an influence in an evolving system.* The interpretation of the meaning of frequency and wavelength that follows is a minimalistic interpretation based solely on empirical evidence and is devoid of unfounded metaphysical assumptions.

3. Using these new understandings of the underlying physical bases of frequency and wavelength, interference patterns will be explained in terms of the way that a causal process can be jointly modified by multiple causal players. We will discuss the potential of such patterns to inform our views about space and the manner in which causal processes evolve in the system.

4. Finally, we will consider other types of two-path experiment. It will be shown how experimental set-ups manifesting complementarity of spin have the same characteristics as "interference" set-ups involving light. In both cases, the apparatus is set up in such a way as to present an ongoing causal event with multiple causal players that can potentially modify the impulse in a joint fashion. The experimenter can set up the apparatus to *prevent* one of the causal players from exerting its influence on the evolving process, or he can choose a set-up that permits *joint* influence.

3.1 Wave or particle?

The view that light exhibits "wave-particle duality" (the notion that electromagnetic radiation behaves as a wave in certain circumstances, like a particle in others, whilst no single physical model can ever account for all the diverse phenomena) has gravely hampered our progress in understanding the nature of electromagnetic radiation. This notion has given rise to a widespread scepticism about the applicability of physical models to the theory of light.

If it is impossible to understand light in non-paradoxical terms, then why bother trying to develop a single model that explains its behaviour in all circumstances? Such fatalism has stifled theoretical developments in this area for the past eighty years.

If the genesis-unit model is to assist in getting theory beyond this impasse, then it needs to say something about the underlying meaning of notions like frequency and wavelength. It will have to clarify the significance of Planck's constant and be able to say exactly what happens when the excess of absorbed light leads to the ionisation of atoms. Most importantly, it must be capable of unlocking the mystery of the so-called "interference" of light.

The last chapter has attempted to describe light in terms of a common-sense physical model. We felt that the structure of the *atom* itself might be difficult to subsume under a physical description where constituent parts interact with each other in intuitively-plausible ways. If the atom is the fundamental constituent of matter, then it is going to be problematic to describe *it* in terms of material constituents. That is why the view was expressed that it might be better to think of the atom in terms of physical *principles* or *processes* rather than physical structures. But light is a different matter. Even if it does not involve a motion of matter through space, it still involves (in our view) pieces of matter (atoms) forming/dissolving relationships of certain sorts with respect to each other. Therefore we should be able to say a good deal about light in terms of the characteristics of these physical structures. The assertion that light does not involve the transmission of material intermediaries will not stop us formulating understandings of frequency and wavelength in terms of the material systems that *incarnate* light.

Say that light is transmitted from hydrogen pair A1-A2 to hydrogen pair T1-T2. At time t, A1 is holding a mini-system of nine atoms to itself (and likewise for A2). At $t+1$, the mini-system disassociates to a mini-system of four atoms. Eighteen impulses of light are thus transmitted and all of them arrive at T1-T2 where the reverse transition takes place. When the nine-atom system held by A1 disassociates, the nine emitted impulses are of two magnitudes: $E(1/2^2-1/3^2)$ and $E(1/1^2-1/3^2)$. In the standard view, each of these magnitudes is associated with a frequency of light. But frequency and wavelength are concepts that seem applicable to a continuous wave motion. What sense can such notions have for impulses of this sort that seem to be momentary events, not spread-out events that endure for a period of time? This question, incidentally, is one that also presents itself for the quantum mechanical account of light. Historically it has led to the wave-particle duality explanation, which is far from satisfactory. It does not explain why light has such a nebulous nature and it does not uncover the mechanism by which it can change its nature in such an apparently discontinuous fashion.

A system that *emits* light continuously over a period of time is typically being *energised* constantly. A light bulb, for example, has electric current passing through it. The gases on the surface of the sun are being constantly irradiated with radiation from within the sun itself. So a mini-system of atoms that disassociates in the source will soon be replaced by another mini-system that will rapidly disassociate in turn. This leads to a constant stream of impulses from the emitter of light. Statistically, given that so many impulses are emitted by a light source in any instant, the electromagnetic radiation will have a highly regular and continuous character. In the common way of visualising the wave pattern of light, we tend to think of it in these terms: a steady stream of radiation being emitted by a source; the steadiness and regularity of the causal activity of the source generates a stream of impulses or waves of a certain frequency. But, if we look closer, this comfortable picture cannot correspond to what we mean by the "frequency" of the light. Sometimes it is radiation that is emitted sparsely and irregularly that demonstrates a very high frequency. Indeed, even a *single* impulse can be attributed a high frequency under the current understanding. What sort of frequency can this be?

The question boils down to the issue of how individual impulses can display properties that are typical of waves. The classic way of illustrating a wave motion is in terms of the oscillating rod in a tub of water that creates water waves. The movement of the rod forms crests and troughs in the water and the motion is propagated through the medium to the tub's edge. During the nineteenth century it was believed that electromagnetic radiation moved in a similar way through the ether, which provided the medium for the motion to be propagated. But now, with the demise of the ether already over a century old, it is difficult to see how a stream of individual photons through a complete void can display the apparently unique marks of a wave motion.

So why do we continue to insist that electromagnetic radiation is a wave? The phenomenon of the interference of light is the principal obstacle that wards off anyone who would deny that the transmission of light involves some kind of wave motion. But apart from interference, there is also the fact that the primary properties attributed to electromagnetic radiation in everyday discourse are its frequency and wavelength. One of the standard practical assignments for students of physics is the exercise of measuring the wavelength of light. If light is not a wave-motion, how can such straightforward experiments, using common-sense principles, yield values for the wavelength and frequency of light?

3.2 The apparent movement of light through space

If light is as the genesis-model describes it, and does not involve the motion of either a wave or a particle through space, then how does it appear to manifest frequency and wavelength? A simple procedure for measuring the frequency/wavelength of light uses a diffraction grating. This is a piece of transparent material with parallel opaque wires placed in it at regular intervals during the process of fabrication. A typical diffraction grating can have many hundreds of lines per millimetre. When light passes through very small slits of this sort, it is diffracted, a phenomenon considered typical of a wave motion. This means that the opening acts as if it were a source of light in itself. Once the light passes through, it spreads out in all directions.

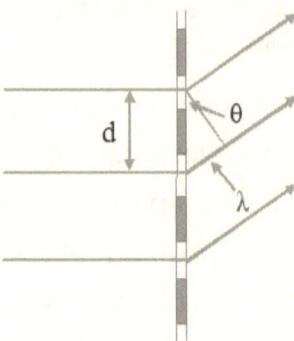

Figure 3.1 Measuring the wavelength of light with a diffraction grating

Consider Figure 3.1. Monochromatic light (i.e., of a single wavelength only) approaches the diffraction grating from the left. The slits are spaced at distance d apart. Once the light passes through the slits, it spreads out and gives rise to a series of light and dark bands on a screen placed to the right of the diffraction grating. The standard explanation for this phenomenon considers light to behave like a wave motion. The wave that approaches the diffraction grating from the left has crests and troughs arriving at each slit on the grating at the same time. However, once the light passes through the slits and diffraction has taken place, the waves that emerge on the right hand side may or may not be in phase - it all depends on the angles at which the rays of light emerge. In the diagram, the middle ray of light has a longer distance to travel than the top ray. The difference in the length of the "wave path" is λ. If this difference corresponds exactly to the wavelength of the light, then the crests and troughs of the two rays will again be in phase and constructive interference will take place between them.

In other words they will give rise to a bright area on the screen placed to the right of the diffraction grating. If they are out of phase, destructive interference takes place, and a dark band will appear at the screen.

A simple classroom procedure can measure the wavelength of light using this arrangement. The distance d between the slits is determined at the time the grating is manufactured. Monochromatic light is passed through the apparatus and the diffraction angle that leads to the first band of constructive interference is measured. This can be used to yield a value for the angle θ. The expression $d\sin\theta$ will then give us the wavelength of light, λ.

How can this so-called "interference" be explained without having recourse to the notion of wavelength? Later we will outline an approach to understanding this most mysterious of empirical phenomena. Beforehand, we will discuss the properties of light transmission that resemble the characteristics of frequency and wavelength.

When a mini-system disassociates and releases the tension that we call light, this causal impulse has definite properties that are determined by the nature of the mini-system that generated it. The evolving causal impulse does not have a material nature (it is not a wave, nor a particle, nor a field), but when the system arrives at a certain point of causal evolution, this influence will prompt an electrostatic alignment to be reincarnated in another pair (or group) of atoms elsewhere. From the spatial perspective, this evolution will be represented as a transmission of light.

The electrostatic tension that is transmitted will have a particular magnitude, depending on the bond in the source mini-system that has been disassociated. The greater the magnitude, according to the standard approach, then the shorter the "wavelength" and the higher the "frequency" (as we shall discuss later, the frequency and wavelength are also influenced by the relative velocities of source and target). Surprisingly, these notions of wavelength and frequency can be adapted quite well to fit with our model, even though we reject the view that light involves the transmission of a wave or any other intermediary. Let us see how this is the case. In Figure 3.2 there are two diagrams which show the progression of impulses of light emitted by two light sources, one producing light of high electrostatic tension (in the sense that it has the potential to form groups of atoms with a higher net electrostatic attraction between them) and the other of low tension. For convenience we will use conventional language and refer to these as light sources of high and low "frequency". The low frequency source produces light of "wavelength" corresponding to the length x0-x3, whilst the high frequency source produces wavelengths of x0-x1. The sources are set up so that they emit light in one direction only along the line x0-x9. Along this line are positioned nine smoke particles which remain stationary for the duration of the experiment. Therefore the nine particles are depicted on all ten timelines of the diagram in the same positions (the horizontal lines with the grey dots representing the smoke particles).

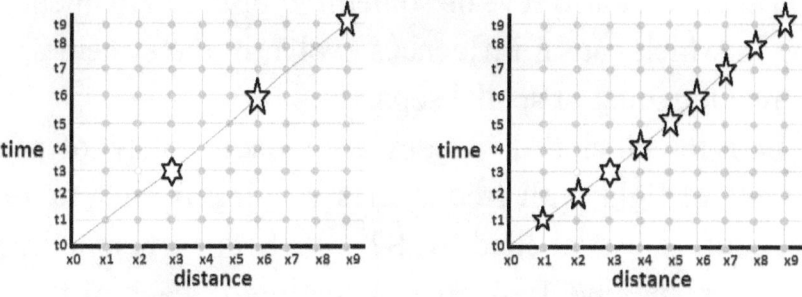

Low frequency source emits impulse of
light from point x0 at time t0

High frequency source emits impulse of
light from point x0 at time t0

Figure 3.2 The frequency of light. *Two light sources emit light of different "wavelengths". The wavelength of the low frequency source corresponds to the length x0-x3, whilst the wavelength from the high frequency source corresponds to x0-x1.*

At time t0, both sources emit multiple simultaneous impulses of light. The impulses emitted by the low frequency source interact with smoke particles at points x3, x6 and x9. The impulses emitted from the more powerful source each interact with one of the nine smoke particles along the line. The point of this exercise is to reflect on the fact that the causal influence emitted by the low frequency source *cannot* interact with the smoke particles positioned at points x1 and x2, and this fact is usually understood in terms of the "frequency" and "wavelength" attributed to the causal influence.

Why cannot a causal influence of this magnitude manifest itself at points x1 and x2? The answer to this question provides the physical underpinning of the notions of frequency and wavelength, and in doing so finally solves the vexing issue of how a causal influence that gives rise to point-like effects can manifest the phenomenon of interference. It also reveals something of the fundamental way in which causal influences evolve in the system and the real meaning of spatial separation.

Imagine a closed system composed solely of ten sources of light with frequencies varying from 1 to 10. The sources are labelled S1, S2, . . ., S10, corresponding to their respective frequencies. At time t0 each of these sources emits an impulse of light towards the other objects in the system. These impulses are not to be thought of as causal intermediaries flying through the space between source and target. Rather, the system as a whole is in a state of evolution. On the ontological level, all ten sources of light have a causal interconnectivity that is not apprehensible from the spatial perspective. The spatial viewpoint is constructed from sense data that is generated by causal processes that play themselves out in the genuine disunities that prevail between objects. The problem with this perspective is that it gives a disproportionate significance to the disunities (which it represents as spatial gaps) and fails to appreciate that there are other ontological connections that have no need of material entities that cross spatial gaps. Indeed the phenomena of gravity and electromagnetism offer ample reason for being open to the existence of such direct connections on the ontological level.

The system is in this state of causal evolution. For simplicity we will only consider the evolution of the causal impulses that were emitted at t0, and we will only consider ten separate "moments" of the evolution from t0 to t9. The challenge is to try to consider the evolution at the ontological level, without being hampered by a purely spatial perspective on things. At t0 the various causal emissions occur and, on the ontological level, these are all part of *a unitary evolution of causal processes*. It is not that each of the ten impulses is a completely autonomous process, flying on its lonely way from source to target. The transmission of light does indeed involve particular influences between sources and targets, but the evolution of these influences occurs in the context of the whole system. At the ontological level, the system is, as it were, "pulsating" as a result of the causal interactions that are taking place in it. And it is the reciprocal interplay of all these causal players that determines the rate at which the system throbs. In our scheme, t0 to t9 represent ten of these successive pulsations (of impulses that were emitted at t0). At t1, the impulses from S1 to S10 are all evolving in the system, but the impulse from S10 (the highest frequency impulse) will already have the potential to effect a causal change at the point to which it has evolved. Its greater potency gives it an increased causal potential in the causal configuration of things, the causal space of which is represented poorly with our intuition of spatial extension. The fact that S10 has twice the "frequency" of S5 means that it has twice the capacity to effect causal change and penetrate the ontological gaps between the source and other objects in the system.

At t2, S5 has the potential to effect causal change, but *S4 has not*. The lesser causal potential of S4 means that it has lesser capacity to make its influence felt in the causal configuration of things. The "wavelength" of these impulses is really an inverse measure of their causal capacity within the configuration of things. The lesser their causal capacity, then the more time elapses before they have the potential to have an influence on an object in the system. The more time elapses, then the more the system as a whole has evolved in the meantime, meaning that the impulse will have its effect at a greater distance from the source. This is the correct understanding of the "wavelength" associated with causal impulses of this kind. It is a minimalistic interpretation based solely on empirical evidence and is devoid of unfounded metaphysical assumptions regarding entities like wavelengths, particles or fields.

The rate of velocity (from the spatial perspective) of these various impulses is perfectly constant because they are not really ten autonomous influences at all but a *single system in evolution* together. The system as a whole is pulsating as a result of its internal causal dynamics. Each pulse is like a ripple that convulses the entire system as the individual causal impulses evolve step in step. The rate of evolution of each impulse is the same because at root there is only one evolution – that of the system - and all the individual processes are just particular manifestations of that unitary progression. In this sense it is quite accurate to speak of the "causal clock" of the system. The rate of ticking of this clock is determined by the total interplay of causal events in the system. In turn, the clock determines the rate at which the effects of each individual event reverberate through the system. But even if their rate of reverberation is the same, the *potency* of particular processes is an individual matter.

Let us examine this view in a little more detail. Earlier we asserted that light does not involve the movement of either a wave or particle through the system. But what we are really saying now is that light is *not movement at all* in the conventional sense of the word. Rather, it is the *reverberation of an influence in an evolving system.* This interpretation of the nature of light has rich potential for the solution of its various mysteries. An event (let's call it A) happens in the system at point P (say A is the disassociation of the electrostatic alignment of a mini-system of atoms). The system evolves. As each moment passes, the influence of A reverberates wider in the system, but it will not manifest itself in another event unless the conditions are right (i.e., unless this particular electrostatic tension that has been released into the system encounters a suitable group of atoms for it to be re-embodied in new mini-system aligned electrostatically). The more time that passes before A makes itself felt elsewhere in the system, the further from P the effect will be eventually manifested. Understandably, this gives the impression that something is moving in the system from P outwards, but such an impression is ultimately misleading. The system as a whole is evolving, and as time passes, any given event will have wider and wider reverberations, resulting in an ever greater "distance" between cause and effect.

3.3 The notions of frequency and wavelength in more detail

With the passage of time, the influence of a causal event reverberates further and further from its source. Whether a particular influence has the potential to be felt at a given intermediate point in the system, however, depends on the potency of the impulse. A causal impulse of double the magnitude has twice the capacity to effect change in its eventual target, but it also has twice the potential to manifest itself at any given point in the causal configuration of things. It is like having two axes; one is twice as sharp as the other and it is wielded by an arm that is twice as strong as the other. Not only does this axe have the capacity to split a tougher block of wood, but it can splinter it more finely. Impulses of light that are greater in magnitude give rise to a greater magnitude of causal influence in their targets, but they also have a proportionately greater capacity to penetrate the "spaces" in the causal configuration of things and *pick out* their targets. This amplified picking-out capacity is usually understood in terms of higher frequency and shorter wavelength radiation.

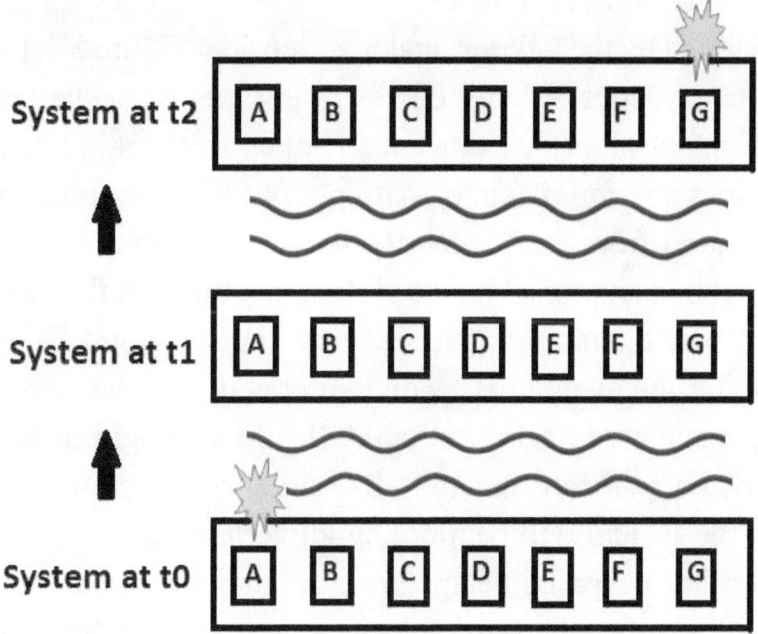

Figure 3.3 The reverberation of causal influence as the system evolves. *At t0 a mini-system of atoms dissociates at point A. Many other causal events occur in the system at the same time and these various events have a mutual influence on each other. It is only at t2 that the event at A finally manifests itself at G. In this sense, the system can be said to "pulsate" with the causal events that are evolving in it. Low "frequency" events take longer to have their effect because their potency places them lower in the pecking order than high frequency impulses.*

When a balloon is inflated to high pressure and then deflated, its rate of deflation is more rapid than a balloon inflated only slightly. The greater the disequilibrium between the pressure inside the balloon and the air outside the balloon, the quicker will be the discharge restoring equilibrium. The causal impulse we call light is renowned for its constancy of velocity, but, in a very real way, higher frequency light is "quicker" than lower frequency light, as illustrated by the balloon analogy. Consider Figure 3.3. Low frequency light is "emitted" by A at time t0. At the same time many other causal events happen in the system. We can imagine for example, that light of twice the frequency is emitted from B. At time t1, the low frequency light has not yet had an impact on the system, but the high frequency light would manifest itself at that time at point E. The fact that the event at B involves atoms in a higher state of electrostatic equilibrium than A (leading to light of twice the potency) means that the "discharge" of this equilibrium into the system will be more rapidly effected.

From the spatial perspective, the impulses from A and B will manifest the same velocity since they will "traverse" the exact same spatial distance (allowing that the medium they pass through contains the same density of material) in the same interval of time. And this is right and proper. It may have taken longer for the event at A to manifest itself in the system, but events of this sort evolve in unison with the system. The more time that passes, the more the event will have reverberated in the system, and the further from the source will the effect be felt. The events at A and B will have had equal reverberations in the system over the same time interval (because it is the *system* as a whole that has reverberated during this interval) and in this sense they have the same velocity. There is nothing mysterious or counter-intuitive about this once we recognize that it is the *overall evolution of the system* that determines the way in which the events at A and B reverberate through the system.

Where those causal events eventually have their effects will depend on the time that elapses and the density of material in the intervening region of the system. If two events occur at A at time t0, both evolving in opposite directions in the system, then the spatial distance they "traverse" will depend on the density of material in the system. An impulse reverberating through a portion of the system that is packed with more material will cover less distance (from the spatial perspective) because the higher density of material greatly complicates the reciprocal interplay of interactions that determines the rate at which impulses reverberate in that part of the system. Despite appearances, both events that occurred at A at time t0 are in fact reverberating through the material substrate of the system at exactly the same rate. The denser the medium, the more "system" there is to be traversed.

The spatial distance at which the effect occurs depends only on the time that elapses and the density of material in between, but the *potency* of the causal event determines the capacity of the impulse to incarnate itself more rapidly in the system. Again, this is best thought of in terms of the promptness with which objects in a greater state of disequilibrium discharge their influence in the system. Let us consider in detail how this leads to the characteristics of light that are usually understood to be its wavelength and frequency.

At time t0, the mini-system of atoms at S1 disassociates and the released electrostatic tension reverberates with the evolving system. At the same moment there are multiple other causal events in the system. On the ontological level, these various events interact in reciprocal fashion. This particular moment, t0, can be thought of as the beginning of a single pulsation of the system. The various players make their causal input and the system as a whole reacts. It is very loosely analogous to a billiard table with multiple players standing around it. As a clock slowly ticks, all the players simultaneously strike a ball with their cue and the balls impact off each other. On the next tick of the clock, the players strike another ball. The interplay of the balls represents the evolution of the system. In the interval between ticks, a particular ball might not strike any other ball. This corresponds to an influence in the system of lower "frequency" that has not yet had its effect. The balls that are struck more potently are more likely to impact off other balls in any given time interval, corresponding to higher frequency impulses.

Let us return to the electrostatic tension (ET) that has been released from S1 at time t0. It evolves in relation to the multiple other influences in the system that are current in the same moment. As time passes, as we have seen above, ET will potentially have an effect at a greater and greater distance from S1 because this is simply the nature of the "transmission of light": a causal event reverberates in the system; the more the system evolves, the wider will be the effects of reverberation. At some moment subsequent to t0 (let us call it tx), ET will have a causal effect in the system if and only if the following two conditions are met.

The first condition is that *sufficient time must have elapsed* since t0 in order for ET to discharge into the system. The rate at which ET discharges into the system depends on the potency of ET. Like the inflated balloon, the greater the disequilibrium that was released when S1 disassociated, the more rapidly it can potentially have its influence in the system. In this regard, ET must take its place in the pecking order of the system which pulsates in step with the reciprocal interaction of its various causal players. If ET originated in an event that prompted a low state of disequilibrium, then it will have a less rapid capacity to influence other players in the system.

The second condition that must be met is the *presence of suitable target material at the right place at the right time.* Say that at moment tx sufficient time has passed for ET to have an influence in relation to the totality of causal processes that are taking place in the system. During the interval of time t0-tx, ET has reverberated in the system and it can potentially have an effect at a certain causal distance from its source *if* suitable material is present at that point. Suitable material would consist in a group of atoms that could realign themselves with an electrostatic tension of just this magnitude.

If the interval t0-tx is the minimum interval needed for ET to give rise to an effect in the system, then it is a measure of the *frequency* of the impulse. The frequency as we quantify it will consist in the number of these intervals that there are in an appropriate time interval, such as a second. Therefore 1/(t0-tx) seconds, for example, would be a standard way of designating this quantity. From the spatial perspective, the distance traversed by the impulse in this time interval would constitute the *wavelength* of the impulse.

Say that ET is released at t0, but at time tx no suitable material is present at the point in the system to which ET has reverberated. What happens then? Our wandering impulse just needs to be patient, for the players around the billiard table are poised to strike again. Some of the balls that failed to impact during the first interval may have done so because their velocity was too low (corresponding to lower frequency). But others that were of higher frequency will have failed to impact because no target ball was present at the right point on the table. Another interval corresponding to t0-tx must elapse before ET again has the potential to exert an effect on the system in relation to the other causal effects that are evolving in the system. In this way, ET reverberates along with the overall evolution of all the various causal influences in the system. They each plumb the depths of the system, so to speak, in successive moments. If suitable material is not present when their act of plumbing has been completed, then they plumb it again, reverberating farther and farther from their source until a target is found.

3.4 The interference of light – preliminary considerations

In a previous section we saw how a diffraction grating can be used to illustrate the phenomenon of the interference of light. Let us now attempt to describe what is happening in terms of our model. The transmission of light involves no intermediaries on our view, so we must seek to understand the process primarily in terms of the *source* of the causal impulse, the *effect* of the impulse, and the *structural* nature of the intervening apparatus that modifies the evolution of the process. The standard approach *assumes* that a material intermediary is involved in the process and places disproportionate focus on the business of how this intermediary behaves as it passes through the diffraction grating. Hence the common questions: "Which slit did the light pass through? Did it pass through one, both or neither?" These questions are all prompted by the presumption that a material entity of some sort is transmitted from source to target. When we consider light to be a causal process with no intermediaries, by contrast, it is clearer from the outset that the set-up of the *entire* apparatus will influence that process under certain conditions.

A group of atoms in the source form a mini-system around an excited atomic pair, as per our model of the absorption of light. This mini-system disbands and the tensions between the atoms in the group evolve through the system. From the spatial perspective, these tensions can be transmitted over enormous distances before they are reincarnated in matter, giving rise to the formation of new mini-systems. As we have reiterated time and again, the spatial gaps between objects (and the spatial extension of objects) is a feature of our way of apprehending the world. Our sensory apparatus is sensitive to the effects of a particular type of causal interaction between objects. There *are* genuine ontological gaps between objects (and genuine "extensions" or presences of objects), and the type of causal interaction that generates the spatial perspective produces effects in our sensory apparatus that register these gaps and extensions above all else. Indeed these gaps and extensions become the basic framework of the perspective that is generated. But this spatial viewpoint, based though it may be on genuine features of reality, is only a *partial* perspective on reality. And the very lucidity and power of this perspective only serve to cover and obscure the other causal relations that are very much present between objects. When the tensions between the atoms of the mini-system in the source are transmitted elsewhere, it is not that they become independent, autonomous entities cut off from other matter as they pass through "empty space". This empty space is the way that our *sensory apparatus* registers a genuine causal gap between objects, but the same apparatus *fails* to register an enormous network of causal relations between all objects in the universe. It is in *this* network that the tension evolves.

In the case of the interference experiment that we are considering, the tension is eventually transmitted to a group of atoms on the final screen. But first it has to evolve through and with the system, and this evolution is mediated by the intervening diffraction grating. We do not accept that a material intermediary *passes through* the apparatus, but this does not mean that the apparatus cannot *block* the evolution of the causal process. If the diffraction grating were replaced by a completely opaque material, then the tension would be transmitted to atoms in that material (or reflected by those atoms) and would not have any influence on the final screen. The diffraction grating presents the evolving causal process with a combination of opaque areas that block the process from evolving towards the final screen, as well as areas (the apertures) that do not hinder the progression of the process to the final screen (the atoms around the apertures do not block the evolution of the process, but they do *modify* it, as we shall see). Many of the impulses from the source will end up influencing atoms in the opaque wires of the diffraction grating, but these impulses are of no interest to us. We are only concerned with the impulses that will end up having their effect on the final screen.

Take the situation where the diffraction grating has two slits open. A mini-system of atoms in the light source disbands and tension is released. This tension has its effect at the screen, causing a mini-system of atoms to come together with a tension of exactly the same sort. The question is: how did the tension "get" from source to screen? We must resist the inclination to think that it must have passed through one or both of the slits. This would be accurate if a *material* entity passed from source to screen. Instead we must try to gain a perspective on this situation that is not dominated by spatial considerations. Recall that the spatial viewpoint is only a partial perspective on reality. On the level of things as they are in themselves, the process evolves from source to screen via the mediating apparatus. The evolution is *not* simply a spatial progression. There is an authentic *aspect* to the evolution that is registered by the spatial perspective. The gaps and distances represented in the spatial perspective *do* reflect genuine disunities and fragmentations at the ontological level. But it is simply deficient to describe the progression from source to screen as being primarily a progression through *space*. The spatial perspective, in fact, is generated from the effects of processes that have *already* evolved from their source to their target (our sensory apparatus). It makes no sense to think of an evolving process (which has not yet had its effect) as moving through space - the medium generated by sentient beings from evolving processes that have *already* come to fruition.

Instead of describing a causal process as evolving through space, it would be more correct to describe it, as we have been doing for some time now, as an evolution of a particular impulse in relation to the totality of causal relationships between the relevant causal players in the system. If we think of it as a progression of an autonomous intermediary through space, then the particular and narrow spatial relationships between the intermediary and the apparatus at the time of passage become dominant. Which slit did it go through? How did it apparently manage to go through both together? But if we view the causal process as the evolution of the causal relationships between all of the relevant players in the system, then the *overall* structural configuration of the system is seen as having a more telling significance.

Let us recount the story from this broader perspective. A causal process that is initiated in the light source is mediated by the intervening apparatus and eventually has its effect on the screen. The intervening apparatus confronts the process with two paths of equal resistance between source and screen. These paths are much deeper and richer than spatial paths. They are *causal* paths. This means that the process is influenced by an apparatus that has a double-barrelled effect on the evolving process. The process simply *cannot* evolve through just *one* of the paths, no more than a magnetic south pole could choose to be affected by just one of two magnetic north poles placed in front of it at equal distances. These twin apertures have an equally essential role in influencing the evolution of the causal process. The great fallacy of the naïve spatial approach is to believe that *something* must pass through one or the other of the openings. If the something passes through the upper slit, for example, then the lower slit should have no role in mediating the causal impulse from source to screen. Even if that slit had been closed, what difference could it have made? But of course the experiments show that whether a slit is open or closed makes *all* the difference, and this has thrown the naïve spatial approach into a quandary. Once we appreciate that the mediating apparatus confronts the evolving impulse in a *causal* way, then the situation is greatly clarified. Two slits of equal dimensions present the impulse with causal paths of equal significance, so the impulse *must* be influenced by both paths equally. There is no escape from this conclusion. The way that the impulse plays itself out at the final screen will be inevitably marked by the twin causal influence of the mediating apparatus.

What sort of mediating influence is exerted by the atoms around the slits on the evolving causal process? Light transmission, on our view, involves the evolution through the system of a tendency to draw atoms into an electrostatic bond. The atoms at the slits, like all atoms, have the internal polarity that constitutes their very existence. Thus they can exert an influence on an impulse that is evolving through the system. Recall again our billiard table analogy. At each successive moment, the various players in the system exert causal influence on each other, and these influences reverberate through the system, having their effect at a time that is determined by their potency and by their relative position in the causal configuration of things. At a particular moment, the impulse of "light" that is "passing through" the slits will come under the causal influence of the atoms around the slits. These atoms do not constitute the right kind of material for the light to be reincarnated in a changed electrostatic configuration of the atoms. But they can affect the way that the impulse evolves, and in this case they have a determining influence on the *direction* that the impulse takes in the causal configuration of things.

How do they modify the direction in which the impulse evolves? As we shall see later, the transmission of light is different to the transmission of an "electron". What is normally understood as an "electron" involves the evolution of an influence that gives rise to an electrostatic *imbalance* in an atom. Such an evolution will be dramatically influenced by the presence of an electric field. Light, by contrast, is an electrostatically *balanced* influence whose evolution is not modified by the presence of electric or magnetic fields. But its direction of evolution can be modified by the close proximity of atoms that are *already* in electrostatic equilibrium. Such atomic bonds have the potential to draw other atoms into electrostatic tension, which is what happens when light is absorbed by a group of atoms that makes a transition to a larger group. Thus, when light evolves "through" the slits of a diffraction grating, a pair of atoms in electrostatic equilibrium in the slits tends to draw the evolving impulse towards itself. The evolution of the impulse continues on through the slits, but its direction has now been modified by the mediating influence of the pair of atoms in the grating. "Diffraction" has occurred.

Why does a so-called interference pattern appear on the final screen? This is the Gordian Knot of physics, a question that has yet to yield its enormous potential as far as our understanding of the character of space, the nature of light and other causal processes is concerned. The general acceptance of the wave "explanation" has prevented us from reflecting sufficiently deeply on this phenomenon. It would be no exaggeration to say that an enlightened analysis that is no longer hampered by the wave assumption has the potential to renew physics. But even a cursory look at the issue reveals that what is happening is not interference at all.

3.5 The so-called "interference" of light – towards an explanation

Consider Figure 3.4. The slits at A and B provide the ontological void required for a causal influence from the source to make its way to the screen. If the aperture was not present, no impulse could evolve through the causal barrier provided by the intervening diaphragm.

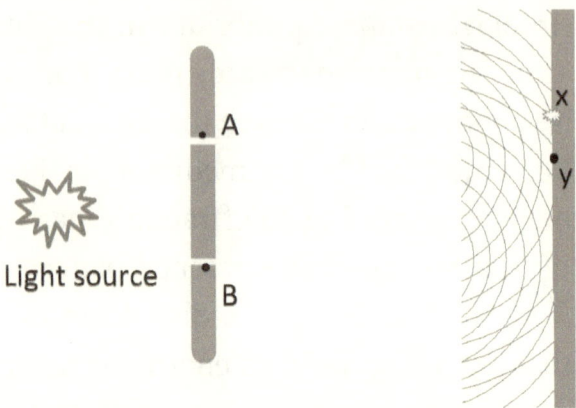

Figure 3.4 The interference of light. A more profitable way of thinking of this phenomenon is not in terms of individual wave trains progressing from the slits and interfering at the screen. Rather a unitary causal impulse from the light source is jointly operated on by atoms at A and B in the slits. The causal impulse has an associated "wavelength" that is inversely determined by its potency. This length entails that the impulse can be jointly guided by A and B to point x on the screen, but point y is inaccessible. The converging "waves" are merely a visual way of calculating the points that are potential targets of such unified action from A and B, as well as showing points (such as y) where such action is impossible.

The narrowness of the slits, however, entails that any impulse that does evolve to the screen will be causally influenced by atoms in the slits. Given the fact that the slits are uniformly identical and positioned at equal distances from the source, it follows that any causal influence from the source that is thus modified by atoms at A will be *equally* influenced by atoms at B. This means that most (if not all) of the impulses that arrive at the screen have been modified in a joint fashion by two separate causal players (or two separate groups of causal players). Consider one such impulse. Though it is a single impulse of a definite magnitude, it will be modified equally by atoms at A and atoms at B. By "modified" we are referring to the fact that the evolution of an impulse of this sort (which originated in the fragmentation of electrostatic equilibrium between atoms) will be susceptible to being influenced by the electrostatic capacities of other atoms that it encounters during its evolution. These influences will help determine where the impulse will eventually have its effect.

In the last section we saw how impulses have a "frequency" and a "wavelength" associated with them. The most potent light impulses are those transmitted when *larger* subsystems disassociate from an excited atom. They represent a greater discharge of disequilibrium into the evolving system and will have a tendency to rapidly embody themselves in suitable target material. The short interval of time required to make their presence felt in the system (relative to other causal influences) gives these potent impulses a higher "frequency". Electromagnetic impulses of lower potency will take relatively longer to have a potential effect on the system. A greater interval of time must elapse before their influence can be made relative to the myriad other influences present. This greater interval of time means that the impulse will have reverberated further from the source, giving rise to a longer associated "wavelength".

This picture of the various causal players reciprocally interacting, thus giving rise to a unified reverberation of the system, highlights the deficiency of representing impulses as moving in *an infinitely-divisible continuous fashion* through space. Consider Figure 3.5. Object A is a group of atoms that disassociates, prompting the reverberation through the system of a causal impulse. Please note that in this diagram the dimensions of the axes do NOT represent the magnitude of the electrostatic tension of the alignment of atoms. Rather, the horizontal axis represents the "wavelength" of the impulse as it reverberates along with the system, whilst the vertical axis represents the fact that this causal impulse *tends to form an alignment of atoms in a particular direction* (this fact can be used to adequately explain the phenomenon of polarisation). In fact, the length of the horizontal axis is inversely proportional to the size of the electrostatic tension released when the impulse was emitted. The greater the electrostatic tension, then the more rapidly the causal impulse will tend to discharge in the system relative to other causal impulses. Therefore, less time will elapse before the impulse will have the capacity to have its effect. The shorter the time interval, then the lesser the reverberation of the effect from its source and the shorter the "wavelength". Hopefully all of this will become clearer as we go along.

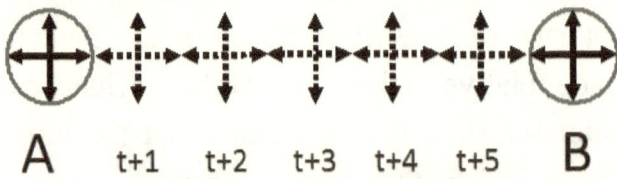

A   t+1   t+2   t+3   t+4   t+5   B

Figure 3.5 The evolution of a causal impulse from A to B. The diagram illustrates the spatial positions where the impulse would form an alignment of atoms if there happened to be suitable matter in that position at that moment in time. But the impulse itself does not move through space. The spatial perspective is the ink in which a causal impulse would be written if suitable matter were present to absorb the influence, thereby producing the effects that generate the spatial viewpoint.

When the mini-system disassociates, the tension reverberates through the causal configuration of things, most likely in a direction orthogonal to the direction of the electrostatic alignment of the T1-T2 pair. Recall once again that this tension is not transmitted through *space*. The spatial perspective is generated from the *effects* that are produced when tensions such as these finally encounter other matter and cause a perceptible alteration in that matter. But - self-evidently - whilst the impulse is *still* evolving, it has not yet produced an effect, so it cannot figure in a perspective that is generated from such effects.

Where then is the impulse evolving? It is our utter dependence on the spatial perspective that prompts us to phrase the question in this way. The impulse is not evolving in a *place* but in the totality of causal connections between all the causal players in the system. From a mathematical point of view, we could call this a "causal configuration" space, and there is no doubt that it would require an enormous number of dimensions to represent this vast network adequately. No such multi-dimensional space exists in reality, of course, but the causal variables and magnitudes that mathematicians would use to construct the space are very real indeed. The distinction between the reality of the *parameters* that mathematicians use to construct multi-dimensional spaces and the *utter non-existence* of such spaces themselves is a distinction that we would do well to keep in mind, lest our account of the physical world degenerate into a form of science fiction.

The impulse under consideration is evolving from A to B in this causal configuration "space". If we were able to represent this evolution accurately from the spatial perspective (consider again Figure 3.5), we would perceive the impulse "jumping" discontinuously in steps across the gap between A and B. The point corresponding to t+1 represents the place (and the time) where the impulse would have had its effect if it had encountered matter there, and so on until the impulse eventually encounters matter at t+6. Needless to say, there is no jumping of this sort going on in reality. The impulse requires an interval of time (relative to the other impulses in the system) to have the potential to effect causal change in the system. By the time the interval has elapsed, the system has evolved, the original causal event has had more time to reverberate in the system, and the impulse will thus have its effect at a greater "distance" from the source. Successive "distances" of this sort correspond to what we normally designate as "wavelength". They effectively mean that a causal impulse has the capacity to effect a change in reality *at a sequence of stepped spatial intervals*, but nowhere in between.

The stubborn belief in the autonomous reality of the spatial void prompts people to assume that the void is perfectly divisible. If it really *were* perfectly divisible, and if it really *were* the substrate through which objects exert their causal influence on each other, then we would expect these causal processes themselves to exhibit a corresponding continuity and divisibility. A causal influence passing through an infinitely divisible void should be potentially able to have its influence at *any* infinitesimal point in the void. However, once we appreciate that the spatial void is *not* the substrate through which causal influences pass, but is the fruit of a perceptual process that is *itself generated by the effects of causal interactions*, then our understanding radically changes. It now becomes apparent that the assumption that the space between objects is continuously divisible is simply a category error. Objects are exerting influences on each other and these interactions give rise to effects that generate the spatial perspective. The spatial void between objects represents a causal gap of some sort between objects, a gap that can be closed by a particular sequence of causal events. We have no reason to believe that the number of such causal events could be infinitely continuous since the separation itself must have been caused by a finite series of events.

Even if we are right that space should *not* be thought of as an autonomously existing, infinitely-divisible substrate, then it still might be claimed that the *causal players* in the world are capable of emitting an infinite spectrum of causal influence. In other words, someone might still hold that *energy* is infinitely divisible and that a causal source, such as an atom, can give rise to a complete spectrum of electromagnetic radiation. The quantized nature of action, however, is one of the best-established empirical findings of modern physics. In our scheme, this empirical discovery arises from the fact that the causal activity of the fundamental units of matter is itself integral in nature. It doesn't follow, however, that the influence of a given atom cannot be *distributed* over a number of other atoms. Typically, when we speak of the electromagnetic interaction, that influence is distributed over a number of atoms of the form n^2, giving rise to the particular mathematical characteristics of such phenomena that are expressed in the Rydberg expression. But whether we are talking about electromagnetism or some other type of interaction, the important point is that the causal interactions involve these indivisible units, each of which has a fixed work potential.

The pulsating nature of the evolution of the system derives from the fact that the fundamental causal players in the system have this integral nature. The reciprocal influences of players of this sort will tend to cause reverberations in the system in which the more potent influences reverberate most rapidly. As a result, the causal configuration space (the abstract space in which all of the causal interactions of the system unfold and evolve) will be characterised by a progression of fixed step-like events. The process by which source A has its influence on target B will unfold over a series of definite, indivisible steps.

It may be helpful to consider an analogy at this point. Say that four people – call them A, B, C and D - are gathered in a room and each one has a collection of six questions that he must direct towards his companions. These questions could include asking for that companion's name, age, profession, etc. While asking the questions, each person must move randomly around the room, directing their questions at the person who happens to be closest to them at that particular moment. The game ends when every person has asked all six questions of each one of his companions. Take two of these people, A and B, and consider the "causal process" (let us call it P) by which A asks B all of the questions. The length of time required for P to be completed will depend on contingent factors arising out of the movement of the people around the room, the initial positions of A and B when the system was set in motion, the total number of people in the room, etc... Despite these contingent features, the process will nevertheless evolve in a finite series of definite steps. That is because the number of causal players is finite and the causal "impulses" they emit are of a definite, integral sort.

With just four people and only six questions, it is likely that there would be an irregular character to the intervals of time between each of the question posed by A to B. If we were to enlarge the scale of this game to thousands of people and a much larger number of questions, then significant regularities would emerge, as always happens in large systems that have many causal players. The rate at which questions in general are posed by players would begin to resemble the ticking of a very regular clock. Say that we alter the rules so that the number of questions asked by each player is now a variable amount depending on the individual player. Some players ask relatively few questions, whilst others ask a colossal number. Now imagine that a mathematician plots the progress of causal process P that is evolving between A and B (this is the process by which A asks B a set of questions of variable number). Say that the mathematician has limited cognitive faculties that does not give him access to the real dynamics at play in the system. His sensory apparatus registers the moment that P begins between A and B, and the moment that it comes to completion, but he does not perceive the nature of the process itself. It would be natural for him to look at the line on the graph and believe that there is a continuous, infinitely-divisible progression of events during the evolution of the causal process. In the same way, the system in which we live is characterised by causal players (atoms) exerting integral influences on the rest of the system. The character of these interactions and the nature of their effects generate the spatial perspective in our sensory apparatus, and we end up with a stunningly lucid spatial representation of the world. It is easy for us to believe in the infinite divisibility and autonomy of this world, but it is illusory. And "interference" phenomena bear elegant testimony to the very different nature of reality.

Armed with this notion of the "length" associated with a causal impulse, let us consider again the two-path experiment (see Figure 3.6). An impulse from the causal source is equally influenced by an atom at each of the slits. The impulse, we remember, is evolving in the causal configuration of things in which the causal interaction between all objects in the world plays itself out. While the impulse is still evolving, nothing is detectable from the spatial perspective. The spatial perspective is generated when a causal evolution comes to fruition, producing a change in a causal target. This change impinges on our perceptual apparatus and contributes towards generating the spatial perspective. No such change has yet occurred, however, in the early phase of the evolution of the process when the impulse is influenced equally by the atoms at A and B. At this stage the process is being jointly guided towards its final destiny by A and B. The *nature and magnitude* of the impulse are determined by the causal source atoms that were the ultimate origin of the impulse. The atoms at A and B are merely "diffracting" the impulse, cooperatively exerting their electrostatic influence on where it will have its effect.

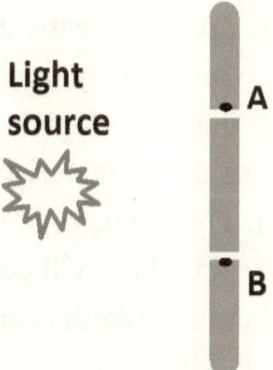

Figure 3.6 An impulse from the source is mediated by atoms at A and B. *The evolving impulse is jointly influenced by the electrostatic capacities of groups of atoms at A at B. Since the group of atoms at any particular slit produces a combined influence, each group behaves like a single causal player.*

The "length" associated with the impulse is dictated by the potency of the event that occurred at the causal source in relation to the other influences that are evolving in the system. The impulse reverberates through the system until such point as its evolution becomes influenced by the electrostatic capacities of atoms at A and B. From this point onwards, the evolution of the impulse is fundamentally changed. The atoms at A and B are not suitable material for the influence to exert an effect in the system, but these atoms *modify* the way that the influence will evolve from now on. Recall our picture of the system as a set of causal players whose impulses reciprocally influence each other, making the entire system pulsate as the totality of these causal processes evolve in unison. Now the influences of the atoms at A and B contribute to the mix. The fact that these atoms are positioned in such a way as to exert an identical effect on the process means that they will tend to exert a *joint* influence on the way that it evolves.

The evolution of the causal process is consequently modified. It still retains its characteristic "wavelength" and "frequency", however. The atoms at A and B do not alter the potency of the impulse, but they will certainly have a say in the direction that it takes in the causal configuration of things. Let us see in more detail how they conjointly determine the points at which the process has the potential to influence the screen. Figure 3.7 depicts the final phase of the evolution of the impulse that is being jointly guided by A and B.

As we have discussed earlier, the impulse has a length associated with it. As the system evolves, the various causal processes that are in reciprocal interaction have the potential to have their effects after a succession of finite intervals of time. The more potent the influence, the shorter the interval of time that needs to elapse before it has the potential to exert its influence in the system. But a suitable target must be present at the position corresponding to the extent to which the process has reverberated in the causal configuration of things, and it must be present at the moment the interval elapses. Prior to being influenced by the atoms at A and B, the causal process had the potential to exert its effect at these disjointed spatial intervals (corresponding to the supposed "wavelength" of the influence). Now that the influences of A and B have entered the equation, the points at which the impulse has the potential to have an impact can be geometrically represented precisely by the traditional wave diagrams showing interference, as we have done in Figure 3.7. The simplicity and convenience of this geometrical representation *does not imply* that what is happening is a case of interference of a wave motion. This is yet another case where a mathematically-convenient model risks becoming the justification for a flawed understanding of material reality. Such models are useful for predictive and descriptive empirical purposes, but they can seriously stifle our comprehension of the physical mechanism underlying the empirical data.

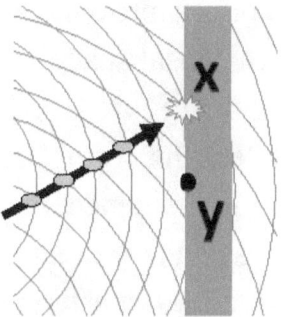

Figure 3.7 Detail of interference. *The point x is conjointly picked out by A and B, whilst the point y is simply inaccessible to the joint influence of A and B for a causal impulse of this "length".*

The evolving process is jointly guided by both A and B. That means that the influence can have its effect at points on the screen (or in the intervening space) that are a whole number of these "wavelengths" from *both* A and B. If it were to have its effect a whole number of lengths from A and a fractional number of lengths from B, then that would constitute sure evidence that B was *not* jointly modifying the influence at all. If B is contributing to the evolution of the process, then the impact will occur at a distance from

B that is consonant with the way that such an impulse reverberates through the system. This stands to reason. The evolving process has come under the influence of B and, in this sense, is reverberating from B (as it is from A). The final point of influence will thus be an integral number of lengths from both.

3.6 The potential of interference phenomena for informing our views about space

The original causal event determines the so-called *wavelength* of the evolving process, whilst A and B jointly govern the *position* in causal space where the process will come to completion. In Figure 3.7, the atom at y is not a whole number of lengths from both slits, so it cannot be affected by a joint influence for an impulse of this length. The blank areas on the screen should be a source of puzzlement for the standard interference explanation. How can causal impulses which have a particle-like or wavelike nature be able to interfere destructively and leave *no effect at all* at these points? Even the destructive interference of water waves leads to a certain minimal disturbance because a crest and a trough must first *converge* before they exert a negative influence on each other. But *nothing at all* seems to converge at the blank areas of an interference screen. It is as if the causal process simply has no influence in this region. There can be no doubt that naïve attitudes about the nature of space have hampered our ability to account for phenomena of this sort.

The first serious question is the issue of the reality of space apart from the causal processes that generate our mode of representing reality. This is a searching question that presents itself for all ontologies, regardless of whether we believe in the possibility of infinitesimally small units of action or not. The ontology that has been presented here (which holds that causal players have causal capacities of particular integral magnitudes) does not conflict in an obvious way with the belief of mainstream physics in the quantum of action. Causal processes, in the genesis-unit model, always have a magnitude of work potential that stands in a particular relation to the basic work impulse that constitutes the atom. In a later chapter we will consider how this work magnitude stands in relation to Planck's constant. For the moment what is important is the claim that causal processes have a certain magnitude and that this magnitude is associated with a particular length (from the spatial perspective) as the process evolves through the system.

The length that is associated with a process has a certain negativity of character: it indicates the successive areas of the causal configuration of things where this particular influence *cannot* have an effect. That is because the process requires an interval of time (relative to the rest of the processes that are working themselves out in the system) in order to be capable of having an effect in the world. During this interval the system as a whole continues to reverberate and, as a result, the particular process in question ends up having its effect at a point ever more distant from the source. The "wavelength" represents the distance that the process has traversed as the system reverberated during that interval of time necessary for the impulse to have its effect relative to the other evolving impulses in the system.

Someone might say: "It is as if the impulse becomes invisible periodically and then reappears at a point that is located one wavelength on from the point where it last appeared. In between these periodic appearances, the impulse travels through another dimension that is not visible to us". This sort of description is useless and misleading. It is not that the impulse vanishes and moves in another "dimension", whatever that might mean. During these intervals, the impulse has no possibility of causing the sort of effects that actually *generate the visible order*. Furthermore, it is the *system* that is evolving during these intervals; nothing is moving in another "dimension". At the end of each periodic interval, the system evolves to a point where that impulse has the potential of effecting causal change, *if matter happens to be present* in the right position in the causal configuration of things. So during those intervals in which the impulse appears absent from the world, it is not that it has disappeared from space; rather, the system is going through an interval of evolution during which this impulse has no possibility of generating the sort of effect that would allow us to experience it from our spatial perspective. Or better, during these intervals, the system is generating various effects that we use to produce the spatial perspective; but none of these effects comes from the impulse under consideration.

Successive areas of the causal configuration of things are inaccessible to causal processes of this sort. The process cannot exert any influence at these points. When a process is being guided by *two mediating influences*, then these influences complicate the pattern of points in the system that are inaccessible to the effects of the process. The standard geometrical depiction showing the interference of concentric waves accurately captures the configuration of points that are accessible/inaccessible to the causal influence. Figure 3.7 shows us points in the intervening space where the process would have had its effect *if* there had been suitable matter in that area. We must be careful not to take the picture at face value and assume that the impulse moved through the intervening space before arriving at x. When causal evolutions come to fruition in the causal configuration of things, then they generate effects that give rise to the spatial viewpoint. In the absence of such effects it makes no sense to speak of an evolving causal impulse moving through space. And once we eliminate talk of entities moving through space, then we pave the way for a solution to the mystery of "interference".

Evidently, this explanation of interference is elementary in nature and requires the kind of elucidation and empirical rationalisation that only experimental scientists can give. But it has potential as a strategy for developing our understanding of a wide range of phenomena. Let us consider just one example. According to our account, the system in which we live consists in a vast range of causal processes reciprocally interacting with each other and giving rise to certain effects in our sense organs. On the basis of these effects, we construct our spatial perspective on reality. The primary shortcoming of our spatial perspective is not the truth or falsity of the impressions that it delivers: the main problem consists in its *limitations*. We have already discussed how we consider the *extent* of spatial voids and magnitudes to be an accurate representation of genuine voids and magnitudes on the ontological level (relative to the motion of the observer). This is where we part with Kant – not that we were ever in his company in any more than a cordial way. Unlike Kant we hold that the human spatial representation of things contains *genuine and significant truths* about the make-up of physical reality. But our perspective is nevertheless limited, and the phenomenon of interference shows us one aspect of this limitation.

Our account does not intend to present a schizophrenic picture of the world, but there is a sense in which the spatial world generated by the effects that impinge on our sense organs resembles a sort of parallel reality alongside the world of real things. It is important that we not describe this state of affairs in terms of parallel "dimensions", lest we fall prey to the kind of wonderland language that has beleaguered quantum mechanics for decades. That being said, let us tolerate this parallel-world view for a moment in order to make a particular point. A causal impulse is evolving in the system in the manner that we have described in this chapter. Until that process gives rise to an effect that impinges on our sense organs, we can have no direct experience of it. In this sense, such effects are the ink in which the spatial perspective is written, but they leave much unwritten. A flash on a screen is taken by us to indicate that light has been transmitted from source to target through the spatial gap that we represent in between. But the spatial perspective is generated from effects, and it is quite wrong to think that the causal impulse must have traversed a space in which that impulse has had no effect whatsoever.

Of course, over in the "parallel" reality of real things, a causal impulse *has indeed* reverberated with the system, but this cannot be simply identified with our spatial world. Reality consists in deep and wide-ranging causal connections that are not accessible to our sense-organs for the simple reason that they do not always generate the kind of processes that impinge directly on our sensory apparatus. Furthermore, even those processes which eventually have the potential to impinge on our sense-apparatus are not apprehensible during the intervals of time that they are in evolution with the rest of the system. "Wavelengths", ultimately, are our way of representing the *potential* of causal processes either to cause or not to cause an effect in the world, an effect that could impinge on our sense-organs. In this sense, returning to our parallel universe picture, wavelengths represent electromagnetic processes as they are evolving in the system, imperceptible to our senses. When each whole wavelength has been "traversed", the process has the capacity to produce an effect. When that effect impinges on our sensory apparatus, it can help to generate the parallel universe of spatial perception.

It is always important to be on our guard against the spectre of idealism. The above discussion might seem to attribute a purely epistemic character to wavelength, as if it existed only in our minds. While it is true that *we* derive the notion of wavelength from our study of the potential of causal processes to have their effect, it simply doesn't follow that wavelength has no basis in real things. As we have seen, the length has a sure foundation at the ontological level, being directly related to the potency of the causal process under investigation. The more potent it is, then the shorter the interval of time required for the process to have the potential to come to fruition relative to the other processes that are reverberating in the system simultaneously. The briefer the interval of time, then the less potential reverberations between cause and effect. Hence, the lesser the evolution in the causal configuration of things and a consequently shorter "wavelength". Thus, the magnitude of a "wavelength" is based on factors that are *utterly independent* of the functioning of our minds or the shape of our conceptual schemes. What we contribute to the notion of wavelength is the intuitive idea of spatial extension (which derives from the natural way we represent reality to ourselves) and the less intuitive concept that it involves an oscillating motion through space.

Interference remains a phenomenon with the useful capacity to perturb our naïve conceptions about space. Imagine a car park attendant who sits in a cabin with no window on the parking lot outside. Whenever a car comes to the entrance/exit of the lot, a light flashes on a screen in front of him. The same happens when a car is parked, and the light is extinguished when the car leaves. The attendant does not see the car move from the entrance to its parking space. His perceptual apparatus can give him no information about that process. We could say that the screen with the lights is the ink with which his view of the world is written, but that does not mean that there is nothing else going on in the world. Processes are continually evolving in the car park, but he only perceives the end point in that process when a causal change is effected in his perceptual apparatus. If it were not for the phenomenon of interference, we might never be aware that our picture of the world is built up from dots in which the lines in between are missing. We substitute those missing lines with the spatial void of our intuition. The picture we generate from the dots is beautiful, but we must never lose awareness of the holes in it: the processes that produce the dots remain inaccessible to our senses and can only be penetrated indirectly by rational reflection.

3.7 Some questions about interference

What is the likelihood that atoms at A and B could exert joint influence on the same causal impulse?

Earlier we used the analogy of a magnet that has two other magnets positioned at equal distances from one end. It is impossible for the influence of the first magnet not to be modified equally by the influence of the other two. Evidently, we would expect the transmission of light to be quite different to magnetic influence. We are inclined to think that the latter must be more "spread out" whilst the former seems to be exactly the contrary. But the usefulness of the magnetism analogy is borne out by the empirical phenomenon of the "interference of light". It is quite clear that the evolution of the light impulse *is* jointly modified by two or more causal players in certain circumstances. In our example, one of the causal impulses from the source is mutually influenced by atoms located in the slits at A and B. The influence exerted by these atoms is usually referred to as diffraction, which really amounts to a form of reflection or deflection of an impulse. The impulse is not absorbed by atoms at A or B, but the path along which its future effect will be felt has been deflected as the impulse evolves towards the final screen.

If the apparatus depicted in Figure 3.4 is set up properly, it is *absolutely inevitable* that a significant proportion of the causal impulses issued by the source (that succeed in making their way to the final screen) will be jointly acted upon in this way. If the slits are correctly positioned with respect to each other, then for every atom (or group of atoms) on the top slit that modifies the evolution of an impulse, there will be an atom (or group of atoms) on the bottom slit that modifies its influence conjointly. This is a statistical likelihood of enormous proportions, given the sheer number of atoms in the material around each aperture.

The fact that atoms at A and B influence the evolution of the process jointly does not mean that they influence it *equally*. This is what gives rise to a pattern of "impacts" that is *distributed* all over the screen. The atoms at A might tend to diffract the impulse in a particular direction. If this influence is dominant then the final point of impact on the screen might lie in this direction from A, but it will still be a whole number of lengths from the atoms at B.

Does this account of interference violate the constancy of the speed of light?

Earlier we cautioned against thinking of this evolution in terms of two separate wave trains from A and B. Such thinking is simply misleading and runs into all kinds of difficulties. For example, point y is closer to A than to B. If an *individual* causal impulse from the source is thought of as a wave (composed of a single crest and a trough) that passes through the slits at A and B and becomes two wave motions, then how can these separate wave motions arrive at y *at the same time* to produce destructive interference? The wave train from A will have arrived *earlier* at y given the absolute constancy of the velocity of light. This is a grave problem for standard approaches to the phenomenon of interference in the case of the transmission of a *single* causal impulse through the apparatus. Even if we think of the impulse as being some sort of particle that is influenced by both slits, we still have to explain how the two interfering influences arrived at the screen at the same time when they had to pass along paths of different lengths.

A more coherent way of thinking of interference is in the terms that we have been advocating. The point at which the influence will have its effect is jointly determined by A and B. The causal influence has an associated "length". This does *not* mean that an entity of this length passes through the apparatus; what it means is that the causal impulse will have its effect in a position in the causal configuration of things that stands a whole number of lengths from A and from B, as well as from the original source. The joint influence of A and B is completely integral in nature, as befits a unitary causal process. And this is so even if the process is the product of the reciprocal interaction of *multiple* influences. This unitary process cannot deliver half an impulse to a point on the final screen. A and B modify the impulse jointly, and this means that certain areas of the screen are completely inaccessible. In order for the causal influence to impinge on a point such as y, it would have to be a whole number of lengths from its source AND be guided *either* by A and B, and that is not the case when the apparatus is set up like this.

It might be objected that the point x on the screen stands at differing distances from A and B. How can A and B have simultaneous joint influences at points that lie at different distances from them? Surely *this* violates the constancy of the speed of light? But the problem does not arise in the account that we have presented. Our approach is very careful about insisting that no impulse moves in the space between A and x, just as no influence moves in the space between B and x. So it is not the case that an impulse along the shorter path is required to move more slowly and vice-versa. The impulse is not taking a particular path but evolving in the causal configuration of things, and this rate of evolution is absolutely constant because it is determined by the playing out of the complex interactions of the entire system. In normal circumstances, when a single impulse of light gives rise to effects from the spatial perspective, this will be measured as having moved through space at a velocity *c*. We identify the point where the light was emitted, the point where it reached its target, measure the distance between and the time taken for transmission. In the case of "interference" phenomena, however, a measurement of this sort makes no sense. The impulse cannot be unambiguously asserted to have taken an individual path because it has evolved in a more complex way in the system under the influence of multiple players.

It is our contention that the causal influence simply evolves with the system, a system that happens to put the impulse in joint interaction with A and B. Eventually the cause has its effect at x. The fact that A is closer than B to x does not mean that A and B cannot influence the atom at x in a joint fashion. Why should it? That argument would only make sense if *intermediaries* travelling at a constant speed left A and B at the same moment. But the transmission of light involves no intermediaries. If it were possible to measure with complete precision the speed of light through an interference apparatus, then the results should be illuminating. In a case such as that represented in Figure 3.4., where the distances from the slits are of unequal length, it could be expected that neither the short path nor the long path would empirically yield a velocity c. To maintain a velocity of c we would have to attribute to the light a pathway *intermediate* to those from the two slits. Such an empirical finding would help to support the contention that the constancy of the velocity of light is due to the fact that its rate is tied to the evolution of causal interactions in the entire system.

How appropriate are terms like "wave-particle duality" and "interference pattern"?

The classic two-path experiment involving light has untold potential for unlocking the secrets of the true character of space and the nature of electromagnetic radiation. There can be little doubt that the notion of wave-particle duality has hampered this potential dramatically. "Interference" phenomena challenge science to face in a more radical way the evident fact that no intermediary moving through space can account for the empirical data. But wave-particle duality has historically provided a means of escape, a way to hold on to entities that cannot cope with the experimental results. It would be of significant pedagogical value to dispose of the notions of wave and particle altogether when we speak about light, and it would also be beneficial to jettison the term "interference". What is taking place in these situations is simply not interference because it is not the case that separate influences are emanating from different slits and interfering with each other. Perhaps a suitable alternative to the term "interference pattern" might be "joint-influence pattern".

Summary description of joint-influence patterns.

Consider Figure 3.8. The paths along which the causal impulse will potentially have its effect are determined by the magnitude of the impulse AND the joint influences of atoms in either slit, such as those at A and B. The arrow represents one such path. The impulse could have an effect at any of the points along the arrow, if there were a physical object positioned at any of those points. But there are no objects in those positions and it would be highly misleading to state that the impulse "passes through" that empty space until it encounters the screen at x. The evolution of an impulse through the causal configuration of things does not involve the transmission of anything through space; not a wave, nor a particle, nor a perturbation in a field of any sort. Furthermore, the process that is jointly influenced by the atoms at A and B is an integral thing. From the time that the initial process has its beginning in the source and begins to evolve in the system, until such time as it has its effect in the target, it cannot be divided or fragmented. It does not split itself up and take both paths through the apparatus, converging at the screen. The persistent error in approaching the two slit experiment is to view it as involving a causal player that divides itself up and takes two paths, effectively becoming two causal players along separate paths that eventually re-converge. This runs into all kinds of difficulties when confronted with the empirical evidence. A better approach is to understand the phenomenon as involving a *single* causal impulse that is *jointly modified* by two causal players (A and B) located in the apertures.

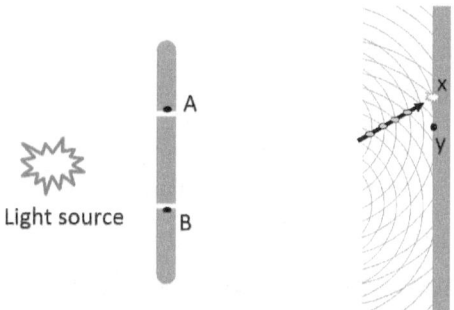

Figure 3.8 Joint-influence pattern

3.8 Other two-path experiments

What is the relation between the classic "interference" experiment and other two-path experiments that do not produce an "interference" pattern, but which involve the disturbance of so-called "complementary" properties? It has long been recognized that the same mystery is at the root of what is happening in both of these types of experiments. In 1922, Otto Stern and Walther Gerlach directed silver atoms through an inhomogeneous magnetic field (see Figure 3.9), a relatively simple experiment that has had far reaching consequences for the development of atomic theory. If the atoms had been magnetised in the classical sense, then they should be deflected by varying degrees depending on the orientation of their magnetic poles when they entered the apparatus. Atoms with their north poles facing up would be expected to be deflected downwards. The opposite should occur for those with their south poles facing upwards. The degree of deflection should depend on the extent to which the magnetic poles within the atom are in alignment or out of alignment with the magnetic field. Thus the final screen ought to manifest a full range of impacts from top to bottom. But when the experiment was performed, it was found that *all* the atoms were deflected either up or down. Experiments of this sort have been extended to other "particles" and are the empirical basis for the hypothesis that elementary particles possess a form of angular momentum called "spin". The idea behind the hypothesis is that a spinning electric charge should give rise to a magnetic field, but it is far from clear why such spin should not give rise to a full spectrum of deflections when the particle is placed in a magnetic field.

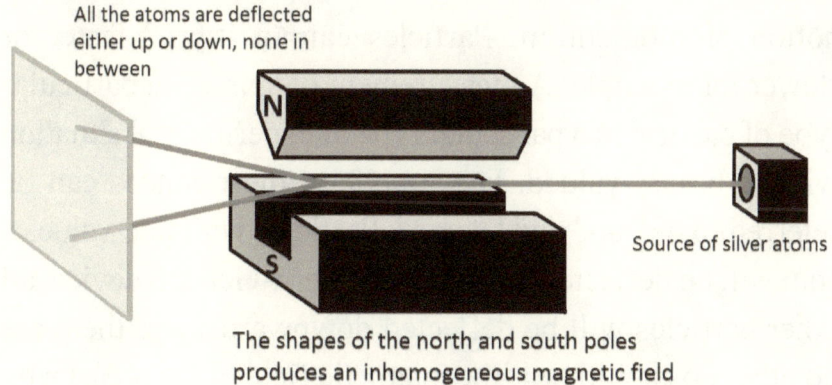

All the atoms are deflected either up or down, none in between

Source of silver atoms

The shapes of the north and south poles produces an inhomogeneous magnetic field

Figure 3.9 The Stern-Gerlach experiment

Figure 3.9 depicts a source emitting atoms towards an inhomogeneous magnetic field. Objects that are magnetised in a classical way have a north pole and a south pole. If the north pole is facing up, then the object ought to be deflected downwards when it passes through the above apparatus. The opposite should occur if the south pole is facing upwards. A large number of objects will have their magnetic poles oriented in random directions. These objects ought to be deflected downwards or upwards to various degrees depending on the extent to which their poles are in alignment with the magnetic field. Thus the screen should show a fairly uniform pattern of impacts of the atoms in a vertical direction from top to bottom. When the experiment is done, however, all of the impacts appear at either the top or bottom of the screen, with none at all in between.

Angular momentum is a straightforward concept in itself, but the property known as "spin" is counter-intuitive in many respects and does not cohere with our everyday notion of momentum. Particles cannot "spin" faster or slower for example. A measurement of spin for a particular type of particle in a particular type of experimental situation will be binary-valued. For simplicity these values can be referred to as "up" and "down". Particles with one value of spin will be deflected upwards in a Stern-Gerlach device; all other particles will be deflected downwards; and there are no other possibilities. If the magnetic field of the measuring apparatus is aligned in a different direction, then we can obtain values for the spin in that direction. For example we can measure the spin of particles along mutually orthogonal x-axes, y-axes and z-axes. Later in the book we will consider this property in more detail and argue that it is in fact a characteristic of the fundamental unit of matter, related to the "protomagnetism" that we have discussed earlier.

We have no need to discuss the physical basis of spin in detail in order to show how it can manifest empirical patterns that correspond to the so-called "interference" of light. The particular characteristic that interests us is the "complementarity" displayed by this property when spin measurements are taken along different axes. For a particular type of spin that an "electron" exhibits (let's call it "spin-type x"), there is a complementary type of spin along a different axis (call it "spin-type y") whose value will be disturbed by the experimental conditions that are necessary to measure the value of the first type of spin. Say that a measurement has already been made of the electron's value for spin-type x, and the result was discovered to be "up". If the electron is then "passed through" a deflector set up to measure the *complementary* type of spin, spin-type y, the previously measured value of spin-type x will be disturbed. What this means is that a subsequent measurement of spin-type x will discover that the electron has a fifty-fifty chance of exhibiting the binary value "up" and a fifty-fifty chance of exhibiting the value "down".

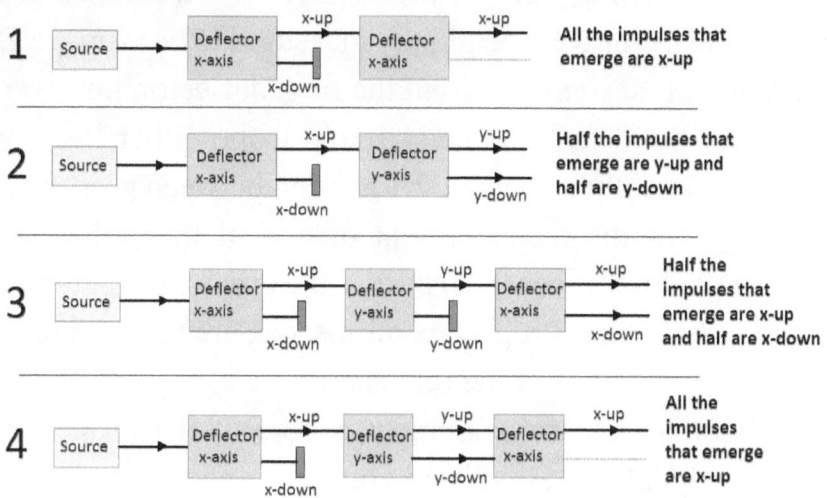

Figure 3.10 Variations of the Stern-Gerlach experiment which together demonstrate "complementarity"

Consider Figure 3.10. The first scenario deals with an "electron" that is passed through a deflector that measures spin-type x. Electrons that emerge from the deflector on the x-down route are blocked, meaning that the electrons that continue into the second deflector are all of the x-up type. The same electrons emerge from the second deflector on the ex-up route, indicating that this property is a genuine and enduring characteristic of the causal impulses themselves.

In set-up 2, the electrons that emerge from the x-deflector are then passed through a y-deflector. Fifty percent of the electrons that emerge from the y-deflector are spin-down and fifty percent are spin-up, showing that there is no correlation between properties x and y. In scenario 3, the electrons that emerge along the y-down route are blocked, meaning that all of the electrons that enter the final x-deflector were originally measured to be x-up AND y-up. When they emerge from the final deflector, however, they are found to be equally likely to be either x-up or x-down. Not only is there no correlation between properties x and y, but the measurement of one of the properties (property y in this case) "disturbs" the previously obtained value of the other. Properties of this sort are referred to as "complementary" in quantum mechanics. The conditions necessary for the measurement of one of the properties necessarily interferes with the conditions necessary for the measurement of the complement.

Scenario 4 presents us with empirical data that is hard to explain in confrontation with the results of the previous three experiments. The barrier along the y-down path is removed from set-up 3. This means that double the number of electrons will enter the final x-deflector as did in the preceding experimental arrangement - all the y-up electrons that were allowed to pass in set-up 3, plus all the y-down electrons that would have been blocked. Now we find that *all* of the electrons emerge from the final x-deflector along the x-up path! This seems to contradict the previous finding that the apparatus for measuring spin-type y disturbs the value of spin-type x. Here in set-up 4, each electron *retains* the value of x-spin that it had when it entered the y-deflector.

These results seem even stranger if we perform experiment 4 *before* experiment 3. The electrons in apparatus 4 *all* emerge along the x-up path. Then we insert the barrier along the y-down path and perform experiment 3. This barrier ought to reduce the output of the entire apparatus by half. Therefore we would expect only fifty percent of the electrons to emerge on the right hand side, but we would expect them *all* of these to emerge along the x-up path. This seems like an exercise of simple logic. If one hundred percent of electrons maintain their value of spin-type x when the barrier is open, then the fifty percent of electrons that make their way through the apparatus when the barrier is closed should still *all* emerge with their value of x intact. But that is not what we find when we run the experiment with the barrier closed. Now only half of the emergent electrons (twenty-five percent of the electrons that entered the apparatus) retain their original value for x. The apparatus, in other words, is now functioning like a simple instrument for measuring spin-type x that respects the principle of complementarity: it scrambles the previously obtained value for spin-type y.

How are we to account for these strange findings? Our approach to understanding this phenomenon is substantially the same as for the classic "interference" experiment. Earlier, we saw how an interference pattern still built up over a period when only *one* "electron" passed through the apparatus at a time. With the closure of one of the apertures, however, the interference pattern was destroyed. This raised the question of how the behaviour of an entity could be influenced by the closure of an aperture through which it did *not* pass. In the set-up involving instruments for measuring electron spin, we are presented once again with the exact same mystery. The placing of a barrier on one of the paths influences the properties of "electrons" who manifestly cannot have taken that path in the first place.

The resolution of the mystery consists in accepting that no material entities of any sort pass along either path. An atom in the causal source (located to the left of the apparatus in the diagram) has an influence on an atom in the target (located somewhere to the right of the apparatus). The influence evolves over time and in interaction with the atoms of the material of the apparatus. The result is an effect in the target that we normally refer to as the reception of an "electron" (in the next chapter we will discuss in more detail what this influence consists in). When both paths are open on the device, each half of the apparatus has a joint impact on how the influence will evolve. The very way that an "open" apparatus is set up (the arrangement that re-converges both up-down paths into the successive deflector) ensures that these influences are perfectly equal and opposite. As a result, the apparatus has no net impact on the evolving influence at all. Consequently the value for spin-type x is left intact. Once we erect a barrier along one of the paths, however, the equilibrium of the causal influence of the apparatus is destroyed. The magnetic deflectors within the spin-detection instrument then exert a net magnetic effect of a particular sort on the evolving influence, disturbing the original magnetic characteristic of the influence that we call spin-type x.

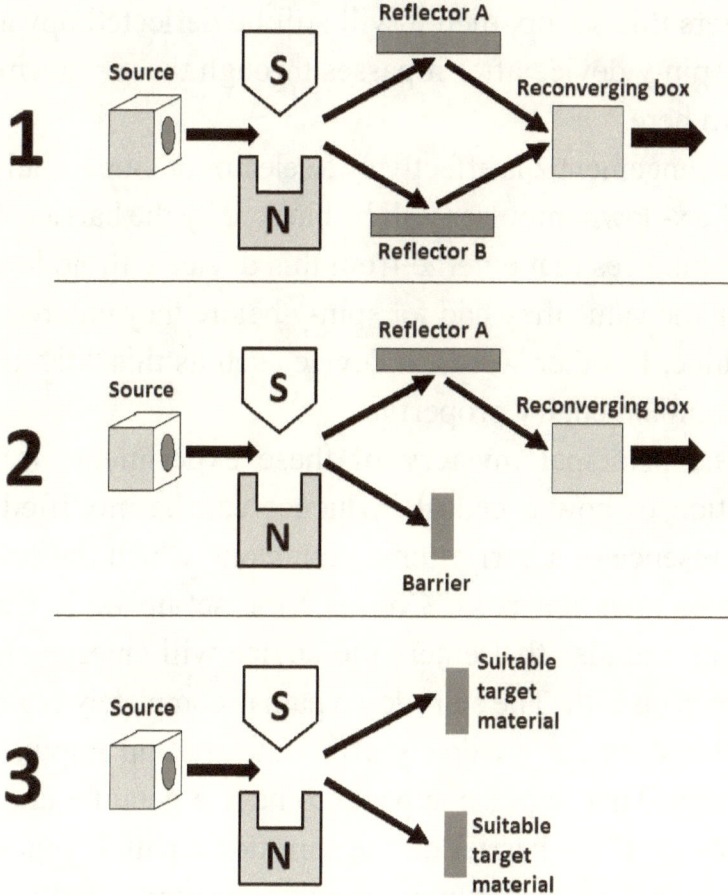

Figure 3.11 Different arrangements of a spin deflection device

Let us look at this in more detail. Figure 3.11 shows the interior of a spin deflection device in schematic form. In the set-up shown in 1, the reflectors ensure that the impulse emerges from the device along the same path regardless of whether it is spin-up or spin-down. As we have seen a moment ago, this arrangement does not disturb the complementary property. Say that the device deflects for spin-type x. If the impulse is known to be spin-y-up before it enters this set-up, then it will still be deflected upwards by a spin-y device after it passes through the arrangement shown here.

Arrangement 2 is effectively a selector of x-up impulses. All the x-down impulses will be blocked by the barrier. The x-up impulses that emerge from this device will no longer retain the value they had for spin-y before they entered the machine. In other words, a device such as this "disturbs" the complementary property.

The principal mystery of these experiments is the question of how a causal influence can be modified by the presence of a barrier on a path along which the causal impulse does not pass. Consider the set-up in 2. Every spin-up impulse that enters the device will emerge along the spin-up path. The spin-down path is completely blocked but this does not adversely affect the spin-up impulse in any way. That impulse appears to have a genuine unitary trajectory. If the barrier on the spin-down path is replaced by a reflector, then future impulses are modified in the sense that their spin-y property is no longer disturbed. How can the modification of the apparatus, on a path along which impulses of this sort did not previously pass, cause an impulse to lose its unitary trajectory? If the impulse is naturally deflected upwards by the magnetic field, how can the presence of either a reflector or a barrier on the *down* path have any influence on the properties of the impulse? These and other questions lose much of their vexing quality when we begin to consider what is happening in terms of the evolution of a non-material causal event in relation to the whole system.

Recall our admonition to refrain from viewing these situations in purely spatial terms. An atom in the source is in a state of disequilibrium. In the next chapter we will consider the sort of disequilibrium that leads to the "transmission of an electron". The disequilibrium arrives at a critical moment and it is discharged into the system. As we shall see later, this disequilibrium will evolve through the system until it encounters suitable target material. Suitable target material for the reception of "light" consisted in *groups* of atoms that could align themselves electrostatically. The target of electron transmission, by contrast, is genesis-units that were neutral and now become ionised, or were positively ionised and now become neutral.

As the causal impulse evolves in the system, it encounters the magnetic field of the Stern-Gerlach instrument. When we consider ionisation and magnetic fields in more detail later, we shall be able to say more about the particular dynamics that is occurring here. What is important for now is the point that the instrument has a tendency to make the impulse evolve along *one of two* separate "paths", but it also has a tendency to change the *character* of the impulse in one of two ways. So there are two types of potential change going on here. The first is a simple change in the geographical *direction* of the target material (upwards or downwards). The second is an alteration in the kind of *internal effect* that the impulse will cause in the target genesis-unit. As we shall see in the next chapter, different types of spin consist simply in different alignments of the B-V axis of the genesis-unit. A Stern-Gerlach instrument will modify the change that will be made to the alignment of the B-V axis of the eventual target atom (thus "destroying" the complementary property) unless the source genesis-unit is already aligned perpendicularly to the magnet field of the device.

The device in 3 modifies *both* the path of the evolving impulse and the kind of effect prompted by the impulse in the target atom. A spin-down impulse will be directed downwards and will give rise to a particular change in alignment in the B-V axis of the target atom. The device in 1, in complete contrast, does not change the impulse in either the geographical location of its target, nor in the particular alignment that is induced in the target atom. The reflectors and reconverging box see to it that the impulse passes straight through the device with the same "trajectory" with which it entered. No effort is required to comprehend this fact. What jars with common-sense is the discovery that a spin-up impulse that emerges from 1 is significantly different to an impulse that emerges from 2.

Let us first consider the magnetic field from the point of view of its capacity to change the *direction* of evolution of the impulse. Later we will consider its capacity to alter the *alignment* of the B-V axis of the target genesis-unit. As the impulse evolves through the magnetic field, there is a twofold tendency on the part of the field to modify the direction of the natural progression of the impulse. A magnetic field of this sort is the joint product of many causal players, but the net influence of the field can be simplified down to two principal tendencies (as far as the path of the impulse is concerned): the tendencies to modify the evolution of the impulse so that its target is in either an upwards or a downwards direction. Say that the impulse is of a sort such that the magnetic field tends to deflect it upwards. This description of the modifying influence of the field is, of course, in spatial terms. But in reality this impulse is still in evolution. It has effected no causal change in any target material. It has produced none of the effects that are necessary for an observer to generate the spatial viewpoint. That being said, it is certainly true that *if* target material were suitably located in an upwards direction, this causal impulse would have its effect there.

For the moment, however, the causal event is still in evolution. Evolutions of systems, we recall, are complex processes in which multiple players contribute their input. In the interference experiment, the impulse of light was produced by the causal source. This evolving event was later influenced by causal players located at either aperture. When the event finally produced its effect at the screen, it was a joint venture on the part of three sets of causal players. In the spin experiment that we are considering, the magnetic field acts on the impulse so as to give it an upwards impetus (speaking in spatial terms, even though no spatial perspective on the event is possible at the moment). But this does not mean that the downward component of the field is absent from the evolving impulse. Even when the impulse of light in the interference experiment finally gave rise to an impact in the upper part of the screen, *this did not mean that the lower slit made no contribution to the evolution of the event. Both* slits jointly guided the impulse to the screen, even if one or other of them had a dominant role in selecting the target atom. In the same way, the downward component of the magnetic field in 1 remains a player in the evolution of the spin impulse whilst the process is still in progress, even if the upward component happens to have a dominant influence in the case of this particular influence.

In 2 (reproduced in Figure 3.12) the barrier placed along the lower path prevents the downward component from continuing to have an influence on the evolution of the process. We must imagine that the event is reverberating through the system, at first encountering both components of the magnetic field. As the event continues to reverberate, the influence of the downward component is blocked by the presence of the barrier. From now on, only the upward component of the field will be able to influence the evolving event. When the impulse later encounters suitable matter, it will give rise to an x-up spin alignment of the target genesis-unit. Alternatively, if the impulse does not encounter target material and is later passed through a spin-y device, there is a fifty-fifty chance that any previously measured value of spin-y will have been disturbed. This is a simple consequence of the fact that spin-x and spin-y involve orthogonal measurements of the alignment of the B-V components of genesis-units. If genesis-unit A1 measures "up" for spin-x, then it is orthogonally aligned to a genesis-unit that shows "up" or "down" for spin-y. If A1 is now passed through an inhomogeneous magnetic field that is aligned in the spin-y direction, its orientation will be changed by that field, and there is equal likelihood that it will be realigned in an "up" or "down" direction for spin-y.

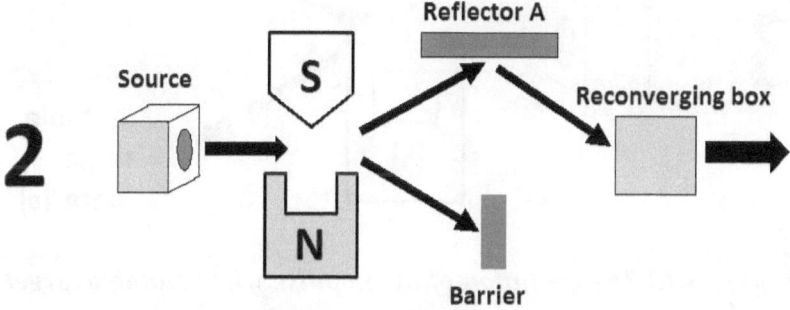

Figure 3.12 Arrangement that "destroys" the complementary property

If, however, the barrier is not present on the lower path, then the downward component of the magnetic field will continue to have influence on the evolution of the impulse. The process will continue to reverberate to the point of encountering the reconverging box. At this juncture, the process develops in an important way. Recall that we are discussing an impulse that would have had its effect in the upwards direction *if* suitable target material had been present. Figure 3.13 shows what happens when appropriate matter is present in the upwards and downwards direction. Every impulse will be dominated by either the upward or downward component of the field.

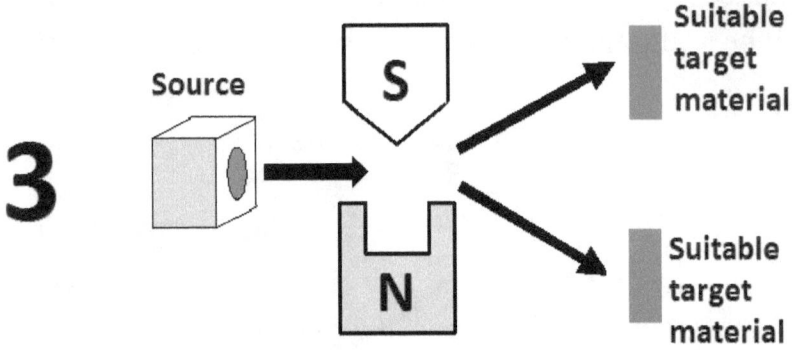

Figure 3.13 The evolution of the impulse when suitable target material is present

But if suitable matter is not available in either direction, and if the process is allowed to evolve under the influence of both components of the magnetic field, then, when it encounters the reconverging box, both components of the magnetic field will begin to exert an *equal* influence on the process.

The empirical evidence summarized in Figure 3.10 indicates clearly that this is what is happening, but we need to explain why it is the case. The impulse we are considering had a natural inclination to be deflected upwards by the magnetic field, but the device did not permit it to do so. If the device had been set up as in scenario 3 (Figure 3.13), then the impulse would have evolved along this natural path, dominated by the influence of the upward component. The device, however, *did not permit* the impulse to be dominated by the contribution of the upward component. Rather, it *facilitated* the ongoing evolution of the process until the point (at the reconverging box) that the downward component gained equal influence over the evolving impulse. Given that the upward and downward components exert equal and opposite influence on the alignment of the B-V axis of a genesis-unit, the device now has a null disturbing effect on the previous potential change in alignment that the impulse would induce in a target atom. Thus the complementary property is not disturbed.

An analogy might provide us with a way of thinking about what is happening here. Imagine a ball that is bobbing along on the top of a water wave. The wave encounters a barrier with two openings in it, A and B, and the ball passes through opening A. For an interval of time, the movement of the ball is dominated by the motion of the water waves issuing from A. As it bobs along on this train of waves, the ball reaches a point where the waves issuing from A reconverge with the train of waves from B. From this point forward, the movement of the ball will be the product of the joint influences of the waves from A and B.

We have no intention of trying to account for what is happening in the two-path experiments using the classic picture of the interference of water waves! No material entities are present in these experiments and we must be very careful not to fall back into attempts to view these situations in terms of material entities following purely spatial trajectories. However, the classic water wave picture helps us to visualize how a causal player (an opening in the barrier) can contribute to the evolution of a causal process, even if that process is *initially* dominated by a different causal player (the other opening in the barrier).

Spin devices that are set up so as to reconverge the upward and downward paths *facilitate the ongoing evolution* of the impulse to the point where the upward and downward components both gain an equal influence on the causal process. This is a simple and uncontroversial explanation of the mysterious "complementarity" phenomena demonstrated by these devices. What made the phenomena seem mysterious was the fact that we believed that *objects* were passing through the devices and that these objects had autonomous properties of either spin-up or spin-down. If an object takes the spin-up path, then its future behaviour should not be influenced in any way by the presence or absence of a barrier on the spin-down path. Evolving processes, on the other hand, are a different matter altogether. Until such point as these processes culminate in an effect (i.e. until they induce a physical change in a material object), they *continue to be influenced* by the various causal players that underlie them. Indeed, their very evolution is a product of the reciprocal interaction of these causal players. Spin device 1 actively coerces the process so that any initial dominating influence (whether it be from the upward or downward component of the magnetic field) is effectively prevented from inducing an effect in a material target. The impulse is then constrained to evolve in such a way that both components eventually exert equal influence on the process. These equal and opposite influences have a null net effect, meaning that the incompatible influences of a y-spin device are left undisturbed.

This solution to spin complementarity has the same form as the solution proposed for the two-path experiment involving light. In both cases, the apparatus is set up in such a way as to present an *ongoing causal event with twin causal players that can potentially modify the impulse in a joint fashion.* The experimenter can set up the apparatus to prevent one of the causal players from exerting its influence on the evolving process, or he can choose a set-up that permits joint influence. Apparent paradoxes arise when the empirical results appear to show that the properties of an object can be determined by the presence/absence of a barrier on a trajectory along which the object did not pass. The mystery is completely diffused when we accept that no material objects are present and that the evolution of causal impulses can be jointly influenced by multiple causal players.

"Interference" phenomena have also been observed in experiments using atoms. In these cases there is no doubt that material objects are passing through the apertures, but their behaviour can still be explained in terms of system-wide processes that guide the movement of the atoms. Such processes initiate in atoms in the source material and are jointly modified by atoms in either aperture. This will become clearer in a later chapter when we consider the non-local manner in which processes can be modified.

Chapter Four

IONISATION AND THE EMISSION OF THE "ELECTRON"

". . . [It] follows on the hypothesis that the cathode rays are charged particles moving with high velocities, that the size of the carriers must be small compared with the dimensions of ordinary atoms or molecules. The assumption of a state of matter more finely subdivided than the atom of an element is a somewhat startling one"

J.J. Thomson

Overview of this chapter and its principal claims

1. Ionisation is described in a way that does not involve the addition or removal of a constituent charged particle. Recall that light is an evolution in the electrostatic equilibrium of the system. If light of a certain magnitude evolves to a pair of atoms, T1-T2, in electrostatic equilibrium, then it can replace the electrostatic influences that T1 and T2 were exerting on each other. This causes the bond to dissolve.

2. Individual atoms would normally possess uncovered B and V poles that would exert influence on other atoms in the vicinity. But this pair of atoms (that has just been dissolved by the incident light) has B and V poles that are "covered" by the electrostatic conservation mechanism that we call the "transmission of light".

246

3. However, for reasons discussed in this chapter, there is a tendency for the V pole to shed its covering almost immediately (although this can depend on the particular electrostatic dynamics within composite atoms). This results in the emission from each atom of an impulse that is called the "electron". The atom is left positively charged - in other words, with a covering on its B pole.

4. We reinterpret the empirical evidence for the postulation of the electron, considering Thomson's classic experiment. The supposed mass of the electron is understood as arising from the fundamental electrostatic capacity of a genesis-unit. When the impulse we call the "electron" arrives at its target, the transmission of this influence gives rise to an associated change of momentum in the target atom. The momentum has a fixed basic magnitude, but this can be augmented by other factors that increase the rate of evolution of the impulse, such as the excess energy of the evolution of light that prompted ionisation. The fact that the causal evolution we call the "electron" gives rise to a change in momentum does not imply that the causal impulse is a particle.

4.1 Preliminary attempt to understand ionisation

This chapter presents a new way of looking at the relationship between light and electrostatic charge. In the standard approach, light and electrons are different sorts of entity altogether – one has a mass and a charge, the other has no mass and is electrically neutral. This raises the issue of how such contrasting entities interact with each other: how does the incidence of light on an atom lead to the ejection of electrons and the ionisation of the atom? The approach advocated in this book tries to develop an integrated framework for matter and light. Light has no existence apart from the electrostatic relationships that prevail between atoms. Once such a relationship of equilibrium is broken, the system tends to reincarnate the same equilibrium in whatever suitable target material is available as the system evolves. If we consider the mechanism by which light is transmitted/absorbed in more detail, then an explanation of ionisation naturally presents itself. As we shall see, if light can be described as an evolution in the system's electrostatic *equilibrium*, then the "electron" can be described as an evolution (or even *resolution*) of a particular type of *disequilibrium*.

The atomic pair, T1-T2, is in natural electrostatic equilibrium. The straightforward nature of this bond is essential if the pair is to become the locus of a system drawn together by the intromission of light. This pair, quite simply, is held in alignment by the mutual attraction between their opposite poles. Thus there is a combined B_{T1}-V_{T2} component and a V_{T1}-B_{T2} component working in unison to maintain the bond. Once the incoming light is absorbed (say that this light is sufficient to draw three new atoms to each side of the T1-T2 pair), then it as if the B and V components of T1 are augmented in capacity so that they can now hold four genesis-units each in equilibrium. In other words, the B of T1 needs the combined V components of four atoms to reciprocate its electrostatic influence (similarly for the V of T1, and similarly for the B and V of T2). It is no harm to recall that the poles of T1 have *not* actually attained any larger electrostatic influence: as we discussed earlier, the augmented electrostatic "energy" of this subsystem is held in place by the *configuration* of the atoms; just as two balloons can develop extra work potential by being compressed together, so too the proximity of like poles can hold electrostatic "energy".

If the groups of atoms held by T1-T2 dissociate, light is transmitted. This consists in the evolution of a tendency to re-embody the same magnitude of electrostatic cohesion elsewhere in the system. There is a perfect symmetry in this tendency arising from the perfect symmetry (a B_{Ti}-V_{Tj} component and a V_{Ti}-B_{Tj} component) in the electrostatic bonds that were dissolved in the source. Say a group of three atoms on either side of T1-T2 dissociates, then this will generate impulses of combined magnitude 6E. Each of the impulses has a dual aspect: it has the tendency to draw the V poles of a certain number of genesis-units into alignment with a pair of atoms in a natural electrostatic bond, *and* it has the tendency to draw the B poles of those same atoms into alignment with the pair. Light always involves this dual component because it arises from groups of atoms disassociating from the B *and* V components of an atom in a natural electrostatic bond. This entails that the causal impulse we call light, as it evolves, is not affected by electrostatic or magnetic *imbalances* in nearby atoms whose properties could otherwise have influenced the evolution of the impulse (or to put it in standard language: "The photon is not deflected by electric or magnetic fields").

Now consider what happens if a T1-T2 pair which exists in isolation becomes the target of light of magnitude E. We are using the expression E to represent the potency of the *attractive* components of the electrostatic bond between two genesis-units when they have converged to the point of equilibrium. The units of E will be the classic gm/s^2 (the disposition of a gramme of the material to accelerate at a rate of a metre per second per second). The Rydberg formula tells us that a pair of atoms in equilibrium can absorb wavelengths that correspond to $1/\lambda = E(1/n_1^2 - 1/n_2^2)$. Each side of the T1-T2 subsystem has just one atom, therefore n_1 is 1. To absorb light and make a transition to a subsystem composed of n_2^2 atoms, the binary subsystem would need to be the target of impulses of magnitude $E-E/n_2^2$.

$1/\lambda$ is an expression of the "energy" associated with the impulse of light that is incident on the atom. In the case under consideration here, this is equal to E. Inserting this in the Rydberg formula yields: $E=E-E/n_2^2$. Therefore $E/n_2^2=0$. But this expression can only produce zero if n_2 is infinite. What this suggests is that light of magnitude E can only be absorbed by T1-T2 if the pair can transition to a subsystem of infinite size, which of course is impossible.

The empirical fact that ionisation occurs when light of magnitude E is incident on the sample indicates that the subsystem is actually dissolved in these circumstances. But why does such dissolution result in the individual atoms becoming ionised, and just what *is* ionisation in the first place? As we shall see, it differs from the transmission of light in one essential respect: light emission and absorption involves the dissociation or formation of *groups* or *subgroups* of atoms in equilibrium; ionisation, by contrast, involves the rupture of an electrostatic bond in such a way that the previously bonded atoms are now roaming free. Light emission *does* require the break-up of a group of atoms into smaller subgroups, but it is a break-up that produces individual, unattached atoms that results in ionisation.

The genesis-unit model is well-suited to describing ionisation in coherent terms. And the fact that light impulses of magnitude E induce ionisation, as the Rydberg formula predicts, fits in very well with our overall understanding of the nature of light. T1-T2 is held together by an electrostatic bond in which the sum total of the attractive impulses (i.e., the influences generated between the unlike poles) amounts to E. When such a pair disassociates, a light impulse will evolve through the system. This impulse will have a dual nature corresponding to the twin V-B components in the original electrostatic bond between T1 and T2.

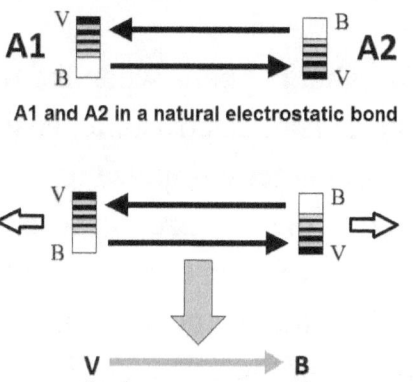

A1 and A2 in a natural electrostatic bond

A1 and A2 disassociate, leading to the evolution of light. This impulse has a dual character arising from the twin B-V components in the original bond

Figure 4.1 Disassociation of an electrostatic bond leading to the transmission of light

To understand the process of ionisation, it is helpful to return to the question of how light is emitted in the first place. Figure 4.1 shows an atomic pair, A1-A2, in electrostatic equilibrium. This bond can be broken by many different means, even by physically striking the object in which the pair resides. When the bond dissolves, an impulse of "electromagnetic radiation" begins to evolve through the system. As the figure tries to show, this impulse has a dual symmetric nature. It evolves though the causal configuration of things with the very particular VB-BV orientation shown in the diagram. This particular orientation – which is really the potential to draw atoms together in *an electrostatic bond aligned in this direction* - is usually referred to as polarisation. When the impulse encounters a target pair, T1-T2, whose electrostatic bond has the same orientation, it draws square groups of other atoms into tension with the core pair, as we have discussed in Chapter Two.

Now consider what happens when the opposite process occurs: a pair of atoms in electrostatic tension becomes the target of an electromagnetic impulse of magnitude E. This impulse is simply the *wrong size* to cause the pair to draw other atoms into electrostatic tension. Remember, in order to draw groups of n_2^2 atoms into tension with a single pair, the impulse must be of the order $E - E/n_2^2$. If the incident light is of magnitude E, then E/n_2^2 would have to equal zero, which is only possible if n_2 is infinite.

So much for what the Rydberg expression says about the situation when a single pair absorbs light of magnitude E. What is actually happening? Why does light of this magnitude cause ionisation? Just what *is* ionisation? Figure 4.2 makes a preliminary attempt to depict an ionisation event. A pair of atoms in natural electrostatic equilibrium,

T1-T2, is the target of the evolution of light of magnitude E. Light is capable of drawing atoms into electrostatic equilibrium with each other, and it is equally capable of *covering* the electrostatic capacities of atomic poles. Indeed, if light can be described as an impulse with the potential to bring atoms into electrostatic tension, it can equally be described as the potential to cover the electrostatic capacities of atomic poles. As represented in the diagram, the incoming light impinges on the bond between T1 and T2, covering that proportion of the V pole of T1 and the B pole of T2 that actually corresponds to the magnitude of electrostatic attraction between them.

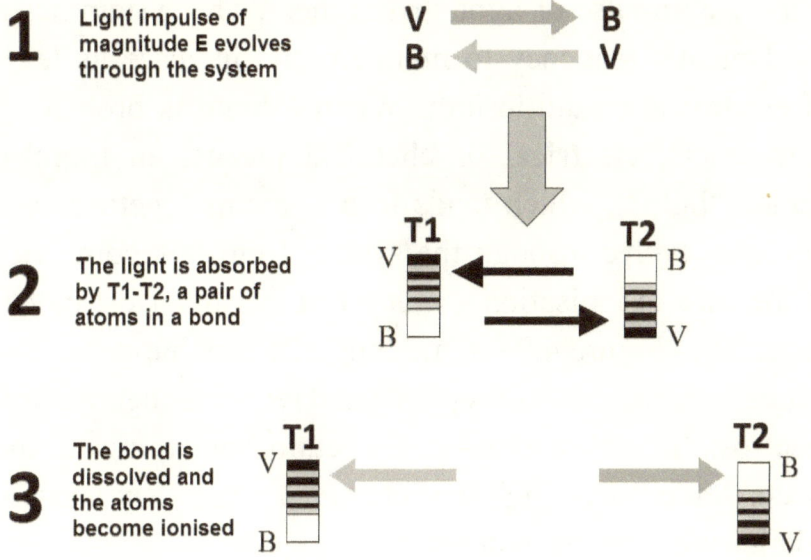

Figure 4.2 A preliminary attempt to understand ionisation

We can imagine how two atoms might be oppositely charged if they have opposite poles "covered". But what does it mean to be covered and how does an atom have the potential to remain in this state for a period of time? According to our model of the transmission of light, the system has a natural tendency to evolve towards greater electrostatic equilibrium. When an impulse of light breaks up an electrostatic bond, *two* separate tendencies conspire together to create a new state of the atom (ionisation) that endures in time. Firstly, the incoming impulse of light originated in a broken electrostatic bond between atoms. This impulse constitutes a tendency to bring atoms into electrostatic tension with each other. Secondly, the target of the light, T1-T2, also had an electrostatic relationship between them which doesn't just dissipate into thin air when the impulse of light strikes them. The system has a fundamental tendency to maintain and increase its level of electrostatic equilibrium. When a bond is broken by mechanical, electrical or chemical means, an impulse (called "light"), which tends to pull atoms together, will typically evolve through the system from this point. But in the case of ionisation – at least in the imperfect model depicted in Figure 4.2 - something different happens. The atoms in the pair do not separate and return to their natural electrostatic states: their poles retain something of the electrostatic "covering" that prevailed when they were in relationship with each other.

The language we are using here can be misleading. It is easy to think of this covering of the poles as some sort of physical entity with a presence in space. But it is really the *system* that is responsible for the covering of the poles after an electrostatic bond is broken. In our scheme, the conservation of electrostatic equilibrium has a position almost akin to the principle of conservation of energy in mainstream science. The system naturally resists changes that occur in the electrostatic status quo. If an electrostatic bond is broken, it is as if the system recoils in a way analogous to Newton's third law of motion. The bond breaks, which means that there is nothing to counterbalance the influence of the B and V poles that were once held in check by the V and B poles of the partner atom; but the vast equilibrium of the system now steps in to the breach. The poles of the atoms are covered by a sort of "inertia" in the overall equilibrium that tends to maintain the status quo.

This approach to ionisation seems plausible but, as it stands, it fails to account for a basic empirical fact: "electrons" from different elements have the exact same charge but different binding energies. If an ion is a genesis-unit which has a pole covered by the very component of light which ionised it, then shouldn't all elements become ionised at the same threshold frequency of incident radiation? Surely the model makes too intimate a connection between the energy of the light that *induces* ionisation and the energy of the impulse that *maintains* the genesis-unit in an ionised state? The Bohr model had no such difficulties. On that approach, charge was an elementary given in nature, and required no explanation. There was no necessary connection between the magnitude of elementary charge and the empirical conditions that bound a charged particle inside a given element. Our model, by contrast, seems to posit a direct connection between the threshold frequency that causes ionisation and the charge that results: an ion just *is* a genesis-unit whose pole had been covered by a component of incident radiation. But the story must be a little different to that if we are to account for the fact that ions of different elements have the *same level of charge*, yet are created by *different frequencies of light*.

4.2 A more complete account of ionisation

A more nuanced treatment of ionisation is evidently required. Earlier we described how two genesis-units, T1 and T2, typically settle into a state of electrostatic equilibrium (see Figure 2.2. in Chapter Two). Beyond a certain proximity, they present to each other as if they were two neutral entities, but, upon convergence, the unlike poles begin to exert an influence that becomes significant. The distance between the unlike poles (e) is now sufficiently smaller than the distance between the like poles (d) for a net attractive influence to be generated. The magnitude of the net attraction (A) exerted by T1 on T2 (and vice-versa) can be expressed as follows: $A = UL - L$, where UL is the sum of the attractive influences exerted by T1 on the unlike poles of T2 at distance e and L is the sum of the repulsive influences exerted by T1 on the like poles of T2 at distance d.

These quantities, UL and L, are appropriate for simple genesis-units (hydrogen atoms) that are not part of larger composite atoms. Now consider the two lithium atoms in Figure 4.3 that have settled into a state of electrostatic equilibrium. In the case of a composite atom like this, there will be many factors that will influence the magnitude of the electrostatic bond. The two genesis-units that are involved in the bond are also directing electrostatic influences to other genesis-units in the structure. This means, inevitably, that they have *less* electrostatic capacity to direct towards each other (the reduced influences of the poles will be designated by UL_1 and L_1). Despite the smaller capacities of the poles, the difference between the attractive influences of the unlike poles will exceed the repulsive influences of the like poles by the same amount, A, as was the case for bonds formed between simple hydrogen atoms.

How can we be confident that the *net* attractive influence exerted by a lithium atom on its partner is the same as the net attraction exerted by each atom in a hydrogen pair? The *total* magnitude of attraction exerted by a lithium atom (UL_1) will be less than that exerted by a hydrogen atom (UL) because of the other mitigating electrostatic influences at play in the composite lithium atom. Say that UL_1 = UL - X. Now consider the repulsive influence, L_1, exerted by the lithium atom on its partner at this distance of separation. This too will differ from the repulsive influence exerted by a hydrogen atom on its partner by the same quantity X. We can be sure of this because the process of formation of atoms evolves in such a way as to maximize electrostatic equilibrium within the composite structure. If the influence of the B pole of a given genesis-unit in the structure is reduced by a certain quantity because of the other electrostatic influences within the atom, then the influence of its V pole will be modified by the same amount. Thus L_1 = L – X. Therefore if UL – L = A then it follows that $UL_1 - L_1 = A$.

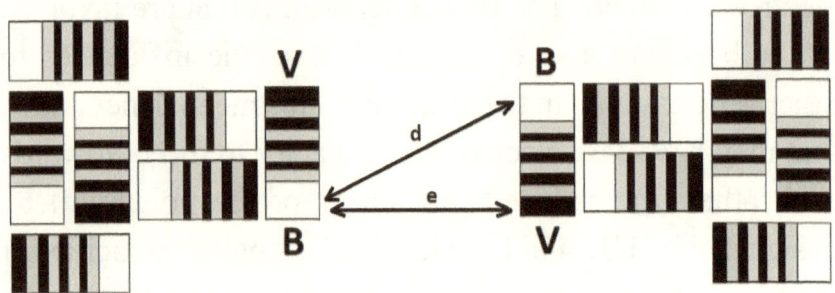

Fig 4.3 Two lithium atoms in electrostatic equilibrium. The lengths d and e are the same for lithium as for simple hydrogen atoms. Closer proximity will lead to net repulsion because of the way the V component is distributed in the genesis-unit.

Now consider the magnitude of the energy of light that would be required to break this electrostatic bond and lead to ionisation. The components of the incoming light would need to be at least equal in magnitude to the diminished influences of the poles if they are to break the bond: therefore the ionising energy of lithium will be lower than that required for hydrogen. In other words, if it takes light comparable to the magnitude UL (or greater) to ionise hydrogen, then light of magnitude UL_1 will be sufficient to ionise lithium.

Thus, we see how the genesis-unit model can account for the variation in the threshold frequencies for the ionisation of different elements. According to this model, a pair of hydrogen atoms in an electrostatic bond should have one of the highest ionisation energies possible since the individual genesis-units are directing their *entire* electrostatic capacities towards each other. And indeed, this is the case – the vast majority of elements have lower ionisation energies than hydrogen. No doubt other factors also contribute to fixing the threshold value. Helium has an even higher value than hydrogen, possibly because the very stability of this atom requires a more dramatic disruption of the bond before ionisation takes place.

We can see how different energies of light can break the bond but we have yet to discuss the type of "covering" of the poles that remains after the break-up which constitutes ionisation. Consider the electrostatic state of two atoms, T1 and T2, in equilibrium. These atoms are exerting an empirically-significant level of influence on each other. When we subtract the repulsive influences of an atom in the pair from its attractive influences, the magnitude of net attraction exerted by that atom is A. The dual components of the incoming light substitute the *attractive* components only (UL) of electrostatic influence between T1 and T2. This light does not impinge directly on the *net* attractive influence, A, exerted by T1 on T2 (and vice-versa). The net attractive influence, after all, is a complex quantity, arrived at by the interplay of the attractive and repulsive forces. The incoming light cannot interact with A directly because this quantity does not have a particular "location" in the causal configuration of things, no more than the combined influence on the tides of lunar gravity and terrestrial winds has a particular location in the causal configuration. But the components of electrostatic attraction between the V and B poles of T1 and T2 have a very definite position in the causal configuration of things and they can be targets of the evolution of light. The light substitutes *these* components and the bond is dissolved.

The relationship of electrostatic tension (of magnitude A) between T1 and T2 does not vanish into thin air. When the atoms go their separate ways, the system's resistance to changes in electrostatic equilibrium ensures that T1 and T2 retain the "mark" of their previous state of bondage. Each atom will have its poles "covered" in the sense that each pole will now exert a diminished influence than was previously the case. Where does the system get the "energy" required to cover the poles in this way? From the light that caused the ionisation. Remember, we do not believe that light is an independent causal intermediary as the photon model would have us believe. The evolution of light is all about the evolution of the electrostatic state of the system. The light that causes the ionisation is really a particular event in the ongoing development of the system. In this case, it causes the breakup of an electrostatic bond, but at the same time it maintains the containment of the electrostatic influence on the poles of the atoms that have been separated.

Let us go through the process of ionisation step by step. The incoming light (of magnitude UL) covers the *attractive* components of the electrostatic relation between T1 and T2. The *repulsive* components therefore become influential, resulting in the dissolution of the bond. As the atoms move away from each other, the inherent tendency of the system to conserve electrostatic equilibrium comes into action. The *net* electrostatic attraction exerted by T1 on T2 (and vice-versa) before the fragmentation of the bond does not simply disappear. It still exists in the system, maintained by the evolution of light, and it is responsible for the fact that the poles of T1 and T2 now exert electrostatically reduced influences.

Thus we see how different frequencies of light can ionise atoms even though the resultant charge remains the same. The light that ionises needs to cover only the *attractive* components of the electrostatic bonds between atoms. But the magnitude of the charge is a measure of the *net* attraction previously exerted by one of the atoms on the other. Before moving on, a few other comments may be timely to help resolve possible misunderstandings. The poles of genesis-units have vastly greater attractive and repulsive capacities than the influences that are exhibited at the distances involved in electrostatic interactions. Earlier we called these enormous influences the "protomagnetic" capacities of the genesis-unit. In the interactions that take place at the minute distances involved in atomic fusion, these greater capacities are manifest. When genesis-units are located at a more significant distance from each other, their mutual protomagnetic influences naturally diminish. The "electrostatic" capacity of a genesis-unit, quite simply, is the influence exerted by the unit on another when they are located at this greater distance.

When we designate the electrostatic attractions between poles with the expressions "UL" or "UL_1", this should not be taken as some sort of measure of the entire attractive or repulsive capacity of that pole: it is simply a measure of the influence of the pole *at the particular distance* where atoms naturally settle into a stable electrostatic bond. A hydrogen atom, T1, in an electrostatic bond with T2, exerts a net attractive influence (A) which is equal to UL – L, where UL is the influences exerted by the poles of T1 on the unlike poles of T2 (at this level of separation where the bond takes place) and L is the influences exerted by T1 on the like poles of T2 at this distance. In the case of a lithium bond, the net attractive influence is the same (A) and it is equal to $UL_1 - L_1$.

When a hydrogen pair is ionised, light of the order UL is incident on the bond. The ionisation of lithium requires incident light of minimum magnitude UL_1. But all of these ions retain the *differential* between the attractive and repulsive influences present in the original bonds. The differential or electrostatic imbalance between the individual poles of any one of these ions will now be A, and this imbalance is what we call "charge". But there is something missing here. If the V pole and the B pole of atom T1 both remain covered by the system's resistance to electrostatic change (a resistance that is embodied in the so-called emission and absorption of "light"), then there *is* no differential in the attractive and repulsive tendencies of the atom. These tendencies have been covered. This question can be answered by considering the difference between negative and positive ions.

4.3 The difference between negative and positive ions

Up to now, charge has been described in terms of an electrostatic imbalance between the poles of a genesis-unit. Negative charge will involve a certain imbalance in the potency of the B and V poles, whereas positive charge will involve the dominance of the other pole. But oppositely charged ions are not simply causal players of the same potency which have opposite effects on the poles of other genesis-units (although that happens to be true). There is a fundamental distinction between negative and positive charge which goes beyond the fact that they have "opposite" signs. The B component is concentrated at one end of the genesis-unit whilst the V component is *dispersed* along a significant portion of its length. Positive and negative ions thus have an essential underlying difference which will lead to significant variations in their empirical behaviour.

When a genesis-unit is ionised, its natural electrostatic capacities are being covered to a certain extent. In the case of a positively charged ion, it is the B component that is covered, whilst a negatively charged ion has its V component suppressed. Why do we associate positive and negative charges with the B and V poles respectively? This conclusion follows from a simple examination of how ionised atoms typically behave. It is relatively easy to induce a negatively charged ion to shed its excess "electron" (shortly we will discuss the nature of the "electron" in more detail), but very difficult to make a positive ion to return to a neutral state. The underlying reason for this distinction is the fact that the V component of an atom is dispersed along the length of the genesis-unit, whereas the B component is located at one end. A concentrated B component will have a more stable relationship with the system; thus the cover that has been applied to it by the electrostatic inertia of the system after the breakup of the original atomic pair will be more enduring. The more dispersed V component would be covered in a less stable way. Indeed, the very dispersion of the V component means that a small portion of the V will be situated close to the massive B pole of the unit. This entails that the B will even resist the covering on the V to some extent.

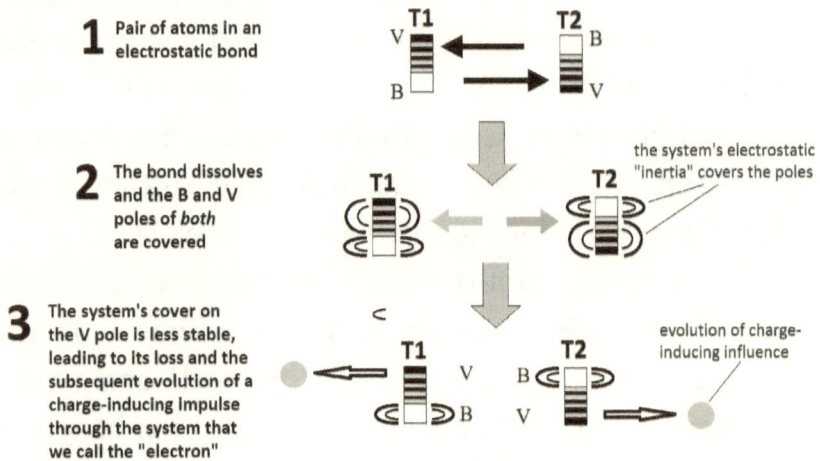

1. Pair of atoms in an electrostatic bond

2. The bond dissolves and the B and V poles of *both* are covered

the system's electrostatic "inertia" covers the poles

3. The system's cover on the V pole is less stable, leading to its loss and the subsequent evolution of a charge-inducing impulse through the system that we call the "electron"

evolution of charge-inducing influence

Figure 4.4 Ionisation and the emission of the "electron"

What determines which atom in the pair becomes negatively or positively charged? Here again we need to develop our account a little, but only focussed empirical study can hope to advance our knowledge in this area. It seems to me that when an electrostatic bond is broken, the two atoms in the pair will have *both* their B and V poles covered momentarily. Nothing else will conserve the sort of symmetry and simplicity that we expect in the physical world. As the atoms separate, the electrostatic inertia of the system, embodied by the very light that has caused the rupture, will see to it that all of the electrostatic influences of the pair are covered. But this situation can only last for an instant. For the reasons just discussed, each separated atom will quickly shed the covering on its V pole. Thus both atoms will end up shedding an electron each and becoming positively charged (see Figure 4.4). It is not just the shedding of the covering on the V pole that causes the positive charge, however. After all, once the V pole completes this shedding, it is just its normal uncovered self. It is the sustained covering of the *B pole* that makes the unit positively charged.

But does this correspond to what actually happens in nature? When salt is placed in water, the chemical bond between the sodium and the chlorine dissolves and we end up with Na^+ and Cl^-. Doesn't this indicate that the atom of one element "pulls" something out of the other element, as the standard electron view holds, leaving one atom with a deficit of that something and the other with a surplus? Our approach to the structure of composite atoms can be developed to account for empirical facts such as these. According to our model, different elements have their genesis-units fused in different ways. This means that the electrostatic influences of the different units within the structure are counterbalanced by the poles of other units to different degrees. When sodium and chlorine bond together to form salt, it is a particular unit in the sodium atom that tends to bond with a particular unit in the chlorine atom. The bond always involves these particular units that happen to be less electrostatically counterbalanced than other units in the atoms. When the bond is broken, *both* the B and the V pole of each unit is covered momentarily, as argued above. Almost immediately, however, the atoms begin to shed their covering. *If* this were a pair of hydrogen atoms, then they would naturally shed the covering on their V poles. But sodium and chlorine have more complex internal electrostatic influences at play. These influences can prompt or inhibit the shedding of the covering on either poles. In the case of sodium, the covering on the V pole of the relevant unit in the atom is shed, which is the most common consequence of bond breakup in atoms in general. With chlorine, this doesn't happen. It is to be imagined that the electrostatic influences of the other units in the atom obstruct the shedding of the V covering and instead expedite the shedding of the B covering, which would give rise to the emission of an impulse similar to the electron but with a positive charge; in other words, a "positron". The fact that positrons are not normally associated with simple chemical reactions may be explicable in terms of the rapidity with which they are absorbed by the natural potency of the concentrated B poles of the atoms in the vicinity. However we have no wish to encourage "explanations" of this sort that have to make unlikely excuses for their own lack of empirical justification. The actual break-up of electrostatic bonds may not always leave both atoms with all poles covered, thus needing to expel a "positron" in order to produce an atom that is negatively charged. It may be that the configuration of the salt molecule is such that the B pole of the relevant unit in the chlorine atom is *already* left uncovered at the moment the bond is broken. In this case no positron would need to be emitted, but a suitable study is needed to understand what actually happens when the atoms in the molecule dissociate.

The so-called "photoelectric effect" can be explained in terms of this general approach to ionisation. This is the phenomenon whereby "electrons" are emitted from a material when light of a certain frequency is incident on it. Increasing the intensity of the light does nothing to increase the number of emissions, so long as the frequency of the light remains below a certain threshold for that particular material. For the genesis-unit model, it will only be light of a certain minimum frequency (which is a measure of the magnitude of an impulse) that can cover the electrostatic attraction between a pair of genesis-units, prompting the fragmentation of the bond. At the moment of fragmentation, both atoms have all their poles covered, but the V poles will quickly shed their coverings, leading to the emission of two impulses that are capable of transmitting a negative electrostatic influence to other atoms. This evolution of the system will look for all the world like the transmission of a negatively charged particle. To the nature and characteristics of this causal impulse we will now turn.

4.4 The causal impulse that we call the "electron"

Figure 4.5 depicts the experimental set-up used by J.J. Thomson for his investigation of what was then known as the "cathode ray". Thomson was the first person to succeed in deflecting these rays in an electric field, thereby corroborating his hypothesis that they consisted in charged particles rather than radiation of an immaterial sort. The beam of rays emitted by the cathode (C) passed through the hole in the anode (A), and were then deflected by the electric field between D and E. If plate D was connected to the negative terminal of the battery, the beam was bent downwards. When the polarity was reversed, the beam curved upwards. Inevitably, the conclusion was drawn that cathode rays consisted in a motion of negatively charged particles. The debate between Hertz and Thomson regarding whether cathode rays should be considered as immaterial radiation (Hertz's view) or particles seemed to have been definitively settled. Ironically, within a few decades there would be ample empirical evidence to demonstrate that the electron *couldn't* be a particle, but the concept of wave-particle duality hindered a radical re-evaluation of Thomson's original conclusion. This notion was sufficiently flexible to render itself resistant to evidence that contradicted either the particle or wave hypothesis.

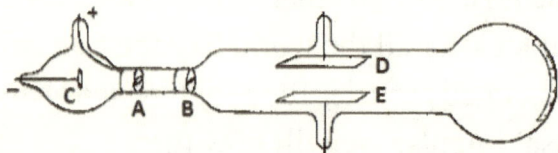

Figure 4.5 The Crookes tube with which Thomson demonstrated that the rays could be deflected by an electric field

Consider what happens when a battery is charged. Various pairs of atoms in the battery are in a natural electrostatic bond. When charging begins, these bonds are broken and the negatively charged atoms are forced to gather at the cathode, with the positively charged atoms congregating at the anode. There will be a natural tendency for the negatively charged atoms (for reasons discussed above) to jettison this impulse and to regain equilibrium with other atoms in the material.

What ensues is normally referred to as the "emission of electrons" from cathode to anode, but it can be equally well explained without having recourse to material intermediaries. Just as the transmission of light was described in terms of the evolution of a process through the system, so the emission of electrons can be described in similar terms. There are also significant differences with the transmission of electrons, as we shall see. To appreciate better how this happens, we must return to our mode of viewing the system as far as possible in non-spatial terms. A1 in the cathode and T1 in the anode are separated according to the spatial perspective. But there is much more to reality than what impacts on us from our very particular viewpoint. On the ontological level there *is* a separation of some profound sort between them, but they are capable of interacting with each other without any causal intermediaries. Before a causal influence emanating from A1 finally has its effect in T1, that influence will have to evolve through a *causal* gap. The gap is not simply a void like the spatial perspective would have us believe, but is a function of the myriad of causal interactions that have to play out in the system before the influence will come to fruition. A1 and T1 are both in internal disequilibrium. A1 has its V component covered, whilst T1 has its B component covered. There is a natural tendency for their respective imbalances to evolve towards a resolution. The evolution of the tension from A1 until its resolution in T1 is what is referred to as the emission and absorption of an electron.

As it turns out, there are also other causal players in the system. The electrical plates at D and E create a "field". What this means is that the electrical circuit creates disequilibrium in individual atoms in D and E. These atoms exert a *combined modifying influence* on the causal impulse emanating from A1. The combined harmonious influence of many atoms working together gives rise to what we call an electric or magnetic "field".

In the interference experiment, we saw how atoms in the slits could jointly modify a causal influence arising in the light source (in that case, it was really the *electrostatic bond* between pairs of atoms in the slits that modified the evolution of light – a modification that is usually described as "diffraction"). By contrast, it is the *electrostatic imbalance* in the atoms of the metal plates that prompts them to modify jointly the causal influence arising in the cathode and naturally evolving towards the anode. The fact that the so-called beam of rays could be deflected by an electrical field was taken by Thomson as a demonstration that the cathode rays were material particles. But this is not the only possible conclusion that can be drawn. In the case of diffraction, we have no difficulty thinking of light in massless terms, but yet we allow that its course of evolution can be modified by other causal players in the system. The fact that electrons are deflected by an electric field does not indicate that they are more material than light is. What it shows rather is that the electron involves an evolution through the system of a tension that is electrostatically imbalanced. The course of evolution of such an impulse in the system will naturally be swayed by the presence of other electrical players. This becomes clearer if we consider the evolution of a particular influence in more detail.

A1 in the cathode is electrostatically imbalanced. At the ontological level, the "gap" between the cathode and the anode does not confer the type of absolute causal insulation that the spatial perspective would appear to indicate. Thus, as the system evolves, there is a tendency for the imbalance in A1 to discharge itself in the opposite imbalance that obtains in the anode. During the course of this evolution, however, other causal players begin to exert a modifying influence on the whole process. Some of these causal players are in the cathode itself. The presence of many other atoms in a similar state of imbalance as A1 could cause the evolution of A1's influence to evolve beyond the anode and to come under the influence of the atoms in the metal plates. We cannot accept the usual description of "electrons" "overshooting" the anode and arriving between the metal plates. Clearly, only a material intermediary can overshoot a target. An account in terms of the evolution of processes in systems can describe this phenomenon in other ways.

The evolving causal process is thus modified by the collective electrostatic imbalance of the atoms in the metal plates. These atoms exert a joint harmonious influence on the evolution of the impulse. The evolving tension that was initiated by A1 consists in the fact that its V component was covered, the cover was shed and is now evolving in the system. What this means is that the impulse will have the tendency to cause similar disequilibrium in a target atom, which means that *it* will end up with its V component covered. Clearly such a causal process cannot come to fruition in a genesis-unit that *already* has an imbalance of this sort. When the D plate is connected to the negative terminal of the battery, atoms in the plate take on a disequilibrium of this sort whilst the atoms in E take on the opposite disequilibrium. Thus the impulse tends to evolve towards the receptive atoms in E, but the impulse is already evolving through the system at a great rate and ends up influencing atoms at the end of the tube in the direction of the E plate.

The evolution of a causal impulse like light cannot be modified by electrically or magnetically imbalanced causal players in the system (such as the presence of an electric or magnetic field). As discussed earlier, the transmission of light involves the disassociation of tension between atoms that are grouped in configurations with *both* the B and the V components in balance. Say that a mini-system of atoms disassociates and the tension evolves through a system that has the presence of negatively and positively charged plates. Not surprisingly, the atoms in these plates have no influence on the evolving impulse. The "trajectory" of the light is determined by the orientation of the source subsystem of atoms at the time of disassociation. If the light were to encounter at very close quarters an apparatus like a diffraction grating, then the trajectory *can* be modified by the diffractive influence of the *electrostatic bonds between pairs of atoms* in the slits, because these *bonds* are the very thing that are relevant to the nature of light and its evolution.

4.5 The mass of the electron

In the last section we mentioned how the capacity to be deflected in an electric field does not indicate that the electron is corporeal. If the impulse we call the electron involves no material intermediary, then how do we account for the fact that measurements of its mass can be made in a fairly straightforward manner? Clearly a distinction needs to be made between the capacity of a causal impulse to impart *momentum* and the question of whether or not it has mass. It is commonplace to speak of the photon as having momentum but no mass. Indeed, the photon is often referred to as a massless particle. Under certain experimental conditions, the photon can be attributed a specific position, and this is considered the classic hallmark of a particle. But the view of causal interaction being promoted in this book argues that causal impulses that are *not* particles can give rise to particular effects located at particular points from the spatial perspective. Once the causal evolution has given rise to an effect at a particular point on a material target, then we tend to label this point as the precise position of the hypothetical causal intermediary. In reality it is simply the point on the target where the evolving process *has had its effect*.

The fact that a causal process gives rise to an effect at a specific *position* on a material target does not mean that the causal process itself has a position or involves a particle. All the evidence attributing precise positions to photons and electrons involves this unfounded assumption. Similarly, the fact that a causal process can induce a change in *momentum* to a material target does not imply that the causal process itself has momentum or is a particle. We are accustomed to speaking of the photon as a massless object that imparts momentum to a target atom. In a similar vein, we wish to describe the electron as a causal influence with no material component but which has the capacity to impart momentum. A source atom, A1, offloads an electrostatic disequilibrium onto a target atom (an event usually described as the "emission of an electron"). The target atom can be in either of two states: it can be electrostatically stable, in which case the causal influence is going to replicate the disequilibrium that previously existed in A1; or it can be positively ionised, in which case the causal impulse will restore it to equilibrium. In either case, it is easy to imagine how the impulse might have very definite and precise consequences for the momentum of the target atom. Just as in the case of the transmission of light, the transmission of the electron involves work being done by the source on the target. A causal change is transmitted and this has "energy" associated with it. In our scheme, energy must be understood - not as a formless, infinitely-divisible entity – but as the specific work potential of an atom having its influence in particular circumstances. In the case of ionisation we are talking about an atom that is in a state of electrostatic disequilibrium. The transmission of the disequilibrium, or the restoration of the corresponding equilibrium, involves work of a very specific magnitude being done on an atom. The upshot of the work is either to restore the atom to equilibrium in the system or knock an atom out of equilibrium. As a result of the work impulse, the atom recoils perfectly in adherence with the laws of classical mechanics. The fact that the target *atom* undergoes a change in momentum doesn't mean that it was "hit" by something which itself possessed momentum. Your pressing of a light switch results in the room being flooded with light, but no one would claim that you are a source of light.

The magnitude of the impulse will be proportional to the magnitude of the net electrostatic B-V tension that holds a pair of hydrogen atoms together in a bond. In our earlier discussion of ionisation, we expressed this magnitude as E. This is a fixed quantity in nature. However, the actual magnitude of the work impulse will also depend on the other causal players that participate in the causal process as it unfolds and on the relative motions of source and target. The standard approach would speak of the way in which the "electron" is energised prior to reaching its target. The circumstances in which the "electron" is emitted by the source will affect its rate of evolution through the system. These circumstances can be more or less violent depending on the potency of the work being done on the source material. Once it has been emitted, electric fields can be used to modify the speed of passage of the impulse through the system so that its rate of evolution is speeded up or slowed down. But whatever the circumstances of the causal transmission, the final momentum imparted to the target will always be the mathematical product of the rate of evolution through the system and the "mass" we ascribe to an electron.

As we see it, the quantity that is usually described as being the "mass" of an electron is in fact a mistaken interpretation of the quantity E. E represents the net electrostatic "pull" exerted by each genesis-unit on the other in a pair (once the atoms have settled into stable electrostatic equilibrium). As such, it is a quantity that expresses the electrostatic capacity of an atom to induce acceleration in another material entity, but E itself is not a material entity. The quantity of matter we attribute to the electron is simply the quantity of matter that is *capable of being displaced* by an atom that has this capacity E.

According to the standard approach, the kinetic energy of an electron emitted from an atom during ionisation is equal to the energy of the photon of light that ejected the electron, minus the ionisation energy of the atom. Imagine a situation where light is incident on a material and "electrons" are being emitted. After emission, they are subject to the influence of an electromagnetic field, which alters the rate of transmission of the "electrons" to their target. Say the incoming light has energy H, the energy required to ionise the atom is UL, and the influence of the electromagnetic field on the kinetic energy of the "electron" is I. Then:

$$K \text{ (the kinetic energy of the emitted electron)} = H - UL + I$$

Our interpretation of the same situation is as follows: UL is the total electrostatic attraction between the unlike poles of the atoms in the bond. Incoming light of this energy will dissolve the bond. After dissolution, the pair of atoms will immediately have a high probability of jettisoning the extraneous impulses that are suppressing their V components. Any one of these impulses has the potential to cover the V component of a target genesis-unit, an event that is usually described as the absorption of an electron. The kinetic energy of this causal influence on the target will be dependent on the extent to which the potency of the ionising light exceeded the ionising energy of the atom in the electron source. Think of the whole process in this way: A mini-system disassociates into smaller subsystems and an impulse of light is transmitted. This light impinges on a pair of atoms in electrostatic equilibrium and dissolves their bond. The cut-loose atoms each have their poles covered as a consequence of the breakup of their previous bond. Now one (or both) of the pair jettisons its cover. The nature of the resultant causal impulse will be very different to the causal influence we call "light". It involves the transmission of electrostatic disequilibrium, whereas the transmission of light involves the tendency to bring atoms into electrostatic balance. The kinetic energy associated with the transmission of electrostatic disequilibrium (K) will be equal to the energy of the incoming light that prompted the ionisation (H) minus the energy required to ionise the atom (E).

Now consider how Thomson interpreted the sort of causal influence that was being transmitted in his experimental set-up. Once he had succeeded in deflecting the cathode rays in an electric field, he worked on the assumption that the causal impulse had to involve a charged particle. The entire apparatus was then placed between the poles of a large electromagnet and the voltage was adjusted until the original deflection induced by the electric field was cancelled out. This gave Thomson the interplay of quantities he needed to come up with a calculation of the ratio of mass to charge of the particles. The deflection induced by the magnetic field (μ) was expressed as μ = Mel/mv, where M is the intensity of the magnetic field, l is the length of the electromagnet producing the field, e is the charge on the "particles", m is their mass and v their velocity.

The intensity of the electric field produced by the metal plates (ø) was ø = Eel/mv^2, where E is the intensity of the electric field and l is the length of the plates. From these equations, it followed that:

$$v = (\mu/ø)(E/M) \text{ and}$$
$$m/e = M^2ø l/E\mu^2$$

The magnetic deflection was identical to the electric deflection previously measured. That meant that μ=ø, which implied in turn that v=E/M, and m/e=M^2l/Eø.

M, l, and E were known quantities, whilst the deflection originally produced by the electric field, ø, could be obtained by observing the scale at the end of the tube.

This impressively simple attainment of a value for the ratio of mass to charge began with the assumption that a particle was moving through the system with a velocity v and a mass m. We can equally interpret the experiment to involve the progression of a causal influence through the system, an influence that has a rate of evolution (from the spatial perspective) equal to v and which imparts a momentum mv to the target atom in which the causal process has its effect. The fundamental quantity mv that occurs in Thomson's calculation *does not* necessarily imply that the causal impulse itself has a mass. The quantity can equally be interpreted as deriving from the fact that a causal influence will impart momentum in its target, and the basic momentum imparted by an "electron" is a function of the electrostatic capacity of a genesis-unit (E). The extent to which the electric and magnetic fields will affect the rate of evolution of this causal influence will naturally be proportional to the magnitude of momentum that the impulse is capable of inducing in its target.

It is a fact of nature that anything that is a causal player in a system is capable of *moving* mass. There is a particularity, however, with the impartations of causal influence that involve what are normally called subatomic "particles". A gravitational causal source will have a *general* influence on all other bodies in the region, and this derives from the nature of the causal influence we call gravity (more on this later in the book). In the same way, a magnetized body will have general influence on other bodies in the vicinity (more on this in the next chapter). Such general influence is *distributed* over all bodies in the region, and we are not so ready to assume that it consists in some sort of exchange of particles (although some adherents of the matter-in-motion project *do* try to explain gravitation and other influences in terms of particles). With the case of the transmission of the "electron", however, the target of causal influence is a *particular atom*. This follows from the nature of the causal influence under consideration. Sometimes ionised atoms are in relatively stable circumstances (as in the case of the atoms on the leaves of a charged electroscope). In this case, the atom will have *general* electrostatic influence, leading to the attraction/repulsion of dissimilar/like charges. But in other more extreme circumstances (such as for example, when a powerful source of electric current is connected to the terminals of a pair of electric plates located close together), the source atoms are impelled to shed their electrostatic disequilibrium altogether. But the target of this particular causal influence is no longer the atoms in the vicinity *in general*. The target of such a causal impulse can only be the *particular* atom that now assumes the disequilibrium of the source, that indivisible disequilibrium which derives from the suppression or uncovering of one of the poles of the atom.

It is important to make a clear distinction between causal influence that will be *distributed* by a source over the bodies in its vicinity in general, and causal influence that is directed towards much more *particular* targets. In the case of the electron, the target is an individual atom. Charge and a momentum change are induced in that target, so it is natural to assume that a particle has been exchanged. The rate of exchange of the influence can be modified by manipulating the electromagnetic fields between source and target, but this does not imply that a particle is involved. The nature and strength of the field will either facilitate or hamper the causal exchange. To the extent that the field facilitates the transmission of the impulse, the greater the rapidity of the causal exchange and the greater the "impact" on the target atom. The causal influence we call the electron will have a ratio of charge to its capacity to induce momentum in a target, but the actual "momentum" of the electron in a particular situation will also depend on other factors that determine the rate of causal interchange.

This distinction between general and particular causal influences seems rather elementary, but it is of fundamental importance. If we do not keep the distinction in mind, then we can easily be led into the tendency of hypothesizing particles as the explanation for certain types of empirical data. Ionized atoms can have *both* general and particular causal consequences. Take the case of a negatively ionised atom, A1. Positively ionised atoms in general will have a tendency to converge on A1, just as material bodies tend to converge on each other gravitationally, whilst negatively ionised atoms will be repelled. This is an example of the general causal influence of an atom that is in a state of electrostatic disequilibrium. But under certain circumstances, that electrostatic disequilibrium can be transferred to another atom (usually referred to as the transmission of an "electron"). In this case, the causal influence can *only* be directed to a single atom. A1 cannot transfer its electrostatic disequilibrium to multiple atoms because the relationship between B and V is something that is fixed and indivisible in nature. While A1 remains ionised, its general influence consists in the tendency to induce *motion* in other atoms. These other atoms may converge on or diverge from A1, but their state of electrostatic equilibrium remains unaltered until the moment when just *one*, and only one, of them becomes the target of the *particular* causal influence of A1 that we call the transmission of the "electron".

4.6 Theoretical reasons for denying the status of particle to the electron

There is a natural inclination to introduce hypothetical *entities* to explain a particular physical phenomenon such as ionisation. There can be no doubt that an unrestrained exercise of this tendency can land physics in theoretical difficulties. If the causal phenomenon under investigation does *not* involve the transmission of a particle, then any account that invokes particles will have difficulties explaining all of the empirical data in a coherent way.

A remedy for this trend is to make a different kind of attitude programmatic in physics, an attitude that makes the hypothesizing of particles or properties a grave matter not to be entered upon lightly. This is the attitude that we have tried to adopt in hypothesizing the existence and properties of the genesis-unit. The vast store of empirical evidence that underlies the construction of the periodic table of elements, not to mention the empirical facts of how elements bond together to form compounds, gives us powerful reasons for thinking that elements are composed of atoms that are themselves formed from successive aggregations of what is contained in the hydrogen atom. That there is a kind of fundamental polarity at the heart of the atom is also testified by an enormous array of evidence. And our attempt at understanding notions such as electric charge and ionisation in terms of a more fundamental "protomagnetic" polarity is in line with the unambiguous evidence that magnetism and electricity are different manifestations of a common condition (in the next chapter we will turn to a proper discussion of the nature of magnetism). But this is where we depart from what has become the standard approach to theoretical physics. Based solely on these minimalist metaphysical assumptions, we seek to understand the pattern of atomic bonding, the structure of the periodic table of elements, the transmission of light, the nature of electrostatic charge and ionisation, and the phenomenon of magnetism. This can be done, we believe, without invoking particles, waves or fields.

This attitude resists the inclination to describe ionisation as an atom that has lost or gained a part. Evidence at this level is always going to be ambiguous so we must tread lightly with our metaphysical suppositions. How can analogical talk advance our understanding of ionisation in a way that the hypothesizing of particles cannot? We admit that such language may have to be modified, or even dramatically changed, as it is confronted with the evidence. But the fact remains that language of this sort tends to capture the salient elements of causal processes without introducing metaphysical assumptions that may have no grounding in reality. We are not sure if ionisation consists in the gaining or losing of a particle, but we can speak reliably of it as involving atoms being thrown out of electrostatic equilibrium with other atoms as the result of causal events. In our account, it involves one of the components of the genesis-unit being covered or uncovered with respect to the rest of the system. Does analogical language run the risk of being so vague that it tells us very little about a physical phenomenon? Yes, there is a real risk of that. But if ionisation does *not* consist in the gaining or losing of a particle, then the negative repercussions of this mistaken ontological supposition on future theory are potentially enormous. In other words, it is better to construct our understanding of ionisation upon foundations that are sure, even if minimalist, than upon more elaborate statements that are false.

The contention is that atomic theory can be taken forward, little by little, using minimalist language of this sort whilst refraining as much as possible from introducing metaphysical entities. Once it was considered to be relatively straightforward to verify or refute the existence of hypothetical entities such as the luminiferous ether. All that was required was a decisive experiment carried out to the correct degree of precision. The irony of modern physics is that its hypothetical entities are so slippery that no single experiment can fish them out. And the experiments that do "corroborate" their existence are themselves so weighed down with theoretical baggage that it is difficult to discern just what they are saying unambiguously.

The treatment given to the electron in this chapter is in line with the epistemological attitude that resists the impulse to hypothesize particles. It is already accepted in physics that causal impulses such as the photon can induce momentum in a target even if they themselves are massless. It is also accepted that photons can be deflected by gravitational influences. What we have done is re-evaluate Thomson's experiment and refused to accept that deflection of a causal influence necessarily implies that it has mass. A causal impulse can have an associated momentum without being a particle. This momentum *is indeed* properly spoken of in relation to particles – the particle that emitted the influence and recoiled as a result, and the particle that absorbed the influence and recoiled as a result. Our mathematical expressions for calculating the momentum associated with the causal impulse can attribute velocity and mass to the influence itself, but this should not be taken as an indication that the *impulse is an intermediary with velocity and mass*. The associated velocity is an expression of the rate of evolution of the causal influence through the system, whilst the mass is an expression of the relative potential of this influence *to do work at its target*. The occurrences of m and v in the mathematical expressions, and the ability of Thomson to obtain a value for m/e, do not corroborate the particle hypothesis in any definitive way.

Perhaps the most positive aspect of our approach is its capacity to explain diverse phenomena in a relatively simple way, beginning from its treatment of the way that matter itself originates. The *protomagnetic* polarity within the atom is generated by the work impulse and it is the separation that constitutes matter itself. The *electrostatic* attraction/repulsion between the B and V components of *different* atoms is a secondary phenomenon that arises naturally from protomagnetism. Electrostatic attraction underpins the bonding behaviour in atoms, explains the structure of the atomic table and is responsible for the alignment of atoms in the mini-systems that give rise to the emission and absorption of light. In certain circumstances, the B or V component of an individual atom may be covered or shielded. In this situation, which is referred to as "*ionisation*", the atom relates in an imbalanced way with other atoms. When the B-V components of the atoms in a material are uniformly aligned, the object is said to be *magnetised* (as we shall argue in the next chapter). All these diverse phenomena are provided with an explanation that rests on a simple and unified foundation, a foundation that would disintegrate if we attributed the status of particle to an electron.

Chapter Five

MAGNETISM, ELECTRICITY AND THE THEORETICAL DERIVATION OF THE VELOCITY OF LIGHT

This velocity is so nearly that of light that it seems we have strong reason to conclude that light itself... is an electromagnetic disturbance in the form of waves propagated through the electromagnetic field according to electromagnetic laws

James Clerk Maxwell in *A Dynamical Theory of the Electromagnetic Field*

Overview of this chapter and its principal claims

1. During the nineteenth century, two separate units were used for quantifying electric current. The first was the electromagnetic unit (the ampere), which measured current in terms of the magnetic effect produced by a current-carrying wire. The greater the current, the greater the magnetic influence emanating from the wire. The second unit was the electrostatic unit (the coulomb per second), and this involved calculating the capacity of a current to induce electrostatic charge in an object. In 1856, Weber and Kohlrausch made the surprising discovery that the ratio of these two units for the same magnitude of electric current equalled the speed of light.

2. We present a thought experiment designed to explore this remarkable finding. If we formulate two analogous types of unit for measuring the flow of *gas in a closed system*, will the ratio of these units also yield a value for the velocity of the gas in the system?

3. A distinction is made between the rate of flow of molecules of gas in a tube and the rate of evolution of electromagnetic radiation through a wire. We discuss why the value for c emerged empirically from the experiment by Weber and Kohlrausch, and why it can be derived theoretically from Maxwell's equations.

4. This prepares the ground for a description of the natures of electrical current, magnetism, electromagnetism, electromagnetic induction and the notion of the field.

5. Magnetism and electrostatic charge are distinguished. Both arise from the basic protomagnetic polarity that constitutes the genesis-unit. The difference is that electrostatic charge concerns the influence emanating from *one or other of the poles of a unit*, especially in cases where the unit is ionised and one of the poles has been covered. Magnetism, by contrast, involves many genesis-units in the same object sharing the *same B-V alignment*. This common alignment has no attractive or repulsive influence on the particular poles of other genesis-units in the vicinity, but it does create a powerful tendency for the units in ferromagnetic objects to *align* in the opposite direction. Once this opposite alignment has occurred, *then* the ferromagnetic object will naturally be drawn to the magnet.

5.1 Introduction: nineteenth century electromagnetic theory

The account of the atom that has been defended in this book considers the basic B-V protomagnetic polarity to be more fundamental than *either* electric charge or magnetism. Following the treatments of ionisation and the nature of the electron that were presented in the last chapter, it might seem that this is an opportune moment to discuss the nature of magnetism. However, it may be more valuable at this point to backtrack a little and approach the whole subject of electrostatic charge, magnetism and electric current with the eyes of the physicists of the nineteenth century.

When an electric current flows through a wire, it gives rise to a magnetic field in the vicinity. This provides a means of quantifying the current, since the magnetic influence caused by the electricity is proportional to the magnitude of flow. A galvanometer registers the amount of current by the extent of deflection of its magnetic needle. The unit of electric current measured by this method is called the *electromagnetic unit.*

But there is another, completely different, means of measuring electric current. Electro*static* charges have attractive and repulsive effects on each other. We can quantify the extent to which an object is electrostatically charged if we bring a similarly charged object to within one centimetre of it and then calculate the mutual forces exerted by the objects on each other. These forces provide a measure of the electrostatic charge on the objects. Let us call the charge on each object q. We then load a Leyden jar (a type of capacitor) with this quantity q and allow the jar to discharge, measuring the time it takes for the jar to be discharged completely. Call this time interval t. The flow of current from the jar will be q/t. This alternative unit for electric current is called the *electrostatic unit.*

In 1856, Wilhelm Eduard Weber and Rudolf Kohlrausch performed a simple experiment that produced a stunning and mysterious result. Using a Leyden jar and a galvanometer, they compared the magnitude of an electric current measured in electromagnetic units with that of the same current measured in electrostatic units. It was found that the ratio of the units equalled *the velocity of light!* In a presentation to the Royal Society of December 8th, 1864, James Clerk Maxwell interpreted this result to signify that light itself was a form of electromagnetic radiation. He proceeded to derive the speed of light theoretically from his equations. But why should the manipulation of equations that deal with electric and magnetic forces yield the value for the speed of light? Or, more starkly, how does the simple comparison of the magnetic deflections produced by an electrostatic unit of charge and an electromagnetic unit of charge produce a value equal to the velocity of light? The accurate empirical measurement of the speed of light was already proving to be a daunting task in the 19th century. It seemed extraordinary that this value should, as it were, accidently pop out of an experiment using a primitive apparatus that was in no way geared to measuring the speed of anything.

Once Maxwell had produced his interpretation, the notion that light was an electromagnetic wave quickly became the standard view. The transmission of light was understood in terms of electric and magnetic fields feeding off each other, generating a wave motion that propagated indefinitely through space. This fitted very well with all that was already known at that time of electricity and magnetism. In 1819 and 1820, Ørsted and Ampere had discovered that electric current gave rise to magnetic effects, whilst a few years later Faraday was able to demonstrate that moving magnets gave rise to an electric current in a nearby conductor. Maxwell built on the work of Ampere, Faraday, Coulomb and others to formulate his series of equations that described the interdependent relationship of magnetic and electrical phenomena. In this new theory of electromagnetic radiation, electricity, magnetism and light were all understood as manifestations of the same underlying phenomenon.

Much has happened in physics in the meantime. Nowadays, light is described as either a transmission of photons of energy or as the propagation of an electromagnetic wave, depending on the circumstances of the description. But whenever we choose the wave model for our account, it is effectively the same model as that presented by Maxwell. Light is viewed in terms of synchronized oscillations of electric and magnetic fields that are perpendicular to each other and to the direction of propagation. At the time of Maxwell, this was a plausible theory based on the knowledge at hand of magnetism and electricity. In our day, however, with the benefit of empirical evidence that was gathered during the twentieth century, we have good reasons for thinking that something very different underlies these phenomena. The first part of this chapter, thus, will offer a critical evaluation of the usual model of electromagnetic radiation. It will do so by providing an explanation of how the value of the speed of light mysteriously drops out of a comparison of the different forces arising from electromagnetic phenomena. Later in the chapter we will present a new account of magnetism and electric current.

5.2 A thought experiment regarding the units of electric current

In this section a thought experiment will be described in which - instead of electric current moving in a circuit - a gas is moving through a closed system. The properties of the gas will be measured using two different types of units. Will a comparison of these units yield the velocity of the gas, just as a comparison of different units led Maxwell to a theoretical derivation of the speed of light?

We are accustomed to using terms like "electric current" and "electric charge" without thinking too much of the causal activity that underlies these phenomena. The units we use in scientific parlance, such as the ampere, the coulomb, or the coulomb per second, influence the way we visualize the evolution of causal activity in the system. For example, if we hear that the current in a wire is x coulombs per second, then we tend to think of this current as involving *electric charge of quantity x passing any point* in the wire per second. When we examine the ways that electric charge and electric current are quantified, however, we soon realize that our familiar ways of visualizing causal events are far from accurate. But these familiar ways were part and parcel of Maxwell's construal of the nature of electromagnetic radiation, as we shall now see.

Imagine the following experimental set-up. A pump is forcing a gas at high pressure to circulate in a tube.

The experimenter comes up with two ways in which to measure the volume of gas flowing in the system. As the gas moves through the system, the material in the tube becomes tauter, although there is no measurable change in the diameter of the tube. The higher the output of the pump, the greater the flow of gas and the tauter the surface of the tube becomes. The tautness of the tube is measured by pushing a probe into its surface. The probe registers how much force is required to indent the surface of the tube at that point by one millimetre. In principle, if the material of the tube is made of the right material, the change in tautness will give an accurate indication of the flow of gas within (once, of course, the gas has attained a certain minimum pressure that leads to the initial increase in the tautness of the material). The experimenter decides to use the term "ampere" for every increase of a certain amount in the tautness of the tube.

The second method for measuring the flow is quite different. A balloon is connected to an offshoot of the tube. The offshoot has a valve connected to a timer that can be opened or closed for any chosen interval of time. The valve is opened for one second and then closed again. In that second, the balloon fills and its tautness is measured. Once again, tautness is quantified by measuring the amount of force required to push a blunt probe one millimetre into the surface of the balloon. The experimenter uses the term "coulomb" for the force required to cause a particular increment in tautness. The "coulomb-per-second" represents an alternative measure of the volume of flow of gas. Whenever the flow of gas is greater, then the tautness of the balloon (after one second of flow) will be greater. Thus, the coulomb-per-second will furnish an accurate reflection of the flow of gas.

In the case of electromagnetism, Weber and Kohlrausch found that a comparison of electrostatic units and electromagnetic units produced a value equal to the velocity of light. Before considering how the ratio of our two units for the flow of gas might be made to yield a value for the *speed of individual molecules* inside the tube, let us reflect for a moment on how the "ampere" and the "coulomb" register flow in radically different ways. The "ampere" in our experiment is measured with a probe that pushes into the surface of the tube. No mechanism for measuring the passage of time is required for the operation, yet it delivers a quantity that does indeed reflect the volume of gas moving inside the tube per interval of time. The measurement of flow using the "coulomb-per-second" unit, by contrast, does involve allowing the gas to flow for one second through the valve into the balloon. Once the balloon has been "charged" in this way, its tautness is measured by calculating the force necessary to indent the surface by one millimetre with the probe. In both of these cases, the unit for flow does not involve a direct mechanism for the measurement of the current of gas through the tube. Rather, both units are generated by different *collateral effects* of the flow.

The real ampere and the real coulomb in the very real world of electricity are generated in a similarly "unreal" fashion. By convention, the ampere is that current in a wire which exerts a magnetic force of 2×10^{-7} newtons per metre on a parallel wire carrying the same current located one metre away. There is no direct attempt here to try to measure what is actually flowing in the wire. Instead we measure a collateral effect of the flow – the extent to which nearby matter is accelerated as a result of the magnetic influence of the current. From this "electromagnetic" unit of current, the "electrostatic" unit of charge can be empirically derived. We simply allow current of one ampere to discharge for one second into an object or instrument of some sort, such as a Leyden jar. Once we have isolated a fixed quantity of charge in this way, we can use it to electrostatically charge other objects at will. Next we bring two of these charged objects to within a certain distance of each other. We measure the acceleration (per gramme of each object) that is mutually exerted by one object on the other as a result of their electrostatic charge. This force, measured in newtons, is called the coulomb. Thus the coulomb is the mutual force exerted (per unit of mass) by objects placed at a certain distance from each other that have been charged by electric current of one ampere flowing for one second.

This electrostatic unit of current gives us a new way of thinking of the electromagnetic unit of current defined earlier. Since the Leyden jar was charged with one coulomb of electricity by allowing one ampere to flow for one second, it is entirely natural to think of the ampere as being the current present in a wire whenever *one coulomb of charge per second is flowing*. But this comfortable picture is misleading. A coulomb is actually the force between two stationary objects that have a particular electrostatic disequilibrium. Do we think that one ampere of current involves particle-like objects of this sort moving inside a wire, each one emanating a certain quantity of force as it goes? We shall see as we continue that electric current is nothing of this sort. A charged object is the *consequence* of the flow of an electric current: current does not involve the flow of charged objects. Our familiar way of visualizing electric current in terms of charge moving along a wire is a product of the way that we quantify the magnitude of the underlying causal process: quantification in terms of collateral *effects*. One of the effects of current is to cause an electrostatic disequilibrium in an object; this disequilibrium causes acceleration in nearby matter – a phenomenon that we call charge; and so we tend to think that current involves a flow of these charges inside a wire.

5.3 Gauss's law

The electric current flowing in the wire gives rise to a magnetic influence on objects in the vicinity, whilst the stationary charges have an electrostatic influence on nearby objects. The range and magnitude of these influences are usually thought of in terms of magnetic and electric *fields* respectively. According to Gauss's Law, if an electric charge is surrounded with a closed surface, such as a sphere, the net electric flux through the surface is proportional to the total electric charge within the closed surface. Say, for example, that a charge of magnitude x is held within a sphere of radius r. If we now quantify the electric influence exerted by the charge at every point on the surface of the sphere and sum all of these quantities together, then the total will be proportional to the charge x within. The constant of proportionality for electrostatics is ε_0.

Gauss's Law can be expressed in vector calculus in integral form as follows:

$$\int E' \cdot dA = q/\varepsilon_0.$$

The mathematical symbols we are using here are not quite right but what the formula means to express is as follows: we imagine the surface of the sphere being covered with a very large number of infinitesimal squares of area dA. At each of these points, the charge q within is exerting an influence of magnitude E' that points outwards at ninety degrees to the surface of the sphere at that point. The integration symbol ($\int$) indicates that all of these influences are to be added together. Their sum equals the total charge contained within (q) divided by the constant of proportionality (ε_0), a constant historically referred to as the electrical permittivity of free space. Intuitively, the idea is very simple: the total charge inside the sphere is spreading its electrical influence evenly all around the surface of the sphere.

The image from our thought experiment of the balloon being "charged" with a flow of gas for one second gives us a fresh way of looking at Gauss's Law. According to the thought experiment, when the level of expansion of the tube showed that one "ampere" of gas was flowing in the system, the valve on the offshoot of the tube was opened for one second in order to fill the balloon. The tautness of the balloon was measured by using the probe to register the tension at any point on the surface of the balloon. Like the electric flux on the surface of the sphere surrounding the electric charge, the vector for tautness will point perpendicularly outwards from the balloon, since the pressure is caused by the gas molecules within. Let us call the tautness vector at any point E', whilst the area of that infinitesimal point on the balloon is dA. If q represents the number of gas molecules inside the balloon, then we can expect the following equality to hold: $\int E' \cdot dA = q/\varepsilon_0$, where ε_0 is a constant of proportionality that relates the number of molecules of gas inside to the tautness of the surface. This constant will vary depending on the empirical characteristics of the material of the balloon itself. In Maxwell's system, ε_0 was considered to derive from the characteristics of the ether through which electrical influence was transmitted. Of course, with a careful choice of units, ε_0 could be made to have the value 1, but in that case the empirical characteristics of the system (that determine the magnitude of the effect that charge exerts on other objects) are really *hidden within the magnitudes of the units selected*. Whatever value is ultimately assigned to ε_0, the opinion of Maxwell and his contemporaries regarding its nature was perfectly valid, and remains so: the constant's *relation to our electromagnetic units* is based on the empirical characteristics of the system, characteristics that determine how charged objects give rise to accelerations in nearby masses.

5.4 Calculating the velocity of the molecules of gas

The ampere is used as a measure of the flow of current, but it is primarily the size of the magnetic force exerted by current of a certain magnitude. The coulomb is used to refer to the electrostatic disequilibrium of an object, but it is really a measure of the acceleration induced by a charged object in nearby matter. In the thought experiment, meanwhile, the rate of flow of gas was measured by the change in the tautness of the surface of the tube, whilst the amount of gas flowing per second was expressed in terms of the tautness of the balloon. The fact that we are using *collateral effects* to quantify the properties of systems leads to some confusion about the real meaning of the units that appear in our descriptions of the system. This confusion is increased when we begin to treat the units as if they were the *actual phenomenon* that, in fact, they only *indirectly* register. We saw how it is easy to begin mixing metaphors, as it were, and describe the ampere as the flow of one coulomb per second of charge. We do this because we visualise the coulomb as a mobile bundle of charge moving along the wire, when in fact it is a measure of the acceleration that charged objects induce in matter that is stationary relative to them.

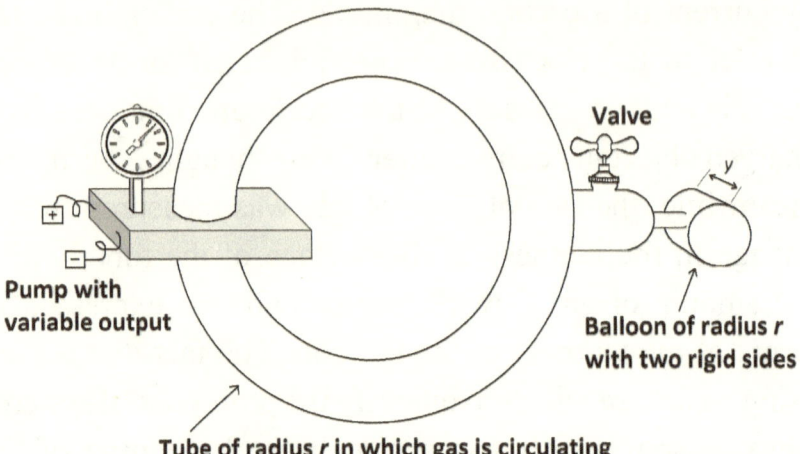

Valve

**Pump with
variable output**

**Balloon of radius *r*
with two rigid sides**

Tube of radius *r* in which gas is circulating

*Figure 5.1 Thought experiment to illustrate the relationship
between electrostatic and electromagnetic units. A pump with
variable output circulates a gas through a tube of radius* r. *An
offshoot of the tube permits a balloon to be filled when a valve is
opened for one second.*

Let us see now how a comparison of the two units in the
thought experiment might be made to yield a value for the
velocity of the individual molecules of gas in the tube. To
achieve this, the experimental set-up will have to be refined a
little. Say that the radius of the tube is equal to *r*. The balloon
that is attached to the offshoot of the tube will also have a
radius *r*, but it will be no ordinary balloon. To simplify our
"derivation" of the speed of the molecules of gas, the balloon
will have two flat rigid sides (see Figure 5.1). These sides
remain flat in response to increased pressure of gas, but the
distance between them (*y*) increases as the balloon fills. This
means that the volume of the balloon is actually a cylinder
and not a sphere. The curved surface of the balloon has radius
r, a radius that does not change as the pressure of the gas

300

increases or decreases. As the pressure increases, the rigid sides move further apart and distance y increases, leading to greater tautness in the curved surface of the balloon. It is this tautness which will provide us with a measure of the amount of "coulombs" of gas contained within the balloon (to make our experiment as closely analogous as possible to the electromagnetic case, we will ignore the fact that a measure of y could also provide an accurate index of the number of gas molecules contained inside the balloon).

According to the empirical law that our experimenter has already derived, the number of coulombs of gas within the balloon (q) is proportional to the tautness of the surface as follows: $\int E' \cdot dA = q/\varepsilon_0$. In this equation, the particular tautness (E') at every infinitesimal point on the surface (dA), when summed together, is proportional to the number of molecules enclosed within. Given that the balloon was filled by opening the valve for one second, we can describe the current of gas in the system as q coulombs per second. This gives the flow in the equivalent of "electrostatic" units.

The current can be written as a differential equation as follows: $I = dq/dt$, where dq/dt represents the instantaneous change in electric charge with respect to time. Now, as we have seen, $\int E' \cdot dA = q/\varepsilon_0$, which means that: $q = \varepsilon_0 \int E' \cdot dA$. If $I = dq/dt$, then, substituting for q, we get:

$$I = \varepsilon_0 \, d(\int E' \cdot dA)/dt \qquad (1)$$

The experimenter has also derived an empirical law that relates the tautness of the tube to the flow of gas in the system: $\oint B' \cdot dl = \mu_0 I$. Again, the mathematical symbols available to us here are not quite right. This time "$\oint$" denotes an integral around a closed *path*, *dl* is an infinitesimal point along that path, and *B'* is the tautness vector of the tube at that point. Thus, the sum total of the tautness at each point along a closed path around the tube in a given instant is proportional to the amount of gas flowing through that particular contour of the tube in that instant. Once again, the magnitude of the constant of proportionality, μ_0, will be dictated by empirical factors such as the extent to which the tube becomes taut as pressure increases. From the above equation we see that:

$$I = \oint B' \cdot dl/\mu_0 \qquad (2)$$

Now we have two formulae for *I*, one of which (1) describes current in terms of "electrostatic" units, and the other (2) in terms of "electromagnetic" units. Putting them together:

$$\varepsilon_0 \, d(\smallint E' \cdot dA)/dt = \oint B' \cdot dl/\mu_0 \qquad (3)$$

Let us see if we can simplify this equation. The radius of both the tube and the balloon is *r*. Therefore the total length of the simplest closed path, *l*, around the tube will be $2\pi r$, whilst the total area of the balloon will be $\pi r^2 y$. Therefore $\smallint E' \cdot dA$ can be rewritten as $E\pi r^2 y$ and $\oint B' \cdot dl$ can be written as $2B\pi r$, where B is the tautness at any point on the balloon and E is the tautness at any point around the tube. Substituting these in (3) we get:

$$\varepsilon_0 \, d(E\pi r^2 y)/dt = 2B\pi r / \mu_0 \qquad (4)$$

302

Say the gas is flowing at a rate of v metres per second. The time it takes (t) to move distance r will be r/v. Therefore $r = vt$. Substituting this for the values of r in (4):

$$\varepsilon_0 \, d(E\pi v^2 t^2 y)/dt = 2B\pi vt/\mu_0 \qquad (5)$$

Differentiating for t gives us: $\quad \varepsilon_0 \, 2E\pi v^2 ty = 2B\pi vt/\mu_0 \quad (6)$

This simplifies to: $\qquad\qquad\qquad \varepsilon_0 Evy = B/\mu_0 \qquad\qquad (7)$

The glaring mathematical error made in the differentiation to give us (6) will be overlooked for the moment. We will return to this error later to examine it more closely and to consider the conditions that would make such a mathematical move justifiable.

To make the thought experiment analogous to electromagnetism we will need to stretch our imaginations a little at this point. Say that the experimenter has realised something similar to what Michael Faraday stumbled upon when he discovered the phenomenon of electromagnetic induction. Faraday found that when a magnet and a coil of wire were moved relatively to each other, electricity was induced in the coil. The electricity was proportional to the rate of change of magnetic flux through any surface bounded by the wire, or:

$$\oint E' \cdot dl = d(\smallint B' \cdot dA)/dt$$

In this formula, the total sum ($\oint$) of the electricity (E') induced in each point (dl) along a coil is equal to the instantaneous change (d/dt) of the net ($\smallint$) magnetic flux (B') at each point in the two dimensional surface (dA)

through which the magnetic source is moving. Sometimes this situation is likened to a loop of wire which has a film of soap spanning it. The magnet moves through the film of soap inducing electricity in the loop of wire. The total magnetic flux over the area of the film of soap will be $\int B' \cdot dA$, whilst the rate of change of magnetic flux as a result of the relative movement of magnet and loop of wire will be $d(\int B' \cdot dA)/dt$.

We imagine that our experimenter has found a corresponding law for the thought experiment. He discovers that by simultaneously pushing many probes one millimetre into the surface of the tube, a flow of gas results. The formula for this relationship is similar to Faraday's:

$$y \oint E' \cdot dl = d(\int B' \cdot dA)/dt.$$

In this case, $d(\int B' \cdot dA)/dt$ gives a measure of the rate at which the probes are pushed into a portion of the tube whose cross section has an area A. The expression $y \oint E' \cdot dl$ gives a measure of the tautness of the balloon when the movement of the probes prompts a flow of this magnitude for one second.

If we once again substitute vt for r, then $y \oint E' \cdot dl$ becomes $2\pi vty E$, whilst $d(\int B' \cdot dA)/dt$ becomes $d(\pi v^2 t^2 B)/dt$, where v is the velocity of the gas, and t is the time required for it to move a distance r. Thus, differentiating for t: $2\pi vty E = 2\pi v^2 t B$

$$\text{Or, } v = Ey/B \qquad (8)$$

This corresponds to the remarkable result achieved by Weber and Kohlrausch (once again we will overlook the recurrent error in the differentiation because later we will see how this mistake reveals something of how the flow of electric current differs from the flow of gas). In formula (8), the velocity of the gas is equal to a simple comparison of the influence of the flow of gas measured in electrostatic units to the influence of the flow of gas measured in electromagnetic units. Or to state it in terms of what is happening in the experiment: the velocity of the gas is equal to the tautness which that particular flow of gas causes in the balloon *divided by* the tautness that the same flow of gas causes in the tube. What can such an expression mean? How does the velocity of the gas emerge from a comparison of these properties of the balloon and the tube?

Before considering these questions, let us "derive" a further result from the experiment. Earlier we had in (7):

$\varepsilon_0 Evy = B/\mu_0$

If we substitute here the value for E from (8) then we get: $v^2 = 1/\varepsilon_0\mu_0$. This is another curious result that exactly parallels a finding from Maxwell's equations. The velocity of light is related in a simple way to the constants of electric permittivity and magnetic permeability. Why should such a simple relationship be the case?

5.5 Why does a value for velocity emerge from these experiments?

We will return now to the errors that were deliberately overlooked when differentiating the expressions for current. In the first instance, we differentiated:

$$\varepsilon_0 \, d(E\pi v^2 t^2 y)/dt$$

and got:

$$\varepsilon_0 \, 2E\pi v^2 ty$$

Here we followed the basic rule for differentiation where a function of the form yx^n becomes nyx^{n-1} when it is differentiated with respect to x. In our formula t^2 became $2t^1$ when operated on with respect to t, which seems legitimate, but the problem is that the t which appears in the expression $E\pi v^2 t^2 y$ is not the same sort of t that we are differentiating. Originally we began with an expression for the current of gas in the system, $I = \varepsilon_0 \, d(\int E' \cdot dA)/dt$. Thus we were considering the rate of *flow* of molecules of gas in the tube. In other words, the differentiation should yield the volume of gas that passes a given point in an instant. But when we made the error, we were differentiating with respect to a t that concerned the *velocity of the gas* in the tube. More precisely, it was the time taken for the gas to move distance r. There is a big difference between the *number of molecules* of gas that pass a given point in time t, and the *distance* that gas travels in time t. Say that the tube has a large diameter. In this case, even if the gas is moving relatively slowly, the flow of current will be high because of the large cross section of the pipe. The volume of gas that passes a given point with respect to time is different to the length of tube that the gas traverses with respect to time. We were supposed to differentiate with respect to the former rate of change but in fact differentiated with respect to the latter.

There is a connection, of course, between the velocity of gas (v) and the magnitude of its flow (I). If the gas is moving faster in a tube of a given cross section, then the current will be greater. If we know the number (n) of molecules of gas that fit in a cross section of the tube, then I = nv. Does *this* connection between current and the velocity of individual molecules of gas reveal anything about the curious emergence of c in the electromagnetic case? Yes, it does, as we shall see shortly. But first let us see what our deliberate error reveals about the connection between electromagnetic and electrostatic units.

Despite the mathematical blunder, it is still a little curious that a value for the speed of the gas should almost emerge from a set-up that has no apparatus whatsoever for measuring speed! In fact, if we knew the number of molecules of gas that fit in the cross-section of the tune at a certain pressure, then we could in principle derive the speed of the gas from the equations that we already have at hand. How this this come to be? The answer is *the process by which we fill the balloon* in order to generate our electrostatic units. Before filling the balloon we have already a value for the flow of the gas in electromagnetic units based on the tautness of the tube ($I = \oint B' \cdot dl/\mu_0$). Then the valve is opened for one second and the balloon is filled. This allows us to generate the electrostatic unit of flow based on the tautness of the balloon ($I = \varepsilon_0 \, d(\int E' \cdot dA)/dt$). The volume of the flow is crucial to the relationship between the electromagnetic and electrostatic unit. If the flow were smaller, then less gas would enter the balloon and its tautness would be reduced. In fact, once purely empirical considerations such as the physical characteristics of the material of the tube and the material of the balloon are taken into account (and these *are* taken into account by μ_0 and ε_0 and the actual choice of the magnitudes of the units), then the volume of flow is *the* determining factor in establishing the relationship between the electromagnetic unit and the electrostatic unit.

As we have said, the volume of flow is itself dependent on two factors: the velocity of the gas and the cross-section of the tube. *This* is how the value for velocity can be made to fall out of an experiment that only measures two types of tautness and compares them with each other! The first type of tautness as a measure of flow is more or less established as a matter of convention. The experimenter does an empirical examination on the tautness of the tube and chooses a certain magnitude of it as the electromagnetic unit of flow. But the second choice of unit is not so arbitrary. In this case, a flow of one electromagnetic unit of gas for one second is used to fill a balloon. There will be a strict mathematical relationship between the unit used to measure the flow of gas in the tube and the unit that measures the state of the balloon as a result of this flow from the tube. And this relationship will depend on the rate of flow of gas into the balloon, which itself depends (in part) on the velocity of each molecule of gas.

We can justifiably say that the electromagnetic unit of gas is used to *generate* the electrostatic unit. The process of generation involves allowing the gas to flow (for 1 second). The magnitude of the resultant electrostatic unit will depend on the velocity of the gas (as well as the cross section of the tube). In formula (8) above, we found (erroneously) that the velocity of the gas, v, was equal to Ey/B (where y was the width of the balloon). This can be corrected to $vn = Ey/B$, where n is the number of molecules that fit in a cross section of the tube at this pressure. Therefore $Ey = Bvn$. The relationship between B, v and E now seems to make much better sense. After all, the electrostatic unit, E (the tautness of the balloon), is generated by B (the flow in the tube measured in terms of the tautness of the tube). How E is generated will be directly proportional to the velocity of the gas, as well as the cross section of the tube. Therefore E will be directly proportional to Bvn.

5.6 Maxwell's derivation of c

We are now in a better position to understand how c emerged from the experiment of Weber and Kohlrausch, and how Maxwell derived it from his equations. The story of how the ampere and the coulomb were historically defined relative to each other is not straightforward, but it can be simplified in terms of the basic pattern of the Weber-Kohlrausch experiment. First, the magnetic influence emanating from a current-carrying wire was noted. An electromagnetic unit of electric current, the ampere, was then defined in terms of a certain magnitude of magnetic force produced by the wire per unit length. Next, two objects were (each) electrostatically charged with this quantity of current. This operation involved allowing the current to flow for one second into each object. The resultant electrostatic force between the two objects was measured and this became the basis of the unit of electrostatic charge, the coulomb. The electrostatic unit of electric current is simply this unit of charge per second.

From the analogy with the gas circuit, we can imagine how the flow of an electromagnetic unit of current *generates* the electrostatic unit, and how this process might depend on the velocity of the charge. There is a strong tendency to visualize the flow of current as involving the movement of electric charges along a wire, just as the movement of gas involved the physical movement of gas molecules inside a tube. The objects mentioned above were electrostatically charged by allowing electricity to flow into them for one second. During that interval, we imagine charges flowing into the objects and accumulating on them. The amount of charges that arrive at each of the objects in that one second will depend on their velocity; hence the link between E and cB.

This cosy picture, however, is simply false. It is well known that charges do not move at velocity c. There is precious little movement of matter of any kind along a current-carrying wire. In fact, the approach that we are taking in this book to electromagnetic activity – namely that electromagnetic influence requires no causal intermediary but involves an evolution of the system – is well-suited to explaining the mystery of the emergence of c from Maxwell's equations. As the system evolves, causal events make themselves felt at other points in the system after an interval of time that corresponds to the distance from the source divided by c. Thus c is really a measure of the speed of causality itself, the rate at which causes have their effects in the system. Let us apply this to what is happening in a current-carrying wire.

Electric current involves the evolution of electrostatic disequilibrium through the atoms of a conductor. The electric generator causes pairs of atoms in the source to go into a state of electrostatic imbalance. Atoms in one type of electric imbalance are forced to the cathode, whilst atoms with the other type of imbalance are forced to the anode. The build-up of atoms with a similar imbalance results in that imbalance being transmitted along the atoms of a wire joined to the poles. The atoms themselves in the conductor do not need to move at all for this imbalance to be transmitted along the wire. Each atom passes its imbalance to the adjacent atom, returns to a state of equilibrium momentarily, before being thrown back into a state of imbalance by the atom on the other side, and so on. The impulses that evolve between adjacent atoms do so with the speed of causal evolution of this sort, namely c.

In order to measure the electrostatic unit of charge, our experimenter allowed one ampere of current to flow for one second into an object or instrument (such as a Leyden jar). Here we must resist the temptation to think of this process as involving a set quantity of particle-like charges (or even charged particles) moving along the wire into the jar and accumulating there. Instead, think of it as a causal process where the source of electricity (whose atoms are in marked electrostatic disequilibrium) exerts a causal influence on a target (ultimately, the atoms of the Leyden jar). The causal evolution happens at the speed of light and the upshot is that the electrostatic disequilibrium of some of the atoms in the source is transmitted to some of the atoms in the jar. Certainly, the Leyden jar ends up with a certain quantity of charged atoms, but the atoms were *already* part of the jar before the wire was even connected to it! It is the disequilibrium that is transmitted along the wire. But even this last statement could be misunderstood. In causal processes such as this one, the end result is generally *greater equilibrium* in the system. The marked disequilibrium in the source of power is lessened by the evolution of this causal process along the wire. The wire, as it were, becomes a conduit for the evolution of the overall equilibrium of the system, even though it will end up charging the Leyden jar!

So this causal evolution occurs for one second, and it happens at the speed of light. During this second, the causal disequilibrium that is responsible for the quantity of magnetic influence that we call the ampere has its effect on the atoms of the metal foil in the jar. In other words, a source whose output is one electromagnetic unit of "current" creates electrostatic disequilibrium in the Leyden jar. The magnitude of this disequilibrium will be the basis for the definition of the unit of electrostatic charge. Thus the electromagnetic unit of current is used to generate the unit of electrostatic charge. What will determine how much charge builds up on the Leyden jar in one second? The primary factor is the extent of causal disequilibrium in the source. Causal sources in relatively high states of disequilibrium have causal effects of greater magnitude (think of a very hot object emanating heat). The extent of causal disequilibrium in this particular source has been measured by means of the magnetic influence emanating per unit length of current-carrying wire, and that magnitude has been found to be one ampere. The next most important factor that will dictate how much charge will build up in the Leyden jar in one second is the *velocity of the causal influence* that evolves from the source to the atoms of the metal foil in the target. We could imagine a very slow causal influence that in one second fails to reach the jar at all. Or we could imagine a swifter influence whose impact on the atoms in the jar will be relatively lessened depending on the length of wire between the causal source and target. In that case, B generates E but its influence will have to be qualified by a factor of *v/d,* where *v* is the velocity of the evolution and *d* is the length of the wire. Or, as is the case with electromagnetic influence, we imagine an extremely rapid evolution, the speed of which is so great that that the length of the wire in a normal laboratory has a negligible influence on the outcome. In that case, B generates E and the principle qualifying factor that dictates how much charge (E) will be deposited on the Leyden jar by one ampere of current (B) will be the speed of the evolution of the influence (*c*) during that second in which the jar is charged. Thus, Bc = E, as Weber and Kohlrausch found empirically, and Maxwell derived theoretically.

In the thought experiment, the derivation of $Bc = E$ was done using a faulty differentiation. We failed to distinguish between the *distance that the gas travelled* in an instant with *how much gas passed a given point* in an instant. The transmission of electromagnetic influence, though, is *not* the same as the circulation of gas in a system. There is a sense in which our thought experiment provides a good model of how we *tend to think* of electric current, but utterly fails to represent what is *really happening* in the case of electric current. Neither electric charges nor charged particles, nor particles of any sort pass a given point in an instant to create electric current. In fact, given that electricity involves the evolution of an influence at the speed of light, and does not involve the movement of causal intermediaries such as charges, the differentiation error we made might not be an error at all in the electromagnetic case. Let us turn to this derivation once again and apply it to electric current.

We had two formulae corresponding to the current in the wire, I, one of which (1) described current in terms of "electrostatic" units, and the other (2) in terms of "electromagnetic" units. Putting them together we got:

$$\varepsilon_0 \, d(\textstyle\int E' \cdot dA)/dt = \oint B' \cdot dl/\mu_0 \qquad (3)$$

Let us apply this to a perfectly circular loop of current-carrying wire of radius r. The total area inside the loop will be πr^2, whilst the total length of the closed path, l, will be $2\pi r$. Therefore $\int E' \cdot dA$ can be rewritten as $E\pi r^2$ and $\oint B' \cdot dl$ can be written as $2B\pi r$. Substituting these in (3) we get:

$$\varepsilon_0 \, d(E\pi r^2)/dt = 2B\pi r/\mu_0 \qquad (4)$$

If the electromagnetic influence is evolving at c metres per second, then the time it takes (t) to move distance r will be r/c. Therefore $r = ct$. Substituting this for the values of r in (4):

$$\varepsilon_0 \, d(E\pi c^2 t^2)/dt = 2B\pi ct/\mu_0 \qquad (5)$$

At the corresponding point in the thought experiment, it was illegitimate to differentiate the first part of this equation as we did, because the t in $c^2 t^2$ referred to the instant in which a molecule of gas moved a certain *distance* in the tube, whilst we were supposed to be differentiating with respect to the instant in which a certain *volume* of gas passed a given point. In the electromagnetic case, this distinction is no longer relevant. There is no movement of a volume of charge in any instant. What is relevant for the magnitude of "current" is the state of the source (and this is represented by B) and the rate at which the causal influence of the source evolves to its target (c). If this reasoning is correct, then the rate of evolution of the electromagnetic influence is the only quantity varying in time that determines the magnitude of current that is flowing, since B remains constant over time. Thus a differentiation that intends to yield the "instantaneous change of electric charge in the wire" can legitimately operate on the t that is derived from the speed of evolution of the electromagnetic influence, c, along the radius of the loop of wire under investigation.

Thus the error of differentiation we made from (5) to (6) in the gas case is not so clearly inappropriate in the electromagnetic case. The result is: $\varepsilon_0 2E\pi c^2 t = 2B\pi ct/\mu_0$ and this simplifies to: $\varepsilon_0 Ec = B/\mu_0$.

Let us try to be as clear as possible about the way in which electromagnetism differs from a system of gas. In the case of the gas, two principal factors are relevant for determining the rate of flow: the pressure created by the pump and the cross section of the tube. These dictate how fast the molecules will flow and how many will pass a given point in an instant. The problem with our differentiation when we introduced vt is that this value of t referred only to the distance travelled by the gas and took no account of the cross section of the tube.

The electromagnetic case is different in a significant sense. When we speak of the magnetic influence of a current-carrying wire, B, then this quantity has no need of a qualifying parameter that would correspond to the cross section of the tube in the thought experiment. B fully expresses *by itself* the magnetic influence of the current. Once we know the magnitude of B exuding from an electric wire then we will be able to state the magnitude (in electromagnetic units) of the current. In the thought experiment we would have needed to know the magnitude of B (the tautness at any point on the tube) and *also* a measure of the cross-section (such as l) in order to calculate the current.

This is a major way in which the thought experiment differs from the real world of electromagnetism. In the real world, $\oint B' \cdot dl$ is an expression that tells us how the magnetic moment of the current is *distributed* in space. The choice of *l* can be arbitrary. We can measure B along a path that begins at ten centimetres from the electric wire or at one metre away, but B itself, the magnetic moment produced by that wire, is a more fundamental quantity than our particular choice of where to measure it. In the thought experiment, $\oint B' \cdot dl$ is quite different. In this case the cross-section of the tube, of which *l* is an expression, actually *dictates* the magnitude of B and the quantity of flow in the system.

A current that has a magnetic influence of B is allowed to flow for one moment in order to charge a Leyden jar and thus generate the electrostatic unit of charge. All we need to know about this current is what magnitude of B it produces! We need no information about the cross section of the wire or any other details in order to be able to state the magnitude of the current in electromagnetic units. B is really telling us something about *the power of the source* in a way that the B in the thought experiment could not without additional information about the cross section of the tube. This fits naturally with the approach to causal influence we have been pursuing in this book: the crucial issue is the magnitude of causal disequilibrium in the source and the way this evolves in the world. In the case of electricity, B gives us a measure of the causal equilibrium in the source, whilst *c* gives us a measure of the way this influence evolves in the system.

5.7 The nature of electrical current

The relationship between B, E and c in Maxwell's equations ultimately originate, of course, in the relationship between electricity and magnetism. Before examining this relationship, we need to clarify the nature of electrical current and its dependence on ionisation. That will be the subject of this section, whereas magnetism will be discussed in the next.

According to our model of ionisation, atoms are ionised when the electrostatic equilibrium of a *pair* is broken. An individual atom standing alone can never be ionised because ionisation is a state of the system that originates solely from the breaking of an electrostatic bond. Let us restate this view briefly. According to the standard view, ionisation usually involves electrons being gained or lost by individual atoms. As such, it is a property of individual atoms. According to the view being pursued in this book, by contrast, the phenomenon of ionisation is always *manifested* in the particular disequilibrium of an individual atom, but it is really the evolving state of the *system* that is responsible for holding the atom in this state. The basic tendency of the system is to maintain or increase its electrostatic equilibrium. A pair of atoms in an electrostatic bond represents a particular state of equilibrium in the system. When the bond is broken, the equilibrium of the system is disturbed and the system resists this change and tends to counteract it. You could say, if you like, that the equilibrium of the system is governed by a sort of Newton's third law that prompts the system to react to disturbances in its electrostatic status quo.

Let us call the two atoms in our pair A1 and A2. After the bond is broken, the V component of one of the genesis-units of A1 no longer balances the B component of one of the genesis-units of A2, and vice-versa. But the tendency of the system to react against the breach in equilibrium means that some of the components remain temporarily "held in check", as if those components were still being balanced by the opposite pole of the unit in the other atom.

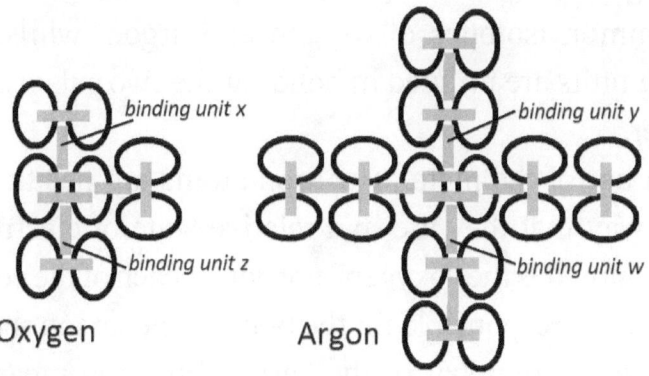

Oxygen

Argon

Figure 5.2 Iron resembles the fusion of oxygen on top of argon. *In reality, it seems likely that the process of forming iron would involve gradual fusion of individual genesis-units to argon until a final structure is formed that resembles oxygen appended on top of argon*

Let us discuss this process with reference to a complex atom like iron. The most common isotope of iron has an atomic mass of 56. According to our model, the final fusion of iron involves a structure that resembles an atom of oxygen fused onto an atom of argon (although the actual fusion process of iron is surely more piecemeal and does not involve individual oxygen and argon atoms being fused together). Figure 5.2 shows the simplified flattened depictions of oxygen and argon. In an atom of iron, the oxygen-like structure would be fused on top of the argon-like structure such that binding unit x would be fused with an additional binding unit to binding unit y, binding unit z would be fused to binding unit w, and so on. In total, four additional binding units would be needed to fuse structures resembling ^{16}O and ^{36}Ar to form ^{56}Fe. But it is more likely that the real pattern of fusion is not so simple; the "oxygen" and "argon" portions may utilize fewer binding units than the common isotopes of oxygen and argon, whilst more binding units are utilized in bonding the two substructures together.

In a bar of the metal, two iron atoms are positioned in such a way that they are in a relative state of electrostatic equilibrium. It is the "oxygen" portion of each of the complex atoms that is responsible for the bonding behaviour because the perfect symmetry of the "argon" portion means that it has little residual electrostatic influence to be counter balanced. Incidentally, this claim that iron has a portion of its sub-structure similar to oxygen, and that *this* part is responsible for its bonding behaviour, finds support in the ease with which iron reacts with oxygen in air, and also the fact that iron is used as the carrier of oxygen in the bloodstream. If iron *already* has a substructure that resembles oxygen, then this would give it a natural affinity to bond with this element, seeing that oxygen atoms already tend to bond with each other in pairs.

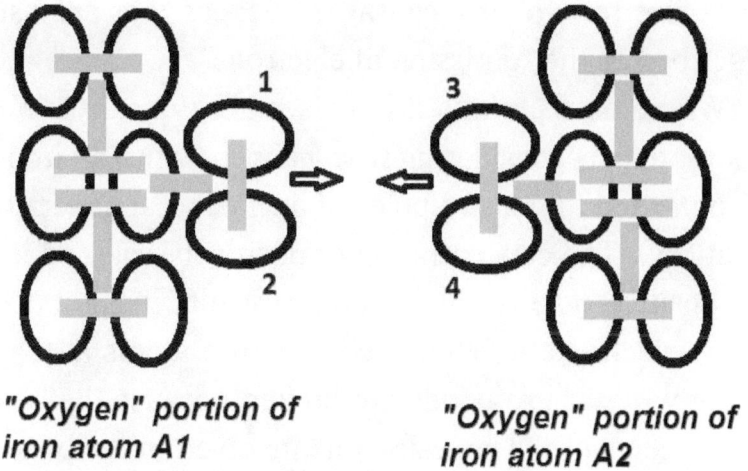

"Oxygen" portion of
iron atom A1

"Oxygen" portion of
iron atom A2

Figure 5.3 The electrostatic bond between two atoms of iron involves the counterbalancing of influences by the "oxygen" portions of the complex atoms

Now consider how the ionisation of iron might occur. We imagine two iron atoms oriented towards each other in different directions so that the "oxygen" portions of each atom enter into a state of relative electrostatic equilibrium (see Figure 5.3). When the bond disassociates, the resistance of the system to the change in electrostatic equilibrium results in the poles of each atom being covered. But this situation will endure for a very short period of time. According to our model of ionisation, when a pair of atoms dissociates, the units involved in the bond (units 1, 2, 3 and 4 in Figure 5.3) almost immediately shed the covering of their V poles. The reason for these differences in stability has to do, we surmised, with the way the B pole is concentrated at one end of the unit whilst the V pole is more dispersed along its length. In any case, the process gives rise to evolving causal influences that are usually described as the "emission of electrons".

Where do the emitted impulses go, typically? We can answer this by considering what happens when an iron wire is connected to the two poles of an electrical generator (or a battery). The concentration of positively-charged ions at the positive pole of the generator results in the breaking of the electrostatic bonds between iron atoms at that end of the wire. These bonds are broken because the positive components of the genesis-units involved in the bonds are counter balanced or repulsed outright by the proximity of the positive ions at that pole of the generator. When a bond is broken, the pair undergoes the process described in the previous paragraph: the covering of the V pole is shed by the "emitting of an electron". Where does the "electron" go? It goes into the positive terminal of the generator and effectively neutralizes the positive charge in one of the imbalanced genesis-units there.

But the process does not end there. Atoms A1 and A2 at the very end of the wire are now positively charged. The covering effects of these positive charges causes the breakup of the electrostatic bond between two atoms, A3 and A4, adjacent to them but located further out along the wire. A3 and A4 break up, emitting "electrons" that neutralize the positive charges in A1 and A2. On the other side they cause a rupture of the electrostatic bond in the next pair of atoms adjacent to them. This chain reaction continues along the wire until it reaches the negative terminal of the generator. If the wire happens *not* to be connected to the negative terminal of the generator, then the atoms in the wire will all quickly become positively charged and remain so, at which point the flow of current will stop. This corresponds to the empirical fact that a tiny electrical current flows in a wire for a short period of time when it is first connected to one end of a battery, even if the other end of the wire remains unattached.

When the other end *is* attached to the negative terminal, current flows freely because the negative terminal now supplies a steady stream of the evolving causal influence we call the "electron". These will neutralise the positively charged units along the wire in a step by step fashion. Thus electrical current involves *mutually sustaining chain reactions moving in opposite directions* along the wire, one of which positively ionises adjacent atoms, whilst the other chain reaction neutralises the same atoms an instant later. Say that the process begins at the left where an atomic pair, A1-A2, has their bond broken by the proximity of the ions in the cathode. A1-A2 consequently have all their poles covered, but almost immediately they shed their negative covering and become positively charged. The negative covers (a pair of "electrons") go from right to left back towards the cathode, whilst the positive charges of A1-A2 breaks the bond of the pair A3-A4 located to their right. They emit electrons which go left and neutralise the ionisations of A1-A2. Positive ionisation is moving from left to right whilst the emission of electrons and the neutralisation of ionisation is moving from right to left. This means that *no net evolution of electrical influence occurs at any point* along the wire except at the very extremities. In the meantime the generator is busy breaking up the electrostatic bonds of atoms in the electrical power source, sending positive ions to the cathode and negative ions to the anode. This serves to maintain the polarities which drive the process.

5.8 Magnetism

During the nineteenth century various causal mechanisms were hypothesized to account for the enigmatic relationship between electricity and magnetism. Poisson's theory that magnetism on the macro-level was caused by primitive pairs of small north and south magnetic poles (see Figure 5.4) gave way to Ampère's view that magnetism arose from perpetually flowing loops of current. At first there seemed to be good reasons for thinking that magnetism was a secondary phenomenon that arose out of particular manifestations of electricity, rather than vice-versa. One of these was the discovery that electricity flowing through a wire generated a magnetic field in the vicinity. But the discovery that a moving magnetic field could induce an electric current in a conductor led to more nuanced views of the relationship.

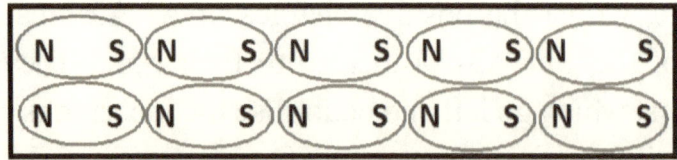

Figure 5.4 Simple approach to magnetism. A magnet is thought of as consisting of many tiny magnetic dipoles with a uniform orientation.

Electric charge is a state of disequilibrium of a genesis-unit that arises as a result of the break-up of an electrostatic bond; when the bond is broken, the system resists this change in electrostatic equilibrium so that the poles of the genesis-unit remain temporarily covered; as a result, the unit no longer presents V and B poles of equal magnitude to the world. Magnetism is a completely different manifestation of the influences emanating from the same B and V polarities in genesis-units. When certain critical genesis-units of a piece of material are aligned in the same direction, then the material will have magnetic effects. We will see in a moment what makes such units critically important for magnetism. All the atoms in the material may well be electrostatically neutral, but the fact that these significant units are *pointing in the same direction* means that they will exert substantial combined influence. The B and V poles of the same genesis-units in a non-magnetised sample of the same material also exert influences, but the random configuration of the units in the material means that these influences will tend to cancel out.

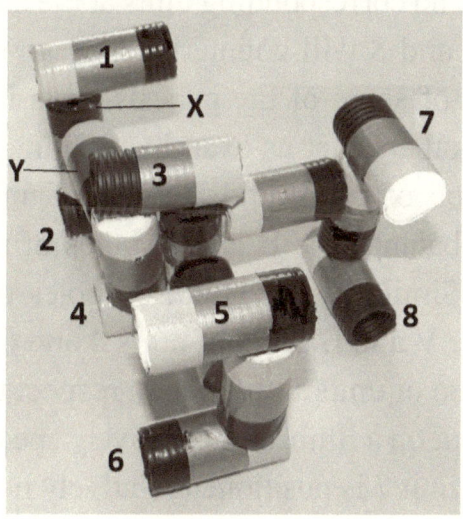

Figure 5.5 The magnetic properties of oxygen

Which units of an element are important as far as magnetism is concerned? Consider the garden hose model of oxygen in Figure 5.5 (we recall that these models were made with convenience in mind; the V components of real genesis-units are not concentrated at one end but are more dispersed along a good portion of the length of the unit). In line with the basic tendency towards electrostatic equilibrium that drives the fusion process, the units are fused in pairs such that 1 and 2 counterbalance each other, 3 and 4 counterbalance each other, and so on. The manner in which any one unit counterbalances another will never be complete, however. Unit 1, for example, is counterbalanced on one side by unit 2, but it will also exert (diminished) influences on its other sides that will affect how the composite atom interacts in the world.

The overall structure of the oxygen atom is not-symmetrical. Units 7 and 8 are located to one side of the structure with no corresponding units at the other side. This means that 7 and 8 will counterbalance, to some degree, the polarities of some of the genesis-units in the oxygen atom but not others. Let us examine this in greater detail. The orientation of the V and B poles of unit 7 are such as to counterbalance the B and V poles of units 3 and 5 respectively. Similarly, the B and V poles of unit 8 will counteract the V and B poles of units 4 and 6 respectively. The orientation of units 1 and 2 with respect to units 7 and 8 puts them out on a limb, comparatively speaking. In fact, the V pole of unit 7 is positioned relatively near the V pole of unit 1. Thus it serves to *increase* the V influence in that region rather than counterbalance it. The B pole of unit 2 is similarly bolstered by the proximity of the B pole of unit 8.

Units 1 and 2 will consequently have greater influence when it comes to the interaction of this composite atom with the outside world. But units 1 and 2 themselves have a disparity of influence. Binding unit X is used in the fusion of unit 1 to 2, whilst binding unit Y fuses this pair to the binding unit that fuses units 3 and 4. Compared to proactive units, the electrostatic influences of binding units are greatly diminished as a result of their position and role within the atom. But even if they have little direct influence *outside* the atom, they still counterbalance a portion of the electrostatic influences of the proactive units that they serve to fuse together. In Figure 5.4, we see that the V of binding unit X is fused to the B of unit 1, whilst the B of unit Y is fused to the V of unit 2. This means that the influences of the B of 1 and the V of 2 should be reduced somewhat. But in the case of unit 1, this diminishing influence is itself counterbalanced by the B pole of binding unit Y. The upshot of all of this is that unit 1 exerts a greater electrostatic influence towards the outside world than any other unit in the composite atom.

Before going any further it might be no harm to emphasize the obvious fact that this conclusion regarding the influence of unit 1 is based on a speculative model of the structure of the atom, not to mention untested assumptions regarding electrostatic shielding. A proper account can only hope to be developed through careful analysis of the empirical data from spectroscopy. But the basic conclusion that can be derived from our approach is surely a valid one: *certain units inside composite atoms will have a disproportional electrostatic capacity as a result of their particular position inside the structure.*

Even if these units indeed punch above their weight as far as external electrostatic influence is concerned, what has that got to do with magnetism? Magnetism, on our view, is just a particular manifestation of electrostatic influence; namely, the effect that arises when many atoms exercise electrostatic influence *in a harmonious way*. A magnet made of iron is an object in which many of the iron atoms have their unit 1 facing in the same direction. The forces emanating from the other units in each atom are weaker and will tend to cancel each out, but many atoms with a common alignment of unit 1 will exert significant collective influence. Figure 5.6 gives a simplified depiction of a permanent magnet made of iron, and we see immediately that it resembles Poisson's view. Since it is the orientation of unit 1 that determines the magnetic properties of the material, it can be described as the "magnetically-significant" unit of iron.

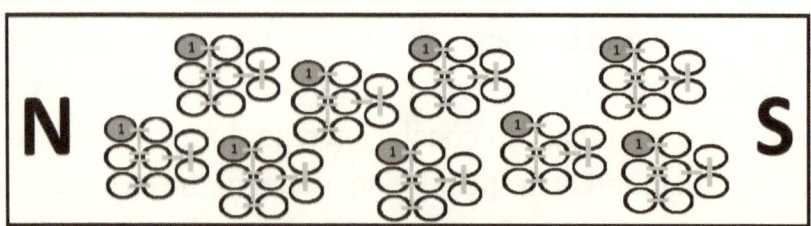

Figure 5.6 The alignment of the "oxygen" portion of iron atoms in a permanent magnet. *The electrostatic influence of unit 1 in each atom stands out above the influences of the rest, so it is its B-V orientation only that is relevant for the external magnetic influences of the bar.*

Say we bring this magnet close to a non-magnetised piece of iron. The combined influence from the magnet of many instances of unit 1 pointing in the same direction will cause some atoms in the non-magnetised bar to realign themselves. The end result is that these atoms in the iron will have their unit 1 facing in the opposite direction to the atoms in the magnet (the V poles of unit 1 in the magnet will repel the V poles of unit 1 in the iron; similarly the B poles will repel, whilst the Bs and Vs will attract; consequently the atoms in the iron will realign themselves in the opposite direction to the atoms of the magnet). Once a certain number of atoms in the iron have realigned themselves in this way, the overall force of attraction between the magnet and the iron will become significant enough to draw the iron bar towards the magnet.

Simple experiments performed by our seven year old son help to support this thesis. For example, when he rubs a piece of metal against the north pole of a magnet, the metal becomes temporarily and weakly magnetised, as demonstrated by the fact that the metal deflects the needle of a compass. If he draws the south pole of the magnet close to the same piece of metal, the metal immediately loses its previous state of magnetism, as shown by the fact that the compass needle is no longer deflected. This seems to indicate that groups of atoms in the metal reorient themselves so that their components are aligned in the *opposite direction* to the orientation of the atoms in the magnet. This leads to a weak form of magnetism being induced in the metal by the north pole of the magnet. When the metal was brought close to the south pole of the magnet, the previously induced alignment of atoms was disturbed.

This understanding of magnetism explains why magnets always *attract*, rather than repel, non-magnetised ferromagnetic objects. Magnets attract because their state of alignment prompts the *opposite* state of alignment in the ferromagnetic material, resulting in net attraction. And ferromagnetism itself is a property of elements whose atoms have a unit (like our unit 1 in iron) whose electrostatic influence stands out above the influences of other units in the atom. Many elements will not have any unit with this characteristic, or they may have two or more units whose influences counterbalance each other, even if they happen to be greater than that of other units in the same atom. Other elements may have a single unit whose electrostatic influence is only very slightly greater, and these materials will be called "weakly ferromagnetic".

What is the relationship between ferromagnetism and electrical conductivity? We might expect that an element that is ferromagnetic ought to be a good conductor of electricity as well, since the presence of a magnetically-significant unit in the material might facilitate electrical conduction also. But many electrical conductors are not ferromagnetic, such as copper and silver. In these atoms, the contrasting orientations of their various constituents means that there is no unit whose influence stands out above the others. They simply lack a magnetically-significant unit, but they can still carry current since all that is necessary to conduct electricity is that these atoms form electrostatic bonds with each other that can be broken. Even elements that lack a magnetically-significant unit can still form electrostatic bonds. Certainly, elements that have a magnetically-significant unit will tend to form electrostatic bonds upon those very units (since they are the ones that dominate electrostatic interactions). But all that is needed to form a bond is the physical proximity of two units from two different atoms in the material. Once the bond is broken by the onset of the electrical current, these units will become ionised. Ionisation involves electrostatic imbalance and any unit that is electrostatically imbalanced *will now* stand out above the other units in the same atom as far as its electrostatic influence is concerned. It is this unit which will realign itself with respect to the power terminals, and it is the collective influence of many of these similarly-aligned units that will give the current-carrying wire its magnetic effects (as we shall see in the next section). This discussion helps to explain why some materials are ferromagnetic only, some are good electrical conductors, some are both and still others are neither.

Charged bodies in general are not attracted to the poles of magnets. But if a magnet is an object with its electrostatically-significant units aligned in the same way, then shouldn't a charged body be attracted to one end of the magnet? Say the charged body is a genesis-unit with its B pole covered. Shouldn't the uncovered B poles at one end of the magnet have an attractive influence on this unit? The thing to remember is that the atoms in a permanent magnet are not ionised. They each have a B pole and a V pole in very close proximity to each other. From a distance, the separation of the B and V pole is negligible and therefore any charged object in the vicinity will be equally attracted and repelled by the twin poles of the atoms in the magnet. *Alignment,* however, is a different story. One million non-ionised atoms have exactly one million B poles and exactly one million V poles: therefore the net attraction and repulsion they exert on another atom that comes into the area will be zero. However, if all of those atoms are *aligned* in a single direction, then we have one million atoms working in concord to change the alignment of that atom that wanders into the area. This is how magnetism works: unlike electrostatics, no net V influence or B influence is exerted; but, out of all the infinite B-V alignments that are possible in space, the atoms of this magnetised object exert a *single* collective influence that prompts other atoms to align in the opposite direction.

5.9 Electromagnetism

Why does the flow of current in a wire give rise to magnetic effects? Recall that electric current is an evolving process of mutually sustaining chain reactions in opposite directions along the wire. Take a pair of atoms, A1 and A2, in electrostatic equilibrium near the positive terminal of the generator. The proximity of the positive ions in the generator counterbalance the positive poles of A1 and A2 so that their bond is broken. Both become positively charged, emitting "electrons" in the process. As soon as they become positively charged, they not only repel each other but also realign themselves so that their electrostatically "significant" units are facing away from the positive terminal of the generator.

Say that the wire is made of iron. Unit 1 in each atom has a disproportionately high electrostatic influence with respect to the other units in the same atom (according to our assumption – if the assumption is wrong, it is still reasonable to hold that *some* unit in the atom will have this heightened role). If this is correct, then each atom will tend to align itself so that the B pole of unit 1 is facing away from the positive terminal – a natural consequence of the electrostatic repulsion of the positive ions in the power source. This business of bond breaking, mutual repulsion, realignment, neutralisation of the positive charges, and finally re-formation of the electrostatic bond, is constantly happening in the atoms along the wire.

When electric current is flowing, a certain proportion of atoms in the wire are repeatedly being aligned oppositely to the terminals. Once an electrostatic bond is broken, the atoms will immediately reposition themselves so that their B-V orientation is contrary to that of the terminals of the electrical source. Is *this* what gives magnetic properties to a current-carrying wire? Does this regular alignment of many atoms in the same direction mean that the electrical conductor becomes a sort of dynamic magnet whose magnetically-significant units are all oriented in the direction of the current, just as in a bar magnet the magnetically-significant units are aligned in an N-S direction? There is one stark difference, however, between the magnetism associated with an electric current and that of a bar magnet: the magnetic influence of an electric wire is *perpendicular* to the direction of current, whilst in a bar magnet the influence is more or less *parallel* to the N-S orientation (apart from at the poles). This is enough to show us that something significantly different is going on where electromagnetism is concerned.

The obvious difference with electromagnetism is that the atoms of iron are already *ionised* at the moment that they align themselves with respect to the electrical terminals of the circuit. Recall the cycle again: the electrostatic bond is broken and the pair of atoms becomes ionised, immediately emitting "electrons" and taking on a positive charge; *then* they align themselves so that their electrostatically-significant units are oriented oppositely to the electrical alignment of the circuit; a moment later the chain reaction coming in the opposite direction neutralises the ionisation and the atoms again form an electrostatic bond. For that instant of common alignment with respect to the circuit, the atoms were positively ionised. Or to be more precise, *unit 1* in each atom was positively ionised. This means that the B pole of these units is covered by the natural resistance of the system to changes in its electrostatic equilibrium. Even though these units have a common orientation, it is clear that they will not give rise to the same magnetic effects that they produce when they are not ionised.

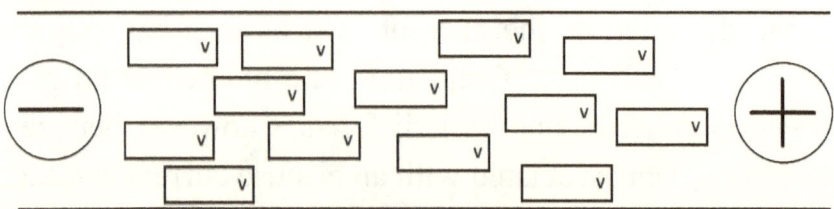

Figure 5.7 Many iron atoms in an electrical conductor will have unit 1 aligned in the opposite direction to the electrical circuit at any instant. *But this cannot lead to magnetic effects in the direction of the alignment because their B poles are covered.*

Figure 5.7 has a simplified view of unit 1 of some of the iron atoms in an electrical conductor (only unit 1 is shown in the diagram since it alone is relevant for magnetism). These atoms have all just been ionised and have realigned themselves against the orientation of the poles of the circuit. This common orientation does not have the collective magnetic effect of the atoms in a bar magnet because each instance of unit 1 only presents its V pole to the world. Magnetic effects, by contrast, arise when many atoms with *a common B-V orientation* have a collective effect on other atoms, prompting their magnetically-significant units to align themselves in the opposite direction to the magnetically-significant units in the magnet, thus giving rise to attraction between the magnet and the ferromagnetic material. Ionised units do not have the B-V influence of other units, as far as the world external to them is concerned, since one of their poles is covered. Thus they can have no magnetic effects in the B-V direction.

How then does a current-carrying wire give rise to a magnetic field *at all*, and why is that field *perpendicular* to the direction of the current? This is part of the same mystery as the strange fact that an electrically charged causal influence, if it enters a magnetic field at right angles to the field, will be always deflected at right angles to the direction of the field and at right angles to its original trajectory. The answer must be sought in the structure of the genesis-unit itself that gives rise to the magnetic field. No other explanation seems possible for this most extraordinary fact.

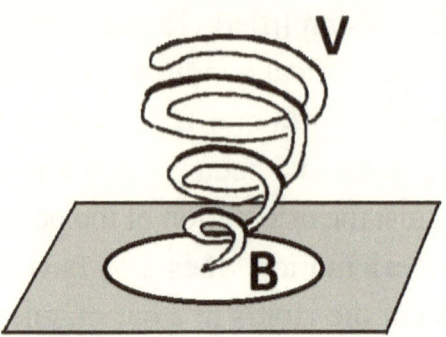

Figure 5.8 Spiral decoration in paper as a means of illustrating the spiral form of the V component of the genesis-unit. When the spiral is cut out of the paper and raised up, a void area is left in the paper. The distinction between the material of the spiral and the emptiness of the void is analogous to the nature of the basic protomagnetic polarity in the atom, but it is analogous in an inverse way. For the genesis-unit model, the spiral V component is actually void, whilst the B component is the contrary.

In the chapter on atomic bonding, we discussed a model of the atom in which the V component was dispersed along a significant portion of the length of the unit, whereas the B component was concentrated at one end. This distinction was sufficient to allow us to develop an account of the structure of the atomic table and explain the pattern of chemical bonding. A more nuanced version of this model will be necessary, however, if we are to provide an explanation of magnetism and the way that ions behave in magnetic fields.

There is explanatory value in describing the B component as being located at one end of the genesis-unit, with the V component being *spiral* in form (see Figure 5.8). It is as if the work-impulse generating the B component moves with a coil like motion through the void in the instant that it gives rise to the B. The final result is that the B is concentrated at one end of the genesis-unit, whilst the V has a spiral form. The spiral is not uniform in nature. It is most dense at the furthest point from the B component of the unit and it progressively lessens in density as it approaches the B end.

This form would assist us in describing the orientation of the magnetic field in a current-carrying wire. The fact that the magnetically-significant units are *ionised* means that their B-V alignment will not determine their external magnetic effects: the B component has been covered by the ionisation and is out of the reckoning as far as magnetism is concerned. Unlike the B component, however, the V component is not uniform. It consists of the "hole" that is left in the void when the B component is generated, and this generation is an ongoing thing. In a sense the "hole" is continually being dug and the B component is still being lifted out of the void. The work impulse at the heart of the unit that is responsible for this activity drags the B component out of the void in a spiral form, we have surmised. When the unit is ionised and the B component is covered, the spiral form of the V unit starts to become significant. This is what causes an electric current to give rise to magnetic influence that is perpendicular to the direction of flow and in concentric rings around the wire.

We will return to the details of this in a moment. First of all, a word about our choice of language in describing the dynamics of the genesis-unit. Evidently, it would be better if this process could be described in a way that does not seem in any sense mystical. The difficulty, as always, is that we are trying to hypothesize the underlying structure of something from empirical effects, and the "something" in this case is the deepest and most hidden of all physical entities. The descriptive language we use will have a certain vagueness and generality about it because our knowledge of this realm is so meagre. If our account does at times seem to be of an ethereal sort, it is only because we wish to draw safe and general conclusions from the evidence without hypothesizing a too-detailed causal structure. Past attempts at a detailed underlying structure, such as the Bohr model of the atom, have only held up progress in our understanding of the causal foundation of reality. In fact, such models continue to coerce our thinking along avenues that are unlikely to lead to the truth about the atom. It is surely better if our descriptions can confine themselves to general principles that are based on clear patterns in the empirical data, avoiding as much as possible detailed hypotheses regarding the underlying structure. That is why we have based our account primarily on a basic polarity at the root of all causal activity, a polarity that is manifest in the most fundamental of natural phenomena. The claims that the V pole is *dispersed* along the unit and is *spiral* in form are two separate assumptions, tentatively made, but at the same time supported by the data, as will become clearer as we go along. Without these assumptions, the polarity by itself lacks explanatory power for some of the empirical data that we seek to understand.

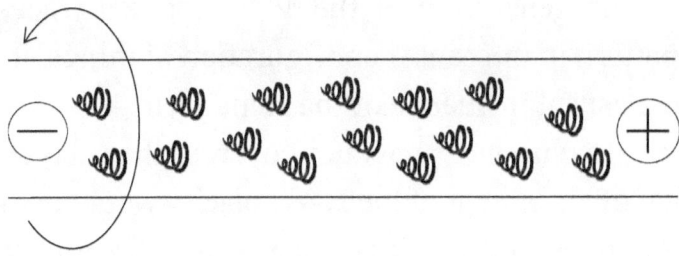

Figure 5.9 The spiral form of the V component gives rise to a magnetic field in the same direction. *Evidently, the real spirals will differ considerably from those shown in the diagram. Empirical study and greater artistic ability are required to improve the model.*

Let us return to the magnetic influence of a current-carrying wire (see Figure 5.9). As the current flows in the wire, pairs of atoms are continually being ionised, aligning themselves oppositely to the poles, and then returning to neutrality. At any instant, a large number of atoms will have their magnetically-significant units lined up in this manner. The B pole of these units is covered, but the work impulse is still active in the atom, generating the B from the V. Thus the spirals that we see in Figure 5.9 should not be thought of as inert "holes" in reality. There is a dynamism in them that derives from the fact that there are being generated continuously. Metaphors here might do more harm than good, but we could liken the unit to a spring that is being stretched. The positive component that is stretching the spring is hidden from view (the B component) but nevertheless the holding of the spring in that position requires ongoing activity, a reality that is usually described as the "rest energy" of the atom. The spring itself has continuous tension in it while it is being extended.

Because the generation of the V component traces this spiral pattern in the causal configuration of things, it gives rises to a causal influence of the same form.

If the B component was not covered, then the general direction of alignment of the B-V polarity would dominate all interactions and the influence of the direction of the spiral would be negligible. But in the case of an ionised unit, the spiral form of the V component comes to the fore. Now take the B component of a non-ionised atom, A1, in an iron bar in the vicinity of the current-carrying wire. This component is attracted to the V poles of the ionised atoms in the wire, but the V component of A1 is repulsed by those same V poles. The fact that there is a gradation of the V component *within* each of those poles in the wire, however, will cause the B-V of A1 to align itself in a particular way with respect to the spiral form of the units in the wire. The quantity of "V" diminishes along the spiral as it winds from the point where it begins towards the covered B component. Thus the B of A1 will tend to orient itself towards the end of the spiral where the V component is most concentrated, whereas the V of A1 will orient itself away from this end. As a consequence A1 will end up with its B-V aligned perpendicularly to that of the magnetically-significant units in the wire. Of course, if the units in the wire were not ionised, the influence of their B components would be such as to not facilitate perpendicular alignment of this sort.

The ionised units are all aligned in the direction of the wire which means that the field of influence of the spirals of their V components will be more or less in a circle around the wire as depicted in Figure 5.9. The direction in which the spirals are wound will not be exactly perpendicular to the wire, but the individual units are so small and there is such an enormous number of them in the wire that their combined field of influence will be in these circles whose plane is perpendicular to the direction of current.

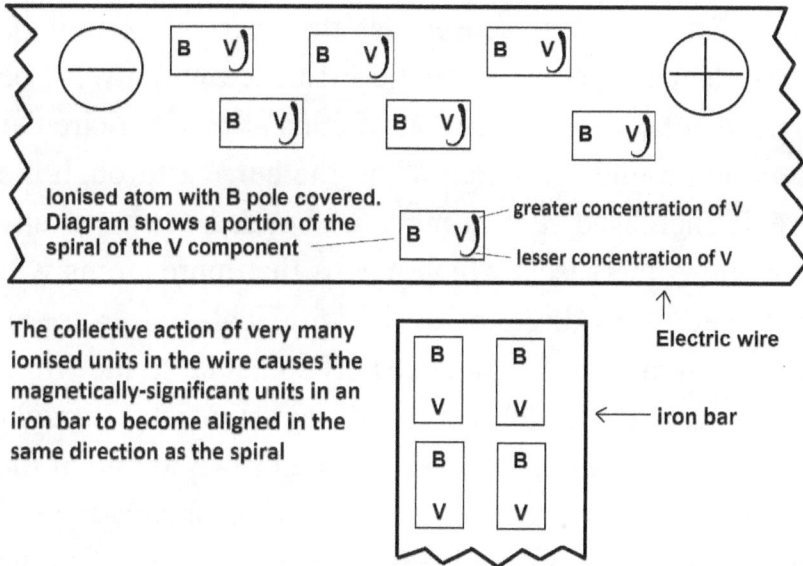

Figure 5.10 The collective action of the ionised atoms in a current-carrying wire. *The figure shows the ionised units (with all B poles covered) at the instant of ionisation and consequent realignment of the units with the terminals of the circuit.*

Figure 5.10 provides a very simple sketch of how ionised units can create a field of magnetic influence that is perpendicular to the direction of current. A portion of the V spiral of a single unit is depicted in the diagram. On the side of the wire nearest to the reader and in the plane of the paper, the V component diminishes as we move down the paper. When very many ionised units are aligned in the same direction for an instant, the combined influence of these steadily diminishing V components can be sufficient to cause the realignment of the units in the bar of iron placed nearby. If this portion of the spiral is in the plane of the paper, then we must imagine that the bar of iron in the diagram is in a parallel plane directly above that of the paper. Because the form of the units is spiral, their collective magnetic influence will always be perpendicular to the direction of current but in a circular direction around the wire. When the current through the wire is increased, the number of ionised atoms at any instant also increases, which means that more atoms will be aligned in the direction of the wire. This will cause an increase in magnetic influence perpendicular to the wire.

If a charged particle enters a magnetic field, it might be thought that the particle should be drawn to one of the poles of the magnet. Say the particle is a positively ionised atom of hydrogen. Its B pole is covered, so shouldn't it be drawn to the pole where all the magnetically-significant units have their B poles facing away from the end of the magnet? To answer this, let us consider the situation where the field is caused by a permanent bar magnet. In this kind of magnet, the atoms are not ionised. Each magnetically-significant unit is exerting electrostatic influence from both its V and its B pole. Given that the length of each of these units is so short, the difference in distance from the charged particle to the B pole compared to the V pole will be negligible. They will both exert equal influence on the particle and consequently it will not be drawn to any particular point on the magnet. But the sheer number of these atoms in the magnet, as a result of their common alignment, *would* have sufficient collective influence to *realign* a non-ionised particle that enters the field. The particle is realigned because each one of the atoms in the magnet that is influencing it has a *common* alignment in the opposite direction. Thus, the particle is realigned but is not normally drawn to any particular point on the magnet because the B and V poles of the many atoms in the magnet are more or less equidistant and cancel each other out. This discussion helps us to understand how the collective influence of commonly aligned atoms can be sufficient to *realign* other atoms in the vicinity but might not necessarily *draw* these atoms towards a particular point on the magnet. The alignment is *collective* and thus gains strength from the sheer numbers involved, but the *position* of each atom in the magnet is *particular* to that atom and has no great causal influence on its own. Thus the target atom is *realigned but not drawn* to a particular position. In the case of a bar of ferromagnetic material in the vicinity of the magnet, by contrast, many atoms become realigned oppositely to the relevant units in the magnet. Here, once again, sheer numbers come into play and the attraction between so many atoms in the bar and in the magnet is sufficient to lead to movement. A slightly different but similar discussion can account for the same phenomena in the case of electromagnets.

These considerations can be used to develop a full account of the movement of charged objects in a magnetic field, which involves constant realignment of these objects (movement along a curved trajectory) but does not result in the object being drawn to a magnetic pole. When physicists talk about charged objects they are usually referring to one of two distinct types of reality: the first is a genuine piece of matter composed of one or more atoms, some of which have been ionised; the second possibility is a non-material causal impulse such as that usually referred to as the "electron". In both cases, when such an entity enters an orthogonal magnetic field, the causal evolution of the "object" is the same: the so-called particle moves at right angles to the field and at right angles to its original trajectory. Furthermore, there are *two* possible directions in space that are orthogonal to the original trajectory and orthogonal to the magnetic field. If the magnetic field lies in the horizontal plane and the direction of movement is also in the horizontal plane, then the particle can move either up or down and still be orthogonal to both. But the direction pursued by the particle is not random; it always goes either up or down depending on its charge.

This extraordinary fact can be accounted for in terms of the spiral form of the V component of the atom. In cases where the charged object is a genuine piece of matter, the object's trajectory follows the curve of the many spirals in the causal source (or goes in the opposite direction when the object entering the field is positively charged). In the case of a non-material causal impulse, the trajectory of evolution of the impulse will be determined by the same influences. An impulse such as the so-called electron, which is negatively charged, will progressively evolve towards the most concentrated portion in the spiral of the V component. When the magnet producing the field is a permanent bar magnet, the atoms in the magnet are not ionised (as they are in the case of an electromagnet), but the spiral form of the V component still determines the trajectory of the causal influence. The contrary influences emanating from the B and V components of the atoms at the poles of the magnet are identical in magnitude. This means that there ought to be no net tendency for the charged "particle" to evolve in either direction. But now the spiral form of the B components of the many atoms at the north pole of the magnet will start to come into play. They will exert an influence in this spiral direction because there is no corresponding influence in the opposite direction coming from the south pole of the magnet (because the B component is not spiral in form).

5.10 Electromagnetic induction

A changing magnetic field will induce electric current in a nearby wire. This process is essentially the reverse of that by which an electric current gives rise to magnetic effects. Consider a magnet that is in a stationary position relative to a coil of wire. The common alignment of the atoms in the magnet will naturally have a collective influence of some sort on the atoms in the conductor. Even if the conductor is not made of ferromagnetic material, the proximity of the magnet will have an effect, and various atoms in the material will realign themselves with respect to the magnet. Now we move the magnet towards the coil of wire. This will give rise to a shift in the relationship between the atoms of the magnet and those in the wire. If the magnet is moved briskly enough, it is inevitable that these shifting relationships will give rise to the breaking of electrostatic bonds. Once a bond is broken, "electrons" are emitted and the pair becomes positively ionised. The continued movement of the magnet causes the rupture of other bonds, leading to more "electrons" being emitted, thus neutralising some of the positive ions just created, and the restoration of electrostatic bonds. In this way a twin chain reaction is set in motion with positive ionisation occurring in a step by step fashion in one direction, with the repeated neutralisation of the positive ions evolving in the opposite direction.

In the case of electromagnetism, we saw how the flow of current causes the ionisation of atoms; these atoms realign in the opposite direction to the terminals; and the common alignment gives rise to magnetic effects. Electromagnetic induction involves the opposite sort of progression: a moving magnet prompts the realignment of atoms in the conductor; this realignment breaks electrostatic bonds; and the ionised atoms give rise to the twin chain reaction in opposite directions that we call the flow of electric current.

At this point, let us return to Faraday's law, which concerns precisely this phenomenon of electromagnetic induction: $\oint E' \cdot dl = d(\int B' \cdot dA)/dt$. The first part of the expression refers to the difference in electrical voltage around a closed loop of wire of length l. We must imagine that the magnet is moving inside this loop and it is the changing magnetic influence that is inducing the difference in voltage in the loop of wire. The second part of this expression, $d(\int B' \cdot dA)/dt$, refers to the instantaneous change in the magnetic field over area A, which is the area enclosed by the loop of wire. As commented earlier, this situation is sometimes likened to a film of soap spanning a ring of wire. The ring is of circumference *l* and has the voltage of magnitude E induced in it by the movement of the magnet. The area of the film of soap is A, and the magnet is moving perpendicularly to the plane of this loop.

Say that the radius of the ring of wire is *r*. This means that $A = \pi r^2$ and $l = 2\pi r$. If the evolution of magnetic influence travels at velocity *c*, and if $r = ct$, then $A = \pi c^2 t^2$ and $l = 2\pi ct$. Consequently, $\oint E' \cdot dl$ becomes $2\pi ctE$, and $d(\int B' \cdot dA)/dt$ becomes $d(\pi c^2 t^2 B)/dt$. Therefore:

$$2\pi ctE = d(\pi c^2 t^2 B)/dt$$

It was at this point in our thought experiment that we did the second illegitimate instance of differentiation with the conclusion: $2\pi ctE = 2\pi c^2 tB$, or $E = cB$. What a stunning result! Now the problem with the differentiation was similar to the previous case. The differential expression is supposed to give us the instantaneous change in the magnetic field, and this is supposed to depend on the rate at which we move the magnet relative to the loop of wire. But the *t* that we operated on was the interval it took for the *magnetic influence* to evolve from the centre of the loop to the wire, not the *t* related to the rate at which the *magnet* moved relative to the wire.

Yet, there is a sense in which this operation of differentiation can be appropriate. As our discussion in this section has unfolded, we have claimed that a change in magnetic influence induces current because of the reshuffling it prompts in the way that atoms in the wire are aligned relative to the magnet. As the magnet moves, the relationships between the atoms in the magnet and the wire change, prompting the rupture of electrostatic bonds. If the magnet is moved systematically enough, the broken bonds, consequent ionisation and immediate release of "electrons" will ensure that significant current will begin to flow.

There are two separate velocities that are relevant for the magnitude of flow of current. The first is the rate at which the magnetic field is *changing*, and the second is the rate at which the magnetic influence *evolves*. At time *t1*, say that the atoms in the magnet are in a particular group of relationships with the atoms in the wire. Now the magnet moves and takes up a new position closer to the wire at time *t2*. Then it reverses direction and at time *t3* takes up the position where it began. These changes in the position of the magnet will alter the relationships between the atoms in the magnet and the atoms in the wire, but the reshuffle of relationships is not transmitted instantaneously. The electrostatic influences of the atoms of the magnet from their new position must be transmitted at velocity *c* to the atoms in the wire. Imagine that *c* is much lower than it actually is. If that were the case, then the influence of the magnet emitted at time *t3* could conceivably arrive at the wire before the influence emitted at *t2*. At time *t3*, the magnet is back in the position it was at time *t1*, so there will be no change in the relationships between the atoms in the source and in the target. This means that the movement of the magnet would have less current-inducing effect on the wire than is the case when the value of c is higher. The lower the value of c then the less efficient will be the transmission of the changes in the relative position of the magnetic source. In other words, the higher the value of c, the more current will be induced in the wire by the relative movement of the magnet.

The velocity of light thus acts as a sort of limiting parameter as far as the efficiency of electromagnetic induction is concerned. The second velocity that determines the quantity of current induced is the rate of change of position of the magnet relative to the wire. A more rapid change of position will mean a greater number of shifting electrostatic relationships per second. So we have these two velocities: one quantifies the rate of change of electrostatic relationships and the other measures how fast these changes are transmitted to the target. The magnitude of the resultant current will be proportional to the product of *the rate of change of electrostatic relationships* (let us call this quantity R) and *the rate of evolution* of the changes to the target (c).

Now consider the minimum rate of change of the position of the magnet necessary to give rise to a measurable current. In this case R will be equal to 1 since the minimum number of changes in electrostatic relationships necessary for the induction of a current is 1. The only remaining factor that will determine the efficiency for inducing current of this single change in relationships will be c. We return now to the expression for the instantaneous change in the magnetic field: $d(\int B' \cdot dA)/dt$. Say that the movement of the magnet is the minimum necessary for the induction of a current. The instantaneous change in the magnetic field felt in the wire will therefore depend only on c. As the magnet moves at the centre of the coil, the change is transmitted to the wire located at a distance r in time t. It will now be legitimate to express $d(\int B' \cdot dA)/dt$ as $d(\pi c^2 t^2 B)/dt$ and to differentiate it as we did before to $2\pi c^2 t B$. The time t measures the time it took for this minimal alteration in interatomic relationships to make it itself felt at its target. Thus, as far as the target atoms are concerned, it is an accurate reflection of the rate of change of magnetic influence. From this we get $E = cB$, a beautiful expression of how the capacity of a magnetic influence to induce electricity is dependent on the rate of evolution of the influence.

5.11 Fields

Under our account, electrical and magnetic fields do not exist as ontological entities that permeate areas of reality. However, it does make perfect sense to speak of a region in which the magnetic or electrical properties of an atom (or the joint influence of groups of atoms) *potentially* have influence if appropriate causal targets are present there.

The origin and nature of what is normally called an "electric field" is relatively straightforward. When an atom is ionised, it exists in a state of electrostatic disequilibrium. As we have discussed, this is a state that naturally seeks "discharge". The evolution of the system as a whole tends towards equilibrium between the various parts. Atoms that are in a state of imbalance will tend to pass this imbalance on to atoms that are in an opposite state of imbalance. Or, in cases where an object (such as a cathode) arrives at a critical number of atoms in the same sort of disequilibrium, the imbalance will be discharged into atoms that were previously in equilibrium. This happens because the target *object* in which these atoms are found is in a state of *overall* lesser disequilibrium that the source object. Thus the extreme disequilibrium of the source object is alleviated by its distribution into the atoms of the target.

What is usually referred to as an "electric field" is the region in space where the electrical disequilibrium of a source object *potentially* has influence. The source object (for example, a charged electric plate) will typically have many atoms in a similar state of disequilibrium. Thus these atoms exert a joint influence, and the influence will wane as the spatial distance from the source increases. But this does not mean that there is a physical *field* permeating space. Spatial separation is a feature of the spatial perspective and merely tells us something about the magnitude of a non-absolute ontological separation of some sort. Despite this ontological separation between source and target, electrical disequilibrium will exert its influence, but the influence will wane as the separation increases. It makes little sense to speak of the effect being due to a *physical* field extending through a spatial expanse that is only, after all, part of our way of representing things. In any case, we have no reason to think that objects require an intermediary such as a field in order to affect each other directly. The concept of an electric field is indeed useful for calculating the magnitude of influence a charged object *would* have if there happened to be a target object at that point in the spatial perspective.

But in the absence of a target object, it makes no sense to talk of the existence of a causal intermediary at that point.

A rather facile analogy may help to make this point. Imagine a situation where there are two rows of tables extending in parallel lines with a wide passageway between them. On one row of tables are placed a series of cakes and sweets whilst the tables on the other row are empty. A child is asked to walk down this passageway right to the end. We can imagine that the row of tables with cakes would exert a certain attractive influence on the child. If many children are made walk down the passageway and some of them give in to temptation, it would even be possible to calculate statistically the strength of the "field of force" that moves children towards the row with cakes. Of course no such field exists in reality. All that exists are *particular* relationships between *concrete* causal players (i.e., children and cakes). But the notion of a field could become a very useful mathematical way of predicting and describing the behaviour of a significant number of particular children who pass through this region of space. Similarly, no field really exists in regions of the world where electric and magnetic causal players are at work. However, once a causal target is placed in such a region, then the consequent influences can be described very usefully in terms of fields.

In this chapter we have tried to show how physical sense can be made of the equations of electromagnetism, whilst rejecting the classical treatment that was couched in terms of fields and lines of force. According to our view, all interactions are to be understood as emanating from the poles of genesis-units. No fields exist in nature, but it is mathematically and pedagogically convenient to describe and quantify the collective action of many atoms in terms of fields.

One of the issues that arose during the late nineteenth century reflection on Maxwell's electrodynamics was the question of the way in which a magnet in motion generated an electric field. If the magnet was stationary with respect to an observer, but in motion relative to the ether, then would the observer detect an electric field or not? The absolute motion relative to the ether should generate an electric field, but the fact that no field was detectable when the observer and the magnet were moving in unison was problematic. Indeed, Albert Einstein cited this bizarre conclusion (that from the viewpoint of absolute rest, a moving magnet generated an electric field, but from the perspective of the moving magnet itself, no field was present) as his decisive motivation for postulating the special principle of relativity.

Our treatment makes all such questions utterly superfluous. There is only one kind of causal influence emanating from genesis-units – the electrostatic influence of its poles. But this influence manifests itself in different ways in different situations. If many atoms have their magnetically-significant units aligned, then their collective influence will give rise to what we call "magnetic" effects. If an object with magnetically-aligned units moves relative to an electrical conductor, then it will begin a process that will lead to the flow of current. But the genesis-units are not giving rise to fields of any sort. There are neither electric nor magnetic fields emanating from atoms, just the same electrostatic influences that give rise to electrical or magnetic effects depending on the circumstances. Therefore whether the magnet is moving relative to the observer or not doesn't matter for our understanding of the reality of the situation: what matters is whether source and target are moving relative to each other, and the other details of their causal interaction.

According to special relativity, the magnetic field perceived by an observer who is stationary with respect to the magnet, will be perceived as an electric field by a moving observer. Both types of field are different aspects of a single electromagnetic field whose particular manifestation depends on the reference frame of the observer. Using field transformations, we can calculate the properties of the electromagnetic field at any point. To perform such calculations we simply need the relative velocities of the causal players in the system.

The unwieldy nature of this framework as a way of describing the world should now be clear. The power of field transformations as a calculating device for describing/predicting electromagnetic phenomena is sufficient motivation for their retention. But our *understanding* of what is actually happening in these situations is greatly improved if we dispose of the notion of the field altogether.

Chapter Six

THE ORIGIN AND MEANING OF THE INVERSE-SQUARE LAW

"The laws of Nature are written in the language of mathematics ... the symbols are triangles, circles and other geometrical figures, without whose help it is impossible to comprehend a single word"

Galileo Galilei

Overview of this chapter and its principal claims

1. Frameworks of explanation that reify space tend to couch their accounts of gravity, electricity and magnetism in terms of forces, fields, waves or particles that allegedly transmit these causal influences *across* spatial gaps. However, if space is merely our way of representing genuine relations between objects, then its reification hampers our development of coherent accounts of these causal relations. The correct starting point for an understanding of gravity (and other phenomena) is to eliminate these causal intermediaries from our descriptions and look on the phenomenon as a natural disposition of the system itself. It is only when we begin to consider gravity in terms of the behaviour of entire systems (rather than as an autonomous property of objects) that it begins to yield its secrets.

2. The gravitational influence of a body is not a property of the body that is exerted equally on *every* other body in the system

depending on its relative distance. Rather it is a total influence that is *distributed* over bodies in the system. If there were more bodies in the system, then the influence would be more greatly dispersed.

3. Based on this assumption, we consider how the inverse-square law would have to be modified if the quantity of matter in our region of the system were different, or if material objects were distributed in a different way. We argue that the value of G is not a measure of the innate gravitational force possessed by matter, but a constant of proportionality determined by system-wide empirical considerations.

4. We reflect on how the empirical behaviour of a heterogeneous portion of a system could easily lead to the functioning of an inverse-square law.

5. Finally, it will be argued that gravitational influence will fall off more dramatically with distance in a more densely populated region of space. This provides an explanation for the surprising speed of rotation of celestial bodies that are located in regions of space that are more sparsely populated.

6.1 Introduction

The law of universal gravitation is the paradigmatic example of an inverse-square law. According to this law, the mutual disposition of two masses to gravitate towards each other has a magnitude of Gm_1m_2/d^2, where G is a universal constant, m_1 and m_2 are masses, and d is the distance between them. Whatever Newton's understanding of this empirical law might have been, there can be no doubt that there was a tendency in subsequent generations to see it as alluding to a *force* that permeated the space around a massive object and that fell away in proportion to the square of the distance around the object. The success of the law in describing celestial phenomena had an enormous influence on the way that scientists approached electricity and magnetism. Therefore, as we hope to show in this chapter, it can be highly enlightening to reflect on the problems and limitations of this pre-relativity approach to gravitation. By doing so, we obtain valuable insights into the underlying nature of any force that has the capacity to act at a distance. We can do this fruitfully without commenting much on the alternative understanding of gravity proposed by the general theory of relativity. Genuine insights into the nature of these forces can be achieved using simple principles that are easily comprehended from within the classical framework.

First, a word about the conceptual backdrop to our approach. For readers who have difficulty concentrating on material of a philosophical nature, it might be possible to skip this part and pick up our conversation from section 6.2. Our assessment of the classic approach to gravitation takes its starting point from a rejection of naive assumptions about the ontological status of space. Even in these post-relativistic days, there is a tendency to look on objects through the same metaphysical lens as the philosophers of the mechanistic age. We see objects separated in space and we consider the space between them to have some sort of independent existence that is relevant for the causal processes that transpire between them. Everyone acknowledges, of course, that gravity, electricity and magnetism are able to make their effects felt by whatever means through this fabric of space (or "space-time"), but the fabric is nevertheless conceived of as real and as holding objects in separation from each other. In confrontation with this world-view, we will assume with fairly minimal philosophical elaboration that the space between objects is a feature of our *mode of representation*. Despite a certain shallow resemblance, this has almost nothing in common with the Kantian conception of space. In contrast to Kant's total idealism about space, our view is that the spatial separations between objects (and the spatial extension of those objects) are based on *genuine* ontological separations and presences in material reality. Let us consider in simple terms what this claim amounts to.

Objects as they are in themselves cannot be either spatially extended or spatially separated from each other. To say otherwise would be to accept naively the objectivity of the content of sensory perception. But we *do* have grounds for saying that material objects must have substance of *some* sort, and that the properties of this substance are responsible for the fact that bodies *appear* as spatially extended under the conditions that make perception possible. Similarly, these objects have some sort of ontological independence or separateness from each other, and it is this separability that makes them appear to occupy a unique position in space under the conditions that make perception possible. If we observe two objects as being separated by a gap of magnitude *x,* then we can be sure that these two objects have *some* sort of ontological disjointedness that is relatively proportional to *x*. But the way that *we* represent these features of the world has a subjective character that does not penetrate to the deepest reality of the material order. Perception itself, in fact, is based on a highly superficial causal interaction between our sensory apparatus and the world.

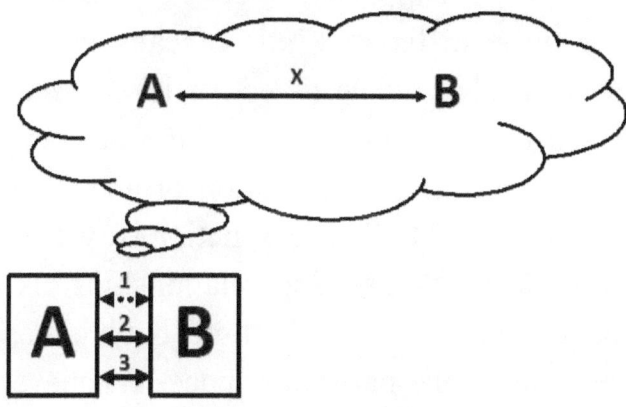

Figure 6.1 Schematic representation of how our spatial perspective fails to penetrate to the deepest core of reality.

When we represent two objects as being separated in space, our mode of representation is such that both objects appear to have a causal insulation from each other that is almost absolute in character. We are consequently startled if these objects appear to be capable of acting on each other "at a distance," failing to appreciate that the definite nature of the distance is a feature of our mode of representation. At the deepest level, these objects have some sort of "distance" between them all right, and that "distance" is represented as spatial separation under the conditions that generate sensory perception. But down there at that innermost level, material objects also have connections and interactions that do not impinge on our sensory apparatus and do not appear at all in our representation of the world. Those fundamental relationships are invisible to us (because they do not impinge on our perceptual apparatus) and we are bewildered when their indirect repercussions become evident in phenomena such as gravitation.

In Figure 6.1, for simplicity, objects A and B are portrayed as having three different kinds of causal relationships between each other at the objective level. Relationships 2 and 3 are relationships that *bind* the objects together, whereas relationship 1 is a relation of an unknown sort that *separates* A and B. This distinction between "binding relationships" and "separating relationships" is a rather crude conceptual tool, as we shall see, but it is a useful one for understanding the nature of space. Whenever we say that a pair of objects has a binding relationship between them, then we are focussing on some sort of equilibrium that exists between them with regard to a certain property. A change in the value of that property for one of the pair will result in immediate repercussions for the value of the property in the other. A simple example would be a bicycle pump that is attached to a tube. When pressure is applied to the pump, the tube inflates. Air pressure is thus a property that binds these two objects in equilibrium together. When we speak of a separating relationship, on the other hand, we are considering ways in which two objects differ from each other, and these differences have *causal* implications. "Having a difference in colour" is not usually a "separating relationship" in our intended sense of the term because such a relationship has no causal repercussions. However in certain circumstances colour differences *can* give rise to causal effects. Say the sun is shining through the window onto two objects, one black and one white. Over time, the black object becomes considerably hotter. This results in the air molecules surrounding the black object gaining kinetic energy leading to the transfer of heat by convection from the area around the black object towards the area around the white object. This is *a typical characteristic of separating relationships*: the "gap" between the objects with respect to that property *is a gap that can give rise to various causal relationships of a secondary kind that need not have any direct connection to the separating property itself.*

Relationship 1 in Figure 6.1 is a separating relationship. It represents a way in which A and B are distinct and separable from each other, even though they are inextricably bound together by relationships 2 and 3. Like any separating relationship, various secondary causal relationships can develop in this "gap" between A and B. Our sensory apparatus is sensitive to one such causal relationship – the interplay of electromagnetic radiation. The gap between A and B is a gap in which this interaction can take place, but like the heat convection in the earlier example, the emission of electromagnetic radiation between A and B is likely to be a *secondary* causal process that develops in the gap. It would be a mistake to think that the *gap itself* can be fundamentally defined in terms of the characteristics of electromagnetic radiation. Indeed, a drastic shortcoming of relativity theory is its tendency to relativise the properties of the gap to secondary interactions such as the transmission of light.

The principal message of Figure 6.1, however, is the way in which the spatial perspective fails to penetrate to the reality of objects. A and B are portrayed as having three relationships between them, one of the separating sort and two of the binding sort. Our sensory apparatus happens to be stimulated by causal impulses that develop in the single "gap" that obtains between A and B. The effects of these impulses accentuate the gap, giving rise to a representation of A and B as occupying separate positions from each other. The separation appears decisive and seems to confer a relative causal insulation on both objects with respect to each other. This impression is further bolstered by the fact that the other two *binding* relationships between A and B are not of the sort that stimulate our sensory apparatus at all. Say that one of these relationships is the gravitational disposition. It will not be directly accessible to our senses, but will become *indirectly* evident when it leads to the closure of that gap that our sensory apparatus happens to be sensitive to, representing it as spatial separation. This will lead to consternation if the observer naively believes that his spatial perspective represents the last word on how objects are causally related to each other. It will drive him to look for hidden mechanisms such as gravitational waves or particles that might explain the closure of the gap.

We do not intend to trivialize the "separating relationship" that lies at the root of our representation of spatial separation. It likely constitutes a fundamental and highly significant ontological disjointedness of objects. The fact that the electromagnetic interaction exists in this gap, and the fact that the gravitational disposition closes this gap, are both indications that it represents some sort of fairly basic carving up of material reality. Our intention is not to underrate this gap but to consider it in a balanced and non-naive way. The fundamental thing is that we must cease to *reify* the form that this gap takes in our spatial representation of things. Perhaps the inappropriateness of reification might be illustrated effectively if we push the colour analogy a little further. Imagine a non-spatial world in which every object has a distinct colour. As it is impossible for us to imagine anything in utterly non-spatial terms, let us simply visualize all of the objects as being arranged contiguously with no empty spaces between them. The "dividing relationship" of colour is, in this case, a way in which reality is carved up, but it would be a mistake to think that it constitutes the way that the world is cut at the very joints. As in the example given earlier, differences in colour can give rise to causal interactions (such as heat exchange) in the colour "gaps" between objects. Now imagine a being whose sensory apparatus is sensitive to one of these types of causal interactions. On the basis of his perception of the divergences of heat emanating from various objects, he represents the objects as spatially separated, occupying distinct positions in space and causally insulated from each other. That is a perfectly natural and useful way of representing the world. We cannot hold the being accountable for the workings of his sensory perception. But if he were to *reify* that spatial gap then he would be making a serious philosophical error.

In our world there can be no doubt that the ontological significance of space has been misinterpreted in the past. Spatial position was absolutized in the classical framework, and it had a central role in the rise to hegemony of the matter-in-motion project. In relativity theory, position in space-time is relativized, and the properties of space-time itself become dependent on the properties of the objects that move in it. But space-time itself, for all its relativity of properties, is still considered to be *a thing that is real*. In the conception that we are defending, space is generated by our perceptual apparatus as a result of a causal interaction that takes place in a "gap" that obtains between objects. This gap is certainly real but in itself it cannot be described as a spatial gap, unless we succumb to naivety of a fairly extreme sort. It is better described as an ontological difference between objects that (with the interplay of light) has causal implications for our sense organs, thus aiding in the generation of the spatial perspective.

When objects are seen to be able to act on each other "at a distance", we naturally look for a mechanism to explain how such action is transmitted. The major barrier to understanding the nature of the mechanisms of gravity, electricity and magnetism is the persistent belief that the space between objects has an ontological character, an objective existence all of its own, even if we allow that its properties depend on the matter it contains. The stubborn conviction that spatial separation is real makes us look for a mechanism that can penetrate space and carry causal influence between source and target. Our attention is dominated by the mystery of transmission through space. So we look for gravity waves or gravity particles that can penetrate the spatial gap, or we speak of a field that is generated by a material object and changes the properties of the space all around. But if space is a feature of our mode of representation, then we are looking for a physical mechanism that penetrates a construction that is ultimately mental in nature. Without knowing it, we are still hot in pursuit of the properties of the ether.

It is time that these ghosts of the nineteenth century were laid to rest. If we acknowledge that spatial separation is not a feature of the world as it is in itself, whilst acknowledging that our representation of the void *is* generated by a genuine fragmentation between objects, then a new approach to understanding gravity soon presents itself. The gravitational disposition of objects is clearly a natural tendency of material reality to close that ontological separation that we represent as space. We ask ourselves: what kind of natural disposition could that be? But now, in forming our answer, we already see beyond the illusion of reified space. We no longer seek to frame our response in terms of forces, fields, waves or particles that fill or penetrate the void. It is our *mode of representation* that gives the impression that these objects are causally insulated from each other, desperately in need of intermediaries or mediums to carry the causal influence between them. Given the superficiality of the process of perception, we have no compelling reason to think that material objects on their deepest level have relationships and interconnections that require *extraneous third parties* in order for their effects to be felt. This is the correct starting point for a treatment of gravity: eliminating third parties such as forces, waves, particles or geometrical changes in the fabric of space-time. We will simply allow material objects to have the natural predisposition to close any ontological fragmentation that obtains between them, a fragmentation that we represent, or *mis*represent, as space. Once the preoccupation with these entities is banished from our quest, some of the salient features of gravitation rise to the surface, features that have been overlooked in treatments of the phenomenon. The most important of these features is the fact that the gravitational disposition can only be spoken of coherently in terms of the behaviour of entire *systems*: it quite simply defies understanding as an autonomous property of individual objects.

6.2 An illustration of the system-wide nature of gravitational properties

What does it mean to approach gravitation from the point of view of the make-up of the entire system? The classical approach agrees that any individual object will exert gravitational influence on all bodies in the system, but it holds that the *magnitude* of the influence is wholly dependent on an autonomous property of the body itself. This magnitude falls off with greater spatial separation and it does not always translate itself into the motion of the other body that is worked on. According to the classical model, other bodies in the vicinity can interfere with the influence exerted by body A on body B, but the magnitude of the particular influence of body A *itself* does not depend on the properties of any other bodies in the system. We wish to challenge this approach and show how the magnitude and nature of the gravitational influence must be considered a system-wide property.

Gravitational influence is the type of causal interaction that bodies naturally enter into with *all* other bodies in a system. It is not a one-to-one type influence such as the transmission of light, which involves an individual source and an individual target. Figure 6.2 gives a simple illustration of this difference between the electromagnetic interaction and the gravitational interaction. Take a body m_1 that exists in an isolated system with a certain number of other objects. If any individual body exerts gravitational influence on multiple other bodies simultaneously, then it makes sense to speak of the *total* gravitational influence, $I(m_1)$, exerted by m_1 on all other bodies in the system. This totality is the sum of all the particular influences, I_i, that m_1 exerts on each other object. Thus

$$I(m_1)=I_1+I_2+I_3+ \ldots +I_n$$

where all the other objects in the system are numbered from 1 to n.

We would need a slightly more complex term to individuate the particular influence exerted by *two* objects, m_i and m_j, on each other. The terms $I(m_i)$ and $I(m_j)$ represent the total influences of these two objects respectively on all other objects. From the totality $I(m_i)$ we would need to subtract the influences exerted on all other objects, bar that exerted on m_j. Similarly, from the totality $I(m_j)$ we would need to subtract the influences exerted on all other objects, bar that exerted on m_i. The sum of the two remainders would represent the total gravitational influence exerted by m_i and m_j on each other.

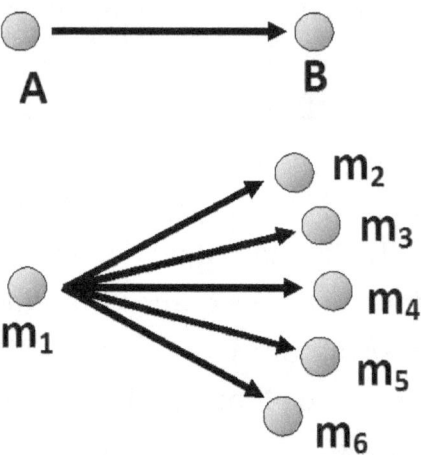

Figure 6.2 Distribution of causal influence. The causal interaction at the top of the diagram represents the transmission of light. An individual impulse of light typically originates in the excitation of a SINGLE source object (such as an atomic pair) and is transmitted to a SINGLE target object. The lower part of the diagram represents the gravitational interaction. Each individual mass will have gravitational influence on ALL other masses in the system

Of course in practice it is impossible to do a calculation of this sort when the system is a universe like ours with an enormous number of individual units of matter. So we must ask the question: how is this impossible task of reckoning by-passed so conveniently with the expression $Gm_i m_j / d^2$, where G is a universal constant and d is the distance between m_i and m_j? The fact that such an onerous calculation *can* be replaced by such a simple expression has contributed greatly to fostering the conviction that objects have an autonomous gravitational force that is utterly dependent on their individual mass alone. To understand the origin of this simple expression we must say more about the way that the gravitational influence of a body is distributed throughout the system.

6.3 The value of G is a measure of the extent of distribution of gravitational influence of an individual object over the system

The principal point made in the previous section was that it makes perfect sense to think of the gravitational power of a body as a *total* influence that is *distributed* over every other body in the system. Now we wish to understand the conditions that determine the way in which the influence is distributed.

In line with our critique of naive assumptions regarding the ontological significance of space, we prefer to speak of the bodies in systems as being arranged in a *causal configuration*. On paper, a causal configuration will resemble the spatial configuration of those same objects, but the "gaps" that exist in a causal configuration will not confer causal *insulation* on the respective objects. Indeed the opposite will be the case. These gaps will represent dynamic causal relationships, although the magnitude of the gap will reflect a relative diminution of some types of causal influence (as we shall now explore).

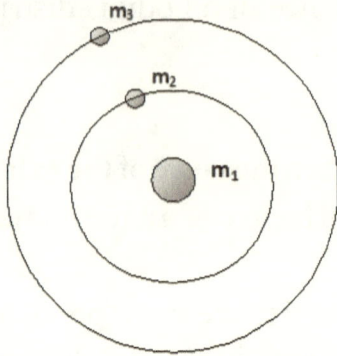

Figure 6.3 A universe with three objects

In Figure 6.3 we imagine a world in which there are only three objects, m_1, m_2 and m_3. The basic puzzle we wish to solve is why the magnitude of the gravitational influence imparted by m_1 on any object m_i should be dependent on the proximity of m_i to m_1 in the causal configuration of things. In particular, we wish to understand the reason why this fall-off of causal influence with distance should follow an inverse-square law. Say that m_1 is the most massive of the objects and that m_2 is located relatively close to m_1 in the causal configuration of things. The object m_3 by contrast, is located at a greater distance. As we have already discussed, it seems right to assert that all of the gravitational influence of m_1 will be distributed among these two objects. In other words,

$$I(m_1)=I_2+I_3$$

where I_2 is the portion of the influence of m_1 exerted on m_2 and I_3 is the influence exerted on m_3. Say that the distribution of influence follows an inverse-square regularity of the sort $Gm_i m_j/d^2$.

Imagine now that m_1, m_2 and m_3 are exactly as before but this time allow that the universe contains three *other* objects as well: m_4, m_5 and m_6 (see Figure 6.4). These three objects are located at an intermediate distance from m_1, between the orbits of m_2 and m_3. The gravitational influence m_1 will now be distributed over m_2, m_3, m_4, m_5 and m_6, which means that proportionally less will be distributed to m_2 and m_3. In other words,

$$I(m_1)=I_2+I_3+I_4+I_5+I_6$$

where I_i is the gravitational influence exerted by m_1 on m_i.

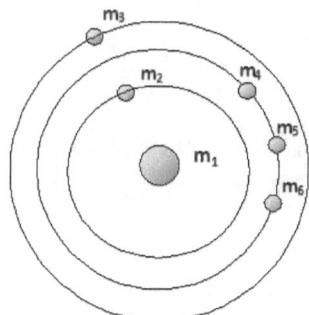

Figure 6.4 A universe with the same three objects as before and three others besides

Any inverse-square regularity with regard to masses and distances that obtained in the first universe with only three objects fails to hold in this second case. In the first universe, the disposition of m_1 and m_2 to gravitate together could be expressed as Gm_1m_2/d^2. In the second universe, the masses and distance of separation of these two objects are identical, but the gravitational influences of both objects are now spread over the three new objects that didn't exist in the first universe. One would expect that the same objects, possessing the exact same numbers of atomic units, would have the same *total* gravitational influence in both universes. That would mean that the influence exerted on each *individual* object would be diminished in the second universe, given that the same starting influence is dispersed over a greater number of objects. If the regularity, Gm_1m_2/d^2, is still to hold in the second universe, then the value of G must be diminished to reflect the increased dispersion of gravitational influence over a greater number of objects. This "calibration" of G will almost certainly be not enough to save the inverse-square law in a universe where the distribution of matter has been altered, as we shall see later. But even if some sort of inverse-square law can be salvaged in the second universe, it is clear that the exact same regularity with the same value of G cannot hold in both universes.

It seems more correct to state that the gravitational interaction between objects depends on the number of objects in the universe and the relative distribution of mass around the causal configuration of things. *Whether* this distribution conforms to an inverse-square law at all would appear to be a purely empirical phenomenon. But we have little reason to doubt that the value of G is determined by the empirical circumstances that prevail in the particular universe. The more objects there are in the universe, the more the gravitational influence of any particular object will be dispersed, and G will have to be modified accordingly.

This is a legitimate interpretation of the meaning of G. It is not a measure of the innate gravitational "force" possessed by matter. Rather, it is a measure of the extent to which the gravitational influence of an individual object is distributed around the entire system.

6.4 Regularities in the distribution of gravitational influence may hold in one part of the universe but fail to hold in others

Up to now we have discussed how the influence of an individual object is distributed over the system. Now let us turn the question on its head and consider how the *entire body of objects* in the system exert their combined influence on any one body. In a universe with a perfectly even distribution of matter, the combined influence of the system on any individual object towards the centre of the system will exert an even "pull" on all sides of the object. In other words, the combined influence of all the objects in the system on this particular object has a null effect – the even and opposite influences cancel each other out (here we are speaking of idealised non-composite objects that do not risk being pulled apart by gravitational influences on all sides).

However, if we consider an object at the periphery, it will be influenced on three sides only, leading to a disposition to move that is markedly different to that of objects at the centre. Say that the inverse-square regularity succeeds in describing the motions of objects towards the centre of the system. It might be expected that the same expression would describe the motion at the periphery. The value for G should be the same at the periphery since the magnitude of this constant (as we have just argued) depends on the combined mass of all the objects in the system. Therefore it should hold as well at the periphery as at the centre. But the problem at the edge of the system is that the *system itself* has a different character here. Objects are no longer surrounded on all sides by an even distribution of matter. The inverse-square regularity was induced from the behaviour of objects at the centre, and the behaviour of objects at the periphery will be markedly different given that their position makes them subject to lop-sided influences from the rest of the system.

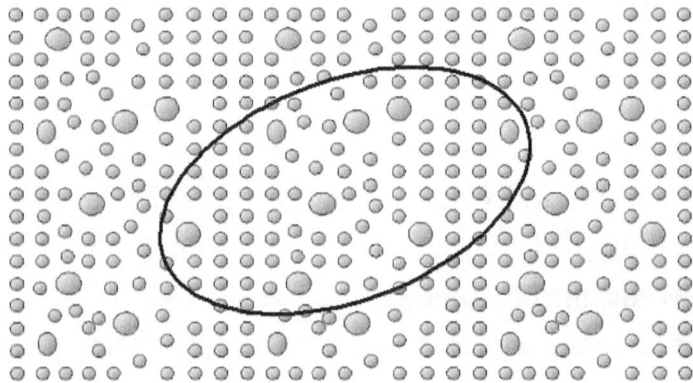

Figure 6.5 The influence exerted by the system on individual bodies within the oval will be different to the influence exerted at the periphery

We can imagine beings living on a planet towards the centre of the oval portion of the universe in Figure 6.5. Their measuring instruments are only able to glean reliable data about gravitational dispositions of the heavenly objects within the oval. On the basis of these measurements, the beings are able to state that the celestial objects have a uniform disposition to move that is proportional only to their mass and relative position. This uniform disposition can be mathematically expressed by invoking a universal constant G. Such an empirical discovery naturally leads to the belief that the gravitational disposition is an autonomous property of individual bits of matter.

A consideration of the wider system leads to a completely different picture. *All* of the objects in the system have a disposition to move towards each other. How this disposition will translate into motion depends on the configuration of the particular part of the system in question. Parts of the system that are more isolated will be disposed to move in radically different way than parts of the system that are uniformly surrounded on all sides by other objects. The fact that an inverse-square law manages to describe the motions within the oval does not mean that it will describe the motions of objects outside the oval that are not so uniformly surrounded by matter. It is the empirical behaviour of a heterogeneous portion of the *system* that gives us the inverse-square law, and we should not be surprised if we find matter in certain isolated parts of the system that does not respect the same regularity.

In case this point is not clear let us restate it in different words. Gravitation involves the distribution of the influence of an individual body over the entire system, and it equally involves the reciprocal influence of the entire system over an individual body. We observe the motions of bodies in a very limited part of the cosmos and derive an inverse-square law from those observations. The beauty of the inverse-square law is that it can tell us the gravitational disposition between any two bodies in our portion of the universe, leaving the rest of the system out of the equation, even though the gravitational influence between any two bodies is utterly dependent on the composition and configuration of the rest of the system! How does this wonderful law manage to give us the influence of the sun on the earth without first having to calculate and subtract the influence of the *entire rest of the system* on the sun and the earth? Because regularities in the composition and configuration of the system in *our* part of the universe are such that the *combined* influences of the rest of the system effectively cancel each other out, AND those same regularities of composition and configuration mean that the influence of bodies on each other happens to conform to an inverse-square law. The classical approach recognizes that the combined influence of the system can sometimes be zero, but it fails to allow that it may be the *particular composition* of the system that permits the inverse-square law to function in certain circumstances. Later we will examine the fairly simple conditions that can lead the gravitational phenomena in a portion of the universe to conform to an inverse-square law. For the moment let us re-iterate the claim that if the matter were configured differently in our part of the universe, then the earth and the sun would not necessarily be gravitationally disposed towards each other according to an inverse-square law. The influence that the *system* exerts on an individual body will be mediated by the local configuration of the system around that body. Let us consider now how this works.

6.5 Identical spatial gaps in different parts of the same system can have different causal implications depending on the composition of that portion of the system

Three principal points have been made so far. The gravitational influence of a body is distributed over the entire system. Therefore the causal sway of any one body over another depends on the composition of the entire system. Secondly, we have argued that the mysterious quantity G is a measure - not of a gravitational force inherent in objects - but of the extent to which the gravitational influence of a particular body is distributed over the entire system. Thirdly, we have argued that the influence exerted by the system on individual bodies cannot be uniform in all areas of a system that is itself only uniform in certain parts. The fact that an inverse-square calculation works in one part of the system is not a guarantee that it will work elsewhere.

It might be objected that these theses are simply qualitative and do not have any quantitative consequences for the motion of bodies that would not be adequately covered by the inverse-square law. In the last example, the objects at the periphery of the system were described as being subject to asymmetrical influences from the rest of the system. Surely the classical approach also accepts that such objects are subjected to lop-sided influences? The individual influences from the objects in the rest of the system can be calculated simply by inserting the masses of each object in turn into the inverse-square expression. Then we perform a vector addition on the results to calculate the total influence of the system on the motion of the object at the periphery. Isn't this classical approach quantitatively identical to our so-called system-wide approach that considers each object to have its influence dispersed over every other object in the system? No. As we shall now see, the magnitude of the gravitational influence of individual objects in the system-wide approach is completely different to that calculated by the inverse-square law.

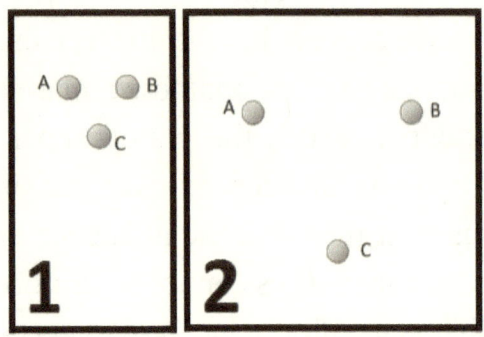

Figure 6.6 Two systems, each with equidistant bodies, but the distances are greater in the second system

According to the law of gravitation, causal influence tapers off in perfect proportion to distance, and this is true whether space is packed with objects or completely devoid of them. But a closer examination suggests that this cannot be accurate for any more than a limited portion of space. The causal implications of spatial separation depends *precisely* on how densely packed it is with objects. The emptier a region of space is, then the less fall-off in causal influence between objects located in that region. Let us illustrate this conclusion with an example (see Figure 6.6). Say that objects A, B and C are the only objects in the universe and they are located in a confined region of space at equal distances from each other. Say that they have a particular gravitational influence on each other. Now imagine a second universe in which the same objects A, B and C are again the only things in existence, but this time they are much more widely dispersed in an enormous region of space at equal distances from each other. This enormous spatial separation, we recall, is a representation of a greater ontological disjointedness than was the case in the first universe. In this second situation, the three objects should exert the *exact same gravitational influence* on each other as was the case in the first universe. Why so? Don't the objects have greater ontological disjointedness in this universe? And shouldn't that lead to a fall-off in causal influence, as per the second thesis? But the second thesis is a relative one, as we remarked earlier. In each universe the objects are all *equally* ontologically disjointed. There is no other object in either universe that is *more closely joined* to any of the three objects A, B and C. If there *were* such an object, then the distinct levels of disjointedness in either universe would become highly significant for the causal influence of A, B and C on each other. The fact that there is no other object entails that the objects exert the same gravitational influence as in the first universe despite their greater separation – a direct violation of the inverse-square law.

Now imagine one of the universes as before but now allow that it has a fourth object, E, located between A and B. The fact that E is closer than B to A will mean that A's causal influence on E will be proportionally greater than on B or on C. That entails that there will be a *tapering off* of influence from A and it will be proportional to the extent to which B and C are more disjointed from A than E is from A. If we admit a further object F between A and E, then the tapering-off of the influence from A towards the more distant objects in the system will become even more marked. From this we can conclude that *a more densely populated region of space will see a greater fall-off in the causal influence of objects over distance than a less densely populated region of space.*

This conclusion is derived from common-sense principles: causal influence such as the gravitational disposition can be legitimately interpreted to be *distributed* over all objects in a system; and the magnitude of such a causal influence is proportional to *relative* ontological separations. Thus we are led to the surprising discovery that the degree to which gravitational influence decreases with spatial separation will *not be constant* if the relative distribution of matter in that region is not constant. In a sparsely populated portion of the universe, widely dispersed objects will exert greater influence on each other than they would do at the same distance in a heavily populated area.

This leads to an entirely new conception of the causal implications of the spatial separation between two objects. The separation *is* symptomatic of an ontological fragmentation within matter, and the magnitude of this ontological fragmentation *can* have implications for the fall-off in causal influence between the two objects. But the magnitude of the fall-off in causal influence *also* depends on whether other bits of matter are more or less ontologically disjoined from the two objects under consideration. In spatial terms, the more densely populated an area of space, the more dramatic the fall-off in the causal influence of an object with spatial separation.

Figure 6.7 shows a universe that has most of its matter scattered fairly evenly throughout a roughly spherical area, with a small number of distant stars in orbit around this sphere at an enormous distance from it. We imagine that the sphere itself contains billions and billions of stars. Star A is separated from the centre of the sphere by a distance x. If we accept naively the "universality" of the inverse-square law, we would expect the gravitational influence of the matter in the sphere to exert an influence of Gm_1m_2/x^2 on A, where m_1 is the estimated mass of A and m_2 is the estimated combined mass of all the objects within the sphere. This particular magnitude of gravitational influence should have consequences for the orbital velocity of A around the central sphere. In particular, the enormous distance, d, between the periphery of the sphere and A will result in the gravitational influence of the sphere being diminished, with negative implications for the orbital velocity of A.

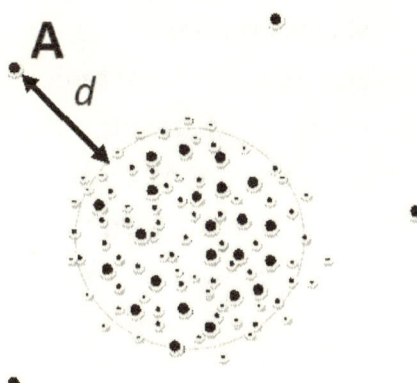

Figure 6.7 A universe with an uneven distribution of matter

According to the view that we are espousing, by contrast, the distance d will have no diminishing effect on the gravitational influence of the combined masses within the sphere. There are *no objects at all* between the periphery of the sphere and A. This means that the combined influence of the sphere is not distributed proportionally over objects between the sphere and A, such that A experiences less influence because of its greater level of disjointedness. If there *were* a few objects between the periphery of the sphere and A, then the total influence of the sphere on A would fall off to *some* degree, but it would not fall-off as rapidly as the inverse-square expression would have us believe. The inverse-square expression would only give an accurate value for the rate of fall-off if the region of space between A and the periphery of the sphere had a similar distribution of matter as the interior of the sphere itself. But in the case above there are no objects at all between the periphery and A. This means that we could still use the inverse-square expression to calculate the velocity of A, but we should treat A as if it were located right on the periphery of the sphere. The adjusted expression would take the form $Gm_1m_2/(x\text{-}d)^2$. In reality the universe is a much more complex place with such local variations in the distribution of matter that the inverse-square regularity cannot always be rescued with such a simple alteration. Let us turn now to those limited conditions that permit the law to work.

6.6 The conditions that favour the functioning of an inverse-square law

The previous section discussed how spatial separation in itself is not directly relevant for the magnitude of causal influence imparted by objects on each other, whilst the *relative* separation of objects is indeed crucially relevant. Causal influence is imparted by an individual object on all objects in the system, but the proportion imparted to any particular object depends on where it lies in the pecking order. This pecking order is determined by the relative ontological disjointedness of the objects in question. In a universe composed of five equal objects, A, B, C, D and E, all of which are represented spatially as being configured along a line, object A will distribute its influence in turn to each of the objects along the line, with the nearest object (B) being the recipient of proportionally the greatest portion of causal influence, and so on down the line. The distribution of influence will be inversely proportional to the distance of the object from A.

$$I = I(B) + I(C) + I(D) + I(E)$$

where I is the total causal influence of A and I(x) is the portion of the influence of A that is distributed to object x.

If the distribution of matter along this line is *relatively uniform* then it will be possible to empirically derive a single expression that would, to a high degree of accuracy, represent the distribution of gravitational influence of A to *each* of the other objects as a function of the spatial separation between them. This should take the form I(x) =

G/d, where G is a constant to be empirically derived and d is the distance of object x from A.

If the objects in the universe are of varying masses, but the overall distribution of matter along the line fulfils certain criteria of uniformity, then the expression for the distribution of A's influence would be expected to conform to an expression of the form $I(x) = Gm_A m_x/d$, where m_A is the mass of object A, m_x is the mass of object x and d is the distance between them.

In a universe where matter is distributed randomly throughout the causal configuration of things, then an observer at a point A in the interior of the universe will represent his world as a sphere centred on A with objects distributed around this sphere at various distances. For a universe that consists in a truly colossal quantity of objects that are scattered relatively evenly throughout the system, the distribution of the gravitational influence of A could be expected to conform to an expression of the form $I(x) = Gm_A m_x/4\pi\, d^2$. A number of factors conspire together to facilitate the functioning of this inverse-square law.

Firstly, the enormous number of objects, allied to their even distribution, will entail that the *combined influence of the system* on any objects in the vicinity of A (which lies well towards the interior of the system) will cancel out. Therefore the influence of A on any particular object can be calculated without recourse to a calculation of the influence of the rest of the system. Secondly, the fact that there are multiple objects in the system located at different distances from each other will entail that the influence of A falls off with distance. Matter is relatively evenly distributed, ensuring that this fall-off will be uniform and will approximate to a mathematical expression that can be empirically derived. Thirdly, if a particular type of causal influence is distributed evenly through the portion of the system in which A is located, then the configuration of the system with respect to that type of causal influence around A can be mathematically represented as a sphere with A at the centre. A's influence (with respect to that type of causal influence) on the rest of the system can be represented as being dispersed evenly on the *surfaces of concentric spheres.* That is why the simple expression for spatial separation in the linear universe is now replaced by the expression for the surface area of a sphere of radius d with A at the centre ($4\pi d^2$). This can be simplified by modifying the value of G so that it now represents the magnitude $G/4\pi$. The new "law" will consequently have the form $Gm_A m_x/d^2$. In the linear universe discussed earlier, the simple distance gave a measure of the dispersal of causal influence. In a universe where the causal configuration is more complex, with the influence of an object being distributed evenly "on all sides" so to speak, the surface area of a sphere will give an accurate representation of the extent of the dispersal at a given point.

If the distribution of matter were not uniform in the universe, however, no such simple expression could accurately describe the gravitational influence of objects in general upon each other. Each object's influence on another depends entirely on the position of those objects in the causal configuration. If the causal configuration does not admit of a uniform distribution of causal influence, then no single empirical expression can be derived that will cover the behaviour of objects in general. But this caveat is also true for regions of our own universe where the distribution of matter is not uniform, or even if uniform, varies to the distribution of matter in *our* portion of the universe. Thus our "universal" law of gravitation does not have a necessary character. It is haplessly dependent on contingent properties of the region of space that we occupy in the world. When we observe the behaviour of more distant parts of the cosmos, we cannot continue to apply without qualification a regularity that is only valid for our own relatively local configuration of matter. If we do then we will be bewildered to observe celestial objects moving at speeds that do not conform to the letter of a supposedly universal law. And what do we do next? As has happened too often in the past, we omit to re-examine the validity of the premises upon which our calculations have been based. Instead we retain those premises by invoking shadowy entities that have not a shred of independent empirical evidence to their name.

A final word on gravitation: the tendency of objects to gravitate towards each other can be reasonably described as a residual effect of the protomagnetic polarity that constitutes matter. Each genesis-unit has a B pole and a V pole, and the magnitude of these poles is identical. Despite this perfect equality, it can nevertheless be argued that a weak residual tendency exists for objects to gravitate towards each other. Take two genesis-units, A1 and A2. The Bs of A1 and A2 repel each other, as do their V components, but the overwhelming natural tendency for each B to collapse into the V of the other unit exceeds the natural repulsion between the like poles, even if only in a very minimal way. Thus, the contention here is that gravity is really a residual attraction that remains because the B-B, V-V, B-V and V-B influences do not quite cancel each other out. When it comes to the interactions between small numbers of genesis-units, these residual effects are often negligible in comparison to normal electrostatic or magnetic influences. But when enormous numbers of genesis-units are involved, then the gravitational influence becomes highly significant. The defence of this claim about the nature of gravity will have to wait for another occasion. Now we turn to the meaning of Planck's constant and the significance of Bell's theorem.

Chapter Seven

PLANCK'S CONSTANT, INDETERMINACY AND BELL'S THEOREM

"'Quantum connectedness', 'passion-at-a-distance', 'incompleteness' – whatever we choose to call it – this feature of the post-'local-realistic' worldview challenges any straightforward notion of the individuation of physical objects. In terms of the Bell-type experimental set-up, for example, we are barred from representing what we ordinarily would call the 'two particles' in a given 'pair' as two distinct entities; . . . Providing an adequate elucidation of this inseparability, and of the measurement process itself, . . . – this is the grand unfinished task"

Jon P. Jarrett

Overview of this chapter and its principal claims

1. Planck's postulate (that the energy E of an impulse of light is equal to $h\upsilon$, where h is a fundamental constant of nature and υ is the frequency of the light) seems to present a congenial picture of tiny oscillators vibrating at a certain frequency, emitting the fundamental quantum of action at each oscillation. As is well known, however, the use of Planck's postulate in the explanation

of the photoelectric effect and for the emission spectra of elements cannot be reconciled with a classic wave picture of light.

2. This presents us with a challenge to articulate the *real physical basis* of the concepts of "frequency" and "wavelength" in a new and intelligible way that is not constrained by the wave model. Our reinterpretation of frequency as signifying a measure of the interval of time it takes an impulse to manifest itself in relation to the evolving system (and this is related inversely to its potency) coheres perfectly with Planck's postulate.

3. Planck's constant is described as a constant of proportionality that relates this manifestation interval to the capacity of the impulse to induce acceleration in matter. We discuss how the constant is *not* a measure of an elementary unit of action but a way of expressing the potency of light in standard units. The quantisation that is apparent in the photoelectric effect and the spectra of elements can be explained without recourse to such a fundamental unit of action. The fact that reality is composed of basic *causal units* that have a fixed work potential explains the quantum character of certain phenomena.

4. Against the notion of a minimum quantity of action, we argue that the causal capacity of a single atom (for example, its electrostatic influence) can be *distributed* over arbitrarily large numbers of atoms. In theory, this would lead to arbitrarily small impulses of light, but there are logistical constraints on the formations of large electrostatic alignments of matter. Thus, arbitrarily small impulses of light are practically impossible.

5. The wave-function is empirically accurate because it manages to capture the structural aspects of systems that determine how processes evolve. Terms in the formalism, such as the conjunction of probabilities that a particle has a certain position, should be reinterpreted to signify the probability of the causal process having *its effect at that position*, if suitable target material were present. It is paradoxical to think of a point-like particle existing in a state where it manifests a combination of multiple positions at the same time. But it makes sense to think of a process as

potentially giving rise to an effect in multiple positions, *if* there happens to be target material in those positions at the right time.

6. We examine the non-locality inherent in quantum systems, a non-locality that has long been manifest in the interference of light. We show how processes *always* evolve in a strictly local manner, but if one of the causal players guiding the process is modified, then this will have an instantaneous effect on the overall manner in which the influence is evolving. This does not mean that a change made at A can evolve to B instantaneously.

7. We discuss how the elimination of particles from the description of quantum systems defuses all the controversy surrounding superposition and collapse, complementarity, indeterminacy of properties and the breakdown of causality.

7.1 The meaning of Planck's constant

In the last years of the nineteenth century, Max Planck was working on a problem posed by his former teacher, Gustav Kirchhoff, on the radiation emitted by blackbodies. A blackbody is a perfect absorber of electromagnetic radiation, regardless of the frequency or angle of incidence of the radiation, reflecting none of the light that falls on it. Blackbodies will eventually re-radiate energy, and Kirchhoff had found that the *energy* of the radiation emitted by a blackbody was related only to two factors: the temperature of the body and the frequency of the emitted radiation. The challenge was to come up with a formula in terms of these three quantities (temperature, frequency and energy) that accurately fitted the empirical data.

In 1900, Planck managed to find a formula that adequately catered for the data. He then set about providing a theoretical foundation for the mathematical formalism. His eventual derivation was based on the postulate that electromagnetic energy could only be emitted in a "quantized" form. In other words, the radiation emitted by blackbodies could only be in multiples of an elementary unit of energy. This postulate takes the following mathematical form: $E = h\upsilon$, where E is the energy emitted, h is a fundamental constant now known as Planck's constant, and υ is the frequency of the radiation.

Planck envisaged the blackbody as consisting of very many "oscillators", each emitting energy in regular fashion. He did not accept the atomic hypothesis, so an "oscillator" was understood to be any tiny piece of matter in the source object. Presumably it would be a dipole, having a positive and negative electrostatic charge. As the blackbody was heated, this piece of matter would be set into more and more rapid motion, in line with the kinetic theory of heat. Every time the oscillator completed a cycle of vibration, it would emit an elementary unit of energy of a magnitude represented by the constant h. Thus, if the frequency of the oscillator was 100 cycles per second, then the energy emitted per second would be represented by the magnitude 100h.

On the basis of this postulate, Planck was able to derive his mathematical formalism describing the patterns of radiation emitted by blackbodies at various temperatures. What was considered unusual about the postulate at the time was the stipulation that an oscillator could only emit packets of energy that corresponded to the magnitude h. Common sense would suggest that different sized oscillators ought to exist in any material object. If these were vibrating at the exact same frequency, then they should produce smaller or larger pulses of energy, depending on the size of the oscillator itself. Developments over the following thirteen years saw Plank's postulate become a pillar of the emerging quantum mechanics. It was put to use in explaining the photoelectric effect and in accounting for the emission spectra of the elements.

Here the significance of the postulate will be reviewed in the context of our account of the frequency of light and the mechanics of its transmission. If we accept that electromagnetic energy comes in bundles that are represented by the magnitude h, then the expression $E = h\upsilon$ makes sense as an expression of the quantity of energy produced by an oscillator in a single period of time. The oscillator completes υ cycles in a period of time, and each vibration generates a quantity of energy represented by h. Thus the total energy generated in every period will be $E = h\upsilon$. This works fine as long as we continue to think of light as a wave motion generated by a pulsating source. It coheres perfectly with the standard way of visualizing a wave motion, which is in terms of an oscillating device in a container of water. Every cycle of the device creates a wave. A more energetic device vibrates more rapidly, generating more waves per second. The energy produced by the causal source can be expressed in terms of the frequency of the wave motion multiplied by the quantity of energy in each wave.

However, as Planck's postulate was put to use in the explanation of the photoelectric effect and in the development of the electron model of the atom, it became increasingly clear that this congenial picture of cyclic oscillators did not represent what was going on at all at the ground level. The explanation of the photoelectric effect speaks of an individual packet of energy – the photon – being *integrally* absorbed by electrons. The energy of *each* photon is represented by hʋ, where ʋ is the frequency of the light. Thus the elementary packet of energy for this frequency of light is not expressed by h at all, but by hʋ, and this is impossible to square with a wave picture. Take two frequencies of light, one of 10 cycles per unit time and the other of 100 cycles. A single photon of the first frequency of light will have energy 10h, and that of the second 100h. These quantities of energy are envisaged as having an indivisible aspect, in complete contrast to the energy of classic waves. If our water oscillator produced one hundred vibrations per unit time, then we could switch it on for half a period of time and generate fifty waves. But a light source that generates light of frequency 100 cannot be made to generate half the magnitude of energy by switching it on for half a unit of time. It is as if the 100 cycles per unit time were compressed into an infinitesimal impulse that behaved like a particle, the photon.

At this point the picture of the oscillator generating the impulse of light breaks down completely. Clearly, we cannot easily make sense of an oscillator that vibrates 100 times in a period of time and only generates a *single* impulse. Does it generate the impulse at the end of its frenetic 100 cycles, or sometime during the process? Despite this breakdown in the wave picture during the early years of the twentieth century, the notions of frequency and wavelength simply couldn't be discarded because they were needed to account for interference and diffraction phenomena. So the wave picture had to be set to one side in some circumstances and given centre stage in others. The result of this has been to introduce a marked scepticism into theoretical physics regarding the possibility of a coherent model of light. And there can no doubt that scepticism of this sort paralyses efforts to understand a phenomenon. But if we *reject* the unfounded philosophical assumption that a coherent physical model of light is impossible, then we are challenged to find a means of articulating the real physical basis of wavelength and frequency in more fruitful ways. The photoelectric effect and other phenomena give us reasons for abandoning the wave picture, so we should simply abandon it, recognising at the same time that hypothetical properties like frequency and wavelength are erroneous ways of visualizing *genuine* characteristics of causal processes.

We have been attempting to construct a new understanding of "frequency" in terms of the interval of time required for a causal event to exert an influence on a target object, if suitable material is available at that moment in the correct region of space. Soon, we shall see how this relates to Planck's work and its consequent repercussions for the meaning of his constant. According to our account, the most potent electromagnetic impulses are those produced by the disassociation of larger mini-systems of atoms. The sizes of the electrostatic components of a mini-system of atoms are proportional to the size of the mini-system itself. Once the system dissolves, impulses are "discharged" into the system. By "discharged" we mean that these impulses begin to evolve in the system in reciprocal interplay with all of the other causal processes that are concurrently active. The more potent the impulse, the shorter the time interval before that impulse has the potential to exert a causal change in matter. If, after that interval, suitable matter is not present at the right place in the system, then the impulse must once again throw in its lot with the other evolving impulses in the system. The same interval of time once again elapses as the whole system reverberates with the interplay of causal impulses. Once that interval has elapsed, then the impulse will have its effect, *if* suitable matter is now present. If not, then the process repeats itself rhythmically until appropriate material is present and the process comes to fruition.

What is the meaning of Planck's constant according to this account of causation? Let us return again to the wave model of transmission of an energetic impulse. In the case of water waves, we can imagine an oscillator producing a wave with energy h at each cycle of vibration. If the oscillator vibrates at frequency υ, then the energy produced over each period of time will be $h\upsilon$. The frequency of the water wave is an indication of the energy of the oscillator, but it doesn't tell us anything about the energy of *each* impulse. It tells us, rather, how many impulses will be emitted over an *interval of time*. On our account of causation, by contrast, the term "frequency" doesn't refer to the number of impulses emitted by a source over a period of time. Rather it is a measure of the *potency* of an individual impulse: the more potent the impulse the shorter the interval of time required for it to make its presence felt in the system amid the many impulses that are evolving concurrently.

We see already how well this account of frequency coheres with Planck's postulate. The meaning of the postulate, E = hυ, changes dramatically once we fully and properly discard the wave model of electromagnetic radiation. We no longer think of this transmission as consisting of a wave in some circumstances and a particle-like photon in others. Rather, the impulse is considered as a unique event in the context of an evolving system. As the event reverberates in the system, there *is* a pulsation associated with it, but this is not the pulsation of a continuous wave emitted by an oscillator with many crests and troughs. Instead it is the pulsation of the *system*. But the impulse under investigation has its particular individuality and location in the causal configuration of things, and the way it reverberates in the system is determined by the potency of the particular event that generated it. We *could* retain the term "frequency" for the time interval required by the impulse to have its effect in the system. In the absence of appropriate target material, the impulse will repeatedly reverberate in the system in this pulse-like fashion, and this gives to the term "frequency" a certain suitability. The problem with the term is that it renders the impression of a wave motion moving through space, and we reject the view that light involves motion of any sort through space. In place of "frequency" perhaps a term like "manifestation interval" could be used, since the term refers to the period of time required for this particular causal event to come to the surface, so to speak, and create an effect amid the other causal events that are evolving in the system. The manifestation interval, of course, is related to the quantity we usually call frequency, but it is actually $1/\upsilon$; the smaller the manifestation interval, the more potent the causal impulse.

If the manifestation interval is a measure of the potency of the causal process under consideration, what then is h? Planck's constant is simply a *constant of proportionality* that relates the manifestation interval to the *capacity of the associated impulse to accelerate matter.* The formulae of classical mechanics describe various capacities of players in the system to induce a standard acceleration in a certain standard quantity of matter. A unit of electric charge, for example, could be expressed in terms of the quantity of charge required to induce an acceleration of 1 metre per second squared in another similar charge weighing 1 gramme placed at a certain distance from the first. Returning to the case of light transmission, if the impulse has energy corresponding to hυ, this means that the impulse has the same energy *as would be* required to induce an acceleration of hυ metres per second squared in a gramme of matter (or some much more minute equivalent). This does not mean that the *light impulse itself* has the capacity to induce motion in matter directly: what is important is that the "energy" of the impulse *corresponds* to the energy required to induce a certain acceleration in matter. The quantity h relates the potency of the impulse (expressed in terms of its manifestation interval) to the capacity required to induce acceleration in matter. The fact that the principal effect of a typical impulse of light may *not* be the induction of acceleration in a portion of matter is not actually relevant.

Energy, or the capacity to do work, is standardly quantified in these terms. Thus we can meaningfully express the energy of light in such terms, even if the principal effect of the impulse is not immediately manifested in this way.

Units of measurement often have this characteristic: they relate *a particular capacity to a second capacity of a more standard sort*, even if the causal player in question could never possess the second capacity! To give an idiosyncratic example: say that I have an instrument in my office that shreds paper. The capacity of the instrument to shred would normally be measured in terms of square metres of paper per hour. But say that I live in a curious world where all the units of measurement are related to my *printer's* capacity to deliver an output of x words per minute. Of course, these two units of measurement *can* be related in a meaningful way: the energy required by my printer to print x words per minute *does* correspond to the energy required by the shredder to shred *some* quantity, y, metres squared of paper in an hour. In our imaginary world, the output of the printer has become the standard of all measurements. Thus even the capacity of the shredder is expressed in terms of words printed per minute. A proportionality constant could be empirically derived that would relate the output of the shredder to the output of the printer for the same input of energy into each machine. Say that the energy required to print 5000 words is equal to the energy required to shred 1 square metre of paper. The proportionality constant, p, that relates the capacity of the shredder to the output of the printer will thus have the value 5000. The units of p will be of an appropriate sort to convert a quantity of metres squared of paper per hour to a quantity of words per minute. If the shredder shreds 10 metres squared per hour then that will be expressed as 50000 words per minute in the "standard units" of this particular world.

In a world where the matter-in-motion project has become dominant, the standard unit of measurement is in terms of motion induced to mass. These units are certainly less counter-intuitive and eccentric than the units in the imaginary world, but they still require constants of proportionality. Sometimes the constants figure in more complex formulae such as the universal law of gravitation or Coulomb's law. It is not immediately obvious in these cases that the constant in each expression plays a role in relating a particular capacity (the size of a mass, the magnitude of a charge) to the capacity to induce matter in motion. But in the case of Planck's postulate we have an explicit expression that *relates the manifestation interval of an impulse to the capacity of an object to induce momentum in matter.* We measure the potency of light in these terms, even though the action of light is more often related to effects that are not directly involved with imparting momentum to matter. Light is reflected, refracted and absorbed by mini-systems of atoms. It is involved in chemical processes such as photosynthesis that are not primarily driven by mechanical motion. Thus the units with which we measure the potency of light are in some ways akin to the units of a printer being applied to a shredder. Despite the differences, their respective magnitudes are perfectly convertible. And how are they converted? It is Planck's constant that expresses the relation between them.

All of this might lead to the suspicion that we are trying to claim that the magnitude of Planck's constant is not a fundamental quantity in nature. It might seem like a mathematical utility that forms a connection between two quantities that are themselves rooted in human conventions regarding the passage of time, quantities of matter and units of length. It is true that our *particular expression* of Planck's constant is the fruit of historically-conditioned factors. Nevertheless it remains perhaps the most fundamental of all physical constants. Our constant for relating the output of the shredder and printer was shrouded in time-dependent conventions regarding the size of paper and the number of printed words. But it still constituted a *purely objective relation* between these two capacities, even if the magnitude of the relation was expressed in terms that depended on conventional units of measurement. In the same way, Planck's constant would be twice as large if our units of matter in motion were twice as small. The *size* of the constant and the *units* in which it is expressed are historically conditioned, but *given these historically-conditioned quantities*, Planck's constant is a completely objective, empirically-derived quantity that authentically relates the manifestation interval of an impulse to the magnitude of work required to induce a standard rate of acceleration in a standard quantity of matter.

There *is* indeed a way in which action is quantised, and we will have a look at that in a moment. Beforehand, however, we wish to offer one more clarification regarding the notion of frequency. In the case of an oscillator in a wave tank, the frequency of the oscillator produces a precise number of wave motions per unit time. If the frequency is five, then the device will produce five sequential wave motions during that interval. We have a tendency to picture light waves in a similar manner. For light that has a known frequency of five, we would be inclined to expect each individual oscillator in the light source to produce five photons a second. When we utilize the wave picture to account for the interference of light, we explicitly invoke this expectation. We visualize the two wave trains as proceeding from the apertures in terms of a series of sequential waves, one wave perfectly positioned after the other. Individual waves that pass through aperture A are not visualized as overlapping with each other. The crest of each wave takes its position exactly one wavelength behind the previous crest that passed through the aperture, and exactly one wavelength in front of the next. Light waves, thus, are visualised as behaving themselves in a very orderly fashion, exactly like water waves produced by a *single* oscillator in a wave tank. But this makes no sense when we consider that a typical light source has *millions* of "oscillators" producing light in any measurable time interval. Therefore there ought to be multiple crests passing through each aperture in the time interval necessary for a single wavelength to pass through. The picture of an orderly train from aperture A converging on an orderly train from aperture B is well wide of the mark. If we replaced our single oscillator in the water tank with just ten oscillators that are out of step with each other, then any interference pattern produced by a single oscillator would soon be destroyed.

The point is that, in a luminous source, the genesis-units that produce impulses of light of frequency υ do *not* produce υ impulses per second. That way of visualizing frequency is a relic of an obsolete strategy for explaining the interference of light. As an explanation of interference it had an intuitive appeal, but it could not cope even with the simple objection that luminous objects consist of multiple individual sources whose causal output simply cannot be in step with each other, even if the light is monochromatic. "Frequency" does *not* entail that the impulse consists in υ crests per second. All it entails is that the potency of an *individual* impulse is such that its manifestation interval will be 1/υ seconds. If we consider a particular genesis-unit that emits an impulse of "frequency" υ within a luminous source, it is extremely unlikely (although not impossible) that it could emit υ impulses per second. Depending on how the source material is being excited by an external source of energy (such as an electric current), A1 may well produce impulses rhythmically, but given the huge number of atoms in any source material, the average rate of production of impulses per atom is likely to be well below υ per second. Our model of frequency has no difficulty accounting for interference regardless of the actual number of impulses passing through the aperture in any instant. The fact is that each *individual* impulse is jointly modified by atoms at both slits, thus excluding certain zones of the final screen from being targeted by the impulse.

In Chapter Eight we will consider another aspect of the manifestation interval: the way that its magnitude varies depending on the *relative motion* of source and target. Say that an impulse is emitted from object O towards object T1 that is receding from O. The manifestation interval of the impulse relative to T1 will be longer than the interval relative to an object T2 that is converging on O. This relative disparity will be discussed in more detail in section 8.2.

7.2 The authentic way that energy is quantised

If our analysis is correct, Planck's constant is not what people often imagine it to be – a measure of an elementary unit of action. Rather, it is a way of expressing the potency of a light impulse in standard units, and these units are in terms of the "energy" required by an agent to induce acceleration in matter. This leads to the question of the existence or non-existence of an elementary unit of action. If such a unit does not exist, then how do we account for the "quantisation" that manifests itself in the photoelectric effect and in the spectra of the elements? The answer to this question, of course, has been the theme of much of this book. There *is* a fundamental unit in nature, but this is not simply a quantity of action, or an amorphous measure of energy. Rather it is a *causal player*, something that has the capacity to act on other causal players, and to be acted on in return. This is the genesis-unit, in terms of which all phenomena - gravitational and electromagnetic - are to be explained. This causal player has a work impulse at its heart that separates B from V, giving rise to the protomagnetic polarity that is the origin of magnetic, electrostatic and gravitational phenomena.

We can speak in the abstract of the magnitude of "energy" that constitutes this work impulse. The secondary dynamics that comes into play between the opposite polarities of different atoms also has a characteristic quantity of "energy" associated with it. But it is to be insisted resolutely that energy *has no existence whatsoever* apart from the relations that prevail between different genesis-units. When we speak of a causal event "discharging" an impulse into the system, and the impulse then evolving with the system until an effect is caused in another genesis-unit in another location, this kind of talk is analogous in character. The evolution of an impulse in the system does not involve the passage of "energy" in some hidden "dimension". The evolution of processes is certainly not accessible to our perceptual apparatus, but we gain nothing by thinking of them in terms of the transition of energy through some sort of substrate. It is better to describe the evolutions of processes in terms of causal sources, causal targets, and other causal players that modify the evolution of the impulse. *All* of these causal players are genesis-units or groups of genesis-units. We have no reason to think of an impulse having an existence apart from the evolving relations that prevail between these material objects.

It is hard to find a suitable analogy to reinforce this point. We can speak in the abstract about the quantity of action contained in the blow of an axe, but only a causal agent wielding a very particular type of instrument can embody the blow of an axe in concrete terms. "Energy" has no existence apart from the causal agents that incarnate this particular capacity to exert influence on each other. On the macro-level we experience many different types of causal influence. All of these have a particular work capacity associated with them, so it is extremely useful to speak of energy in abstract terms and to treat it as if it were arbitrarily divisible. But at root, all energy-exchanging processes involve relations between genesis-units. There is never an instant when a quantity of energy possesses an independent existence apart from the evolving relations between genesis-units.

The genesis-unit has a protomagnetic polarity that has a particular magnitude associated with it. Only a definite quantity of work (or an excess of that quantity) will lead to the ionisation of a unit. This is the fundamental basis for the fact that the photoelectric effect requires incident light of a certain frequency. In our account of the transmission of light, we claimed that the spectra of elements was generated by the formation and dissolution of electrostatic alignments between groups of atoms. Thus the frequencies contained in the spectra were determined by the size of the electrostatic relations between atoms. Neither the photoelectric effect nor the spectra of elements require *energy* to exist in elementary units; they can be explained in terms of an elementary *agent of causal influence* whose electrostatic balance is changed either integrally or not at all. And, in fact, this account of the transmission of light entails that the "energy" of an impulse of light can be arbitrarily small. We can see this more clearly if we return to the distinction between the *impartation* of a particular causal influence and the *distribution* of general causal influence.

This simple distinction is fundamental in physics. Causal activity such as the "absorption of an electron" and consequent ionisation involves a causal source in disequilibrium *imparting* its disequilibrium to a target. The effect in the target may be the restoration of equilibrium if it had been positively ionised previously. An ionised genesis-unit can also *distribute* causal influence over a number of other atoms. In this case, the causal source *remains* ionised, but the disequilibrium in this genesis-unit causes other atoms to move towards or away from it, depending on their state of electrostatic equilibrium.

The transmission of light involves an interplay between impartation *and* distribution of causal influence. Consider a mini-system of atoms that is held in alignment by an "excited" pair, A1-A2. A1-A2 is *distributing* the influence of its heightened electrostatic tension over the mini-system. When the mini-system disassociates, an impulse (or multiple impulses) is discharged into the wider system. This impulse evolves with the system until it is imparted to a target pair, T1-T2, giving rise to a new alignment of atoms around them. Now T1-T2 distributes the influence of its heightened electrostatic tension over the atoms in the newly formed group. If we look more closely at this model of light transmission, then it becomes evident that the impartation of an arbitrarily small impulse is entirely possible as A1-A2's electrostatic components are dispersed over larger and larger subsystems. Of course there will be *practical* limitations on how large a mini-system can be formed in electrostatic disequilibrium around an excited atom. The concrete conditions and logistical circumstances of luminous materials will mean that it is improbable that a single atom could distribute its electrostatic disequilibrium over a mini-system that is arbitrarily large. But there are no obvious theoretical limitations on the size of such systems. Thus there would be no corresponding theoretical limitations on the minimum size of electrostatic impulse that can be imparted between systems.

7.3 A thought experiment: twentieth century quantum mechanics

The following thought experiment is intended to throw light on certain issues that dominated quantum mechanics during the course of the twentieth century. Consider the experimental set-up illustrated in Figure 7.1. The impractical nature of this arrangement has no bearing on the validity of the conclusions that can be drawn from it.

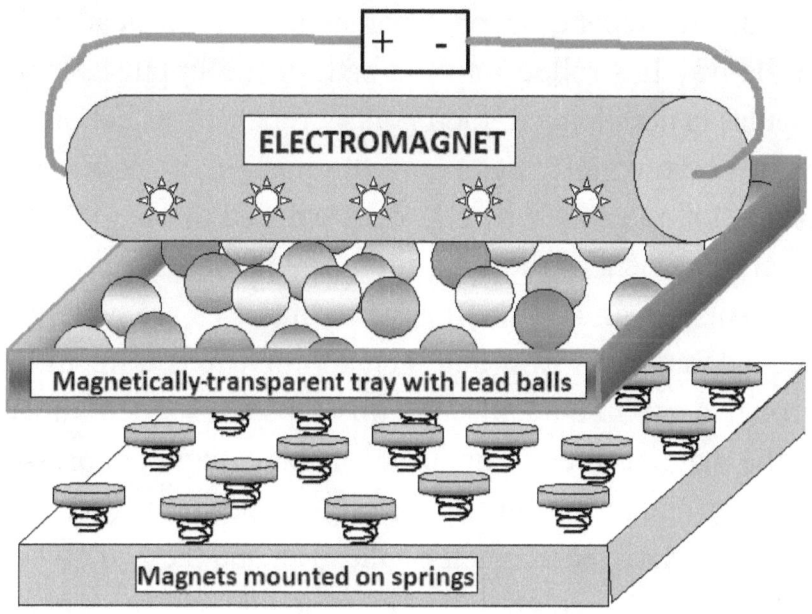

Figure 7.1 Thought experiment apparatus

An electromagnet is placed rigidly in a horizontal position above a tray made of a magnetically-transparent material. On the tray there is a large number of lead balls that can roll around freely. When the balls are set in motion (by inclining the plane of the apparatus), they hit against each other and the side of the tray, rebounding until they eventually come to rest. Just below the tray there is a horizontal grid with twenty cylindrical magnets arranged on it. The magnets are attached to weak springs, one end of which is connected to the magnet and the other to the grid. All of the magnets are oriented with the north pole facing upwards.

When the electromagnet is turned on, it produces a northerly magnetic impulse in the direction of the magnets underneath. This magnetic impulse may or may not be obstructed, depending on whether one of the lead balls on the tray has rolled into the path of the impulse. If the impulse is not blocked, then it has a repelling effect on the magnet below (north against north) pushing it down on its spring, followed by a recoil. We are asked to imagine that the arrangement of the apparatus, and the large number of lead balls, mean that the magnetic impulses are generally blocked, but every now and then an impulse is transmitted through a gap in the balls, and this causes a sudden impact on the magnet directly underneath. The magnet depresses its spring and then returns to its original position. There is a light on each of the magnets on the grid, and these are wired to their own source of electricity. The lights have switches that are activated by a magnetic impulse. Whenever such an impulse makes its way through the lead balls from the electromagnet, the light switches on and off momentarily.

The electromagnet itself also has a series of lights attached to it that are activated when electric current is passing through the electromagnet. The voltage of the current varies randomly, and the lights on the electromagnet glow brighter or dimmer as the voltage is increased or decreased. As the voltage increases, the magnetic moment of the electromagnet will also increase, meaning that any magnetic impulse that makes its way down to a magnet below will impart a greater momentum to the magnet on its spring.

Say that this apparatus is placed on a ship in mildly choppy waters with an observer who has a limited perspective on things. The observer has a simple sensory apparatus that can only apprehend the flashes of light produced by the apparatus and nothing else. He has no notion that the lights he sees at the top of the apparatus are attached to an electromagnet. He cannot perceive the tray with the rolling lead balls, and he does not perceive the magnets on the grid below. What he perceives is the series of lights at the top of the apparatus (changing continuously in luminosity) and intermittent flashes of varying brightness at some distance below.

The system is activated by turning on the electromagnet, and the observer watches the results. After some time monitoring the apparatus, the observer starts to compile tables of empirical correlations. He notices that the lights below are moving during the short interval that they are activated. It seems to him that when the series of lights on the electromagnetic are brighter, then this leads to greater movement below. Our observer does not realize what the magnets are, but he still manages to add a pointer to each one, and a dial fixed rigidly to the apparatus (see Figure 7.2). This allows him to get a value for the momentum that is imparted to the magnets on a spring. Since our observer can only perceive flashes of light, we have to assume that the momentum-measuring device is wired in such a way that a different flash of light is produced for each of the readings on the dial.

After gathering the empirical data, the observer hypothesizes that *particles* are being emitted by the series of lights above. The emission of each particle, he assumes, coincides with a light being illuminated on the electromagnet and is followed shortly by a light appearing on the grid below. Let us call these hypothetical particles "magnetons". The energy of the magnetons, as far as the observer can make out, is proportional to the brightness of the lights of the source. When any such particle impacts on the unknown material below, it causes a recoil with a momentum proportional to that of the particle, and a flash of light is emitted.

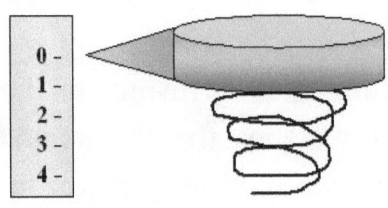

Figure 7.2 The momentum-measuring device attached to each magnet

Our observer is good at maths and it is not long before he comes up an empirical "law" correlating the energy of the causal source with the momentum of the magnetons emitted and the corresponding energy/momentum imparted to the target below. He is also able to come up with a value v for the velocity of the particles (here we have to imagine that the observer has the means of measuring the infinitesimal intervals of time between fluctuations in the brightness of the lights on the electromagnet and variations in the momentum being imparted to the magnets below). Values for the relative masses of the magnetons are also obtained. Since the velocity of the magnetons is constant, their masses are always proportional to readings 1 to 4 on the dial on the momentum-measuring device.

By simple observation over a considerable period of time, the observer is able to state that, on average, a magneton is emitted every t seconds. This, along with the other empirical information gathered, permits the observer to develop mathematical formalisms that describe various properties of a system consisting of a single particle. During time interval t, as he interprets things, a particle is in the system, travelling with velocity v, of relative mass in the range 1 to 4 on his scale. At the end of the interval t, the magneton will be found in one of the twenty positions corresponding to the magnets on springs (no lights exist in the space between the electromagnet and the grid with the springs; therefore the observer cannot say anything with certainty regarding the position of the magnetons as they pass through the system, and the only positions he can attribute with certainty are those that correspond, in his view, with the "impact" of the magneton on the light-emitting device).

Let us consider how the observer might express some of his empirical findings in mathematical terms. Mathematical expressions in terms of vector spaces can be used to represent a property of a system in terms of the vector sum of all the various possible values that the property in question can have. Take, for example, a system composed of a switch that can be in either an "on" or an "off" position. The state of the system at any given time can be expressed as:

$$\Psi = x_1 A_1 + x_2 A_2$$

Where A_1 and A_2 are orthogonal vectors of unit length representing "on" and "off" respectively. In the case where the switch is in the "on" position, then x_1 will have the value one and x_2 will have the value zero. If the state of the system is unknown and both possibilities have equal probability, then the observer's *state of knowledge* of the system can be expressed as:

$$\Psi = \tfrac{1}{2} A_1 + \tfrac{1}{2} A_2$$

In the thought experiment, the observer can express the position of *one* of the magnetons as a twenty-dimensional vector space, in which twenty mutually orthogonal vectors of unit length each represent the state of being in the position of one of the magnets on a spring.

Let us call these vectors (that each represent the state of being in one of the positions corresponding to a magnet on a spring) A_1, A_2, ..., A_{20}. Such mutually orthogonal vectors of unit length form a basis of the position state space of the system (not to be confused with the three dimensional physical space occupied by the apparatus). The position of any given magneton can be expressed as a sum of the following sort:

$$\Psi = x_1 A_1 + x_2 A_2 + \ldots + x_{20} A_{20}$$

where $x_1, x_2 \ldots x_{20}$ are numbers representing the expansion coefficients of the vector Ψ in the position state space that uses A_1, A_2, ... A_{20} as a basis. In the case where the particle is found at spring number 18 on the grid, for example, x_{18} will have the value one, and all of the other coefficients will have the value zero.

Prior to an observation being made, however, the observer cannot know the position of the magneton, and so he cannot attribute the value one to any coefficient in particular. If the empirical data that he has gathered indicates that, in any given interval t, there is an *equal* probability that a particle will be found at any of the twenty locations on the grid, then he can assign the following values to the expansion coefficients of vector Ψ:

$$\Psi = \frac{1}{20} A_1 + \frac{1}{20} A_2 + \ldots + \frac{1}{20} A_{20}$$

The *total* of this sum is of course one. The vector ψ can be written in shorthand as a function (or "wave-function" as it is usually called) of the form $\Psi (x)$, where x represents the coefficients of the basis vectors A_i for the vector representing the state of position. The wave-function $\Psi(x)$ for the state of the magneton being at position 18 has the value one when $x = 18$ and zero in all other cases.

We can summarize the above discussion as follows: The observer hypothesizes that the flashes of light he observes in the system are due to the emission of particles from the target object to the source object. Based on extensive gathering of empirical data, he calculates that in any given time interval t, there is a high probability that there is a single particle in the system. During any such interval, the mathematical formalism that expresses the position of this particle is:

$$\Psi = \frac{1}{20}A_1 + \frac{1}{20}A_2 + \ldots + \frac{1}{20}A_{20}$$

When the observer actually observes the apparatus for an interval t, however, he will invariably "find a particle" (i.e., observe a flash of light) at one of the positions on the grid. Say that this happens to be at position 18. This experimental finding can be mathematically expressed as:

$$\Psi = 0A_1 + 0A_2 + \ldots + 0A_{17} + 1A_{18} + 0A_{19} + 0A_{20}$$

All of the coefficients x_i of A_i are zero, except for x_{18} which is equal to one. In other words, the form that Ψ had prior to the observation now "collapses" to:

$$\Psi = A_{18}$$

The mathematical formalism that predicts the *possible* outcomes of an observation shrinks to the mathematical expression of a *concrete* observation, and a contraction of this sort is known in the quantum mechanical literature as the "collapse of the wave-function". The same literature refers to the particle prior to the collapse as being in a "superposition" of states (in this case, the states in question are the various possible positions the particle could occupy in the system). In the thought experiment just described, superposition and collapse are evidently features of a *formalism* constructed by an observer on the basis of empirical data, not a feature of the *reality* that generated the empirical data. We shall devote some time shortly to exploring the conditions under which an observer might be led to hold the peculiar belief that it is *reality* itself that manifests a superposition of states, and that this superposition "collapses" under the conditions necessary for observation. Before doing that, we wish to consider how the observer arrives at the view that certain properties of his system are "complementary".

The observer realises that the movement of the lights on the grid (movement produced by the recoil of the spring) below precludes the knowledge of the *exact* position of any given particle during this process of compression and recoil. Despite the limitations of his perceptual apparatus, he manages to find a way to clamp any given magnet on the grid in a fixed position, such that it cannot move on its spring when it is subjected to a magnetic impulse. The light on the magnet is still activated by the magnetic moment of the electromagnet. When the magnet flashes (indicating the presence of a "particle") the observer knows its exact position, since the clamped magnet stands in a fixed place in his reference frame. If the observer wishes to know the *momentum* of the hypothetical particle, however, then he must leave the magnet free on its spring and measure the distance moved by the magnet during the "impact" of the magneton. The mutually-exclusive nature of the experimental conditions necessary for the measurements of momentum and position, respectively, impel the observer to term such properties "complementary".

In the thought experiment, it was very clear that the manifestations of superposition, collapse, and complementarity were related to the *state of mind* of the observer, and not to properties of reality itself. There were two major issues with the hypotheses of the observer. Firstly, the observer erroneously postulated the *presence of particles* in the system, accounting for certain empirical data (flashes of lights, movements of pointers) in terms of the position and momentum of such particles. Secondly, he interpreted the absence of empirical data in certain circumstances to indicate the superposition of states, a superposition that would collapse once an event of measurement took place. Evidently, this framework of superposition and collapse becomes simply superfluous once we recognize that there are no particles in the system.

A shortcoming of this analogy is that it provides no motivation for the observer to postulate a superposition of states. Most observers would assume that the magnetons continue to have a definite position and momentum when they are not being observed, even if the mathematical formalism describes the system during these unobserved moments in terms of a superposition of states. The question that must immediately be asked is: what conditions are present in *quantum* versions of such experiments that they could have given rise to such a paradoxical interpretation as a superposition of states? What is it about measurements performed on quantum systems that led such a significant proportion of the scientific community to believe that *reality* itself could be in states of superposition that discontinuously collapsed?

The conditions that lead to such strange interpretations occur generally in quantum experiments and are relatively easy to discern. There are two primary conditions – an epistemological condition (a characteristic of the kind of hypothesis that the observer is willing to entertain), and an ontological condition (a characteristic that pertains to reality itself). Let us see how these two conditions, working together, can lead to the view that entities exist in a superposition of states in atomic systems.

Atomic experiments typically have a source object (S), a target object (T) and a mediating apparatus. S acts as the primary source of causal activity in the experiment. T is acted on by S and usually incorporates some sort of device for measuring a property (such as energy, momentum or position) of the causal influence emitted by S. Between S and T there is often a mediating apparatus, such as a two-path diaphragm.

S gives rise to effects in T (such as flashes on a phosphorescent screen). On the basis of these effects, the observer postulates that particles have been emitted from S and have impacted on T. This almost universal tendency of theorists to postulate *particles* as the bearers of causal influence in atomic experiments is a manifestation of the epistemological condition: *namely, when confronted with situations in which a causal source gives rise to effects on a target object (such as flashes on a phosphorescent screen), there is an overriding tendency to believe that some third object has traversed the space between source and target, thus bearing the causal influence from source to target.*

This condition or attitude has a guiding influence on the kind of explanations that scientists develop to account for material phenomena. When this attitude is confronted with the ontological condition (which we shall treat shortly) it can give rise to explanations that eventually lead to paradoxes. Typically, when an observer has hypothesized the existence of an intermediary that carries causal influence from source to target, he goes on to investigate the properties of these entities. He modifies the experimental set-up. Say that he now introduces a two-path apparatus between S and T, and finds that an interference pattern occurs on T. This interference pattern, which seems archetypical of a wave motion, is a manifestation of the ontological condition: *namely, the phenomenon under investigation involves a causal process that evolves in the system without the aid of intermediaries such as particles, waves or fields. The evolution of this process can be jointly influenced by multiple causal players simultaneously.*

Of course we could state the conjunction of these two conditions in simple terms as follows: *an observer may be prompted by the empirical evidence to conclude that objects exist in a superposition of states if he postulates particles as the carriers of causal influence in cases where the real causal influence involves no particles.* But it is also useful to look at the two conditions separately, particularly because processes of the kind that we are investigating have this capacity to be jointly influenced by multiple causal players.

Consider a causal source, S, that gives rise to a causal process that evolves over a period of time. This causal process is of the sort that requires no intermediaries. A characteristic of such processes is that they are capable of being modified by two or more spatially separated causal players simultaneously. This is not at all as unusual as it might sound. Figure 7.3 presents an analogous situation. Here, two balloons are connected to each other by means of two tubes that can be opened or closed independently. If A has been inflated to a higher state of pressure and then both tubes are opened, a process begins by which A and B attain equal levels of pressure. The process can be said to evolve over both paths simultaneously. In this analogy, of course, material intermediaries (molecules of air) *are* involved in the process by which equilibrium is attained. Our contention in this book has been that the transmission of light (and other types of causal activity) involve systems tending towards equilibrium by evolutions that can be simultaneously influenced by many causal operators *without* the aid of intermediaries.

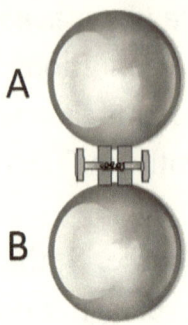

Figure 7.3 Causal activity that evolves over two equal paths simultaneously. *Balloons A and B are connected by two tubes that can be individually opened or closed. When both tubes are open, the process by which A and B attain equilibrium in pressure evolves concurrently over both paths.*

7.4 The collapse of the wave-function

We intend to use this thought experiment as our departure point for a discussion of the contentious issues of quantum mechanics during the twentieth century: complementarity, superposition and the collapse of the wave function. A pillar of the Copenhagen Interpretation of quantum mechanics was the belief that it is impossible to attribute "complementary properties" (such as position and momentum) to quantum mechanical objects simultaneously. The nature of this impossibility was expressed in various ways over different periods and among different writers. At times it seemed like a simple impossibility in the capacity of scientific instruments to measure certain properties simultaneously. In that case, the inability to *measure* both properties simultaneously should not imply that both properties did not *exist* simultaneously. Bohr explicitly invoked Planck's postulate as a foundational element for his claim that complementary properties could not be measured simultaneously. The existence of the elementary quantum of action meant that the influence of the interaction between the measuring instrument and the object could never be arbitrarily reduced; the complementary property would consequently always be disturbed to a greater or lesser degree. On the face of it, this seems to be a claim about the simple *disturbance* of the object by our measuring instruments. However, in its most developed form, the doctrine of complementarity asserted that *it made no sense* to attribute properties to objects in the absence of the empirical conditions that were necessary to measure those properties. Our knowledge is mediated to us by means of scientific instruments that measure certain observables. In the absence of the appropriate instrument for quantifying a particular observable, it was not permissible to assert that reality exhibited that observable. Complementary properties required measurement conditions that were mutually exclusive. Therefore, a situation could never arise in which it was possible to assert unambiguously that an object possessed both properties simultaneously.

Alongside the belief in complementary properties was the claim that there were complementary *models* for visualising quantum systems. The empirical data that generated these models came from sets of mutually-exclusive observational conditions. Neither the wave picture on its own, nor the particle picture on its own, could account for the full range of empirical evidence. Quantum objects exhibited "wave-particle duality", generating effects typical of particles under certain observational conditions and effects typical of waves under incompatible conditions.

The realist philosopher would be inclined to think that objects *possess* both properties simultaneously, regardless of the difficulties of *measuring* both properties simultaneously. Thus, as far as a realist is concerned, complementarity is fundamentally a practical limitation on our ability to *know*, because the experimental set-ups required to gain knowledge of both properties are mutually exclusive. He can choose to measure one or the other, but not both (unless he could come up with ingenious methods of measurement that did not destroy the complementary property). Failure to measure a property, on this view, should have no *ontological* implications for the existence or non-existence of that property.

The adherents to the Copenhagen Interpretation of quantum mechanics (and certain other interpretations), by contrast, believed that the limitations implicit in the notion of complementarity pertained to *reality* itself, because the properties of the objects under examination were *inextricably linked to the set-up of the system* in which they were moving. Atomic objects couldn't possess both properties simultaneously because the conditions that were essential for the presence of both properties (the way that the system was set up) were mutually exclusive. An ingenious method for measuring a property without disturbing the complementary one would not be sufficient for the *existence* of the complementary property. The existence of that property can only be made sense of when the conditions for its measurement are present.

The thought experiment that we have presented in this chapter gives us a way of thinking about complementarity that is completely distinct from the two traditional approaches. The traditional approaches, in fact, share *a common set of assumptions* about what is happening in quantum systems. As far as they are concerned, the causal influence in the system is borne by an *entity*, such as the electron. It is the way that the entity possesses its *properties* that is the source of the dispute between the realist and anti-realist interpretations. Are these properties possessed autonomously, or are they dependent on the set-up of the system in which they are moving?

The thought-experiment challenges the assumption shared by realist and anti-realist alike that there are *entities*, like magnetons, in the system. If it is the case (as it was the case in the thought experiment) that there are *no* entities in the system, then a third form of "complementarity" arises. In the first form of complementarity (the realist version), entities are believed to be present in the system. These entities are considered to possess autonomous complementary properties simultaneously, but they cannot be *known* simultaneously. In the second form of complementarity (enshrined in certain interpretations of quantum mechanics), entities are believed to be present in the system, but these entities are not considered to possess their properties independently of the set-up of the system in which they are moving/measured. In the third form of complementarity, there are *no* entities in the system at all. Some sort of causal process is taking place, and *the way that the process evolves is dependent on the set-up of the system.* Some set-ups of the system are mutually exclusive of each other, thus giving rise to effects in the measuring instruments that are mutually exclusive, and hence "complementary". This very natural description of complementarity gives rise to no paradoxes or difficulties of interpretation.

In the thought experiment, the observer erroneously postulated the presence of magnetons in the system, and he used effects procured from two mutually-exclusive set-ups of the system to quantify "complementary" properties of these fictional particles. The quintessentially realist line that the magnetons *must* possess their properties autonomously is completely inapplicable to what is really happening in this system, because the entities don't exist. By the same token, the quintessentially quantum mechanical line that the magnetons possess their properties *relative* to the system is also inapplicable to what is going on here, because the entities don't exist.

This third form of complementarity has none of the anti-realist connotations associated with the second form. However, as it stands, the thought experiment is not able to replicate in a graphic way the *ontological* condition that occurs universally in quantum experiments and that was responsible for the historical dominance of the second form of complementarity for much of the twentieth century. In quantum experiments, it is not just the erroneous postulation of *particles* that is responsible for the bizarre results, but the ontological fact that in certain circumstances causal processes *can be jointly modified by multiple causal players simultaneously.*

The elements of scepticism and positivism made complementarity a difficult doctrine to confront philosophically. Moreover, the results of the Bell test seemed to support the worldview which asserted that objects *do not* possess their properties in the absence of the appropriate conditions for observing/measuring those properties. This profound question retains a valid place in the philosophy of science. What we wish to focus on right now is a related issue that is still very much alive in physics even if complementarity in its old form is no longer a widely held doctrine: the phenomenon of the "collapse of the wave-function". This is historically linked to the development of the notion of complementarity and to the belief in wave-particle duality.

Some currently-held interpretations of quantum mechanics deny that collapse takes place at all, whilst other interpretations accept collapse in a more nuanced way than was historically the case, denying that collapse occurs in response to a special category of causal event (the act of measurement). Our approach to this matter can be best illustrated by outlining it in contrast to the old Copenhagen account. Even if few people still accept this interpretation in its classic form, the salient elements of that interpretation have been decisive in shaping the current debate and the new interpretations that have since been offered.

What is meant by the "collapse of the wave-function"? Take a hypothetical particle such as the electron and a property such as x-spin. At times when an electron is in a state of transition through a measuring apparatus, the quantum mechanical description of the state of the particle (the wave-function) does not attribute either x-up or x-down to the electron. Instead, the formalism suggests that the particle exists in some kind of indeterminate state with respect to this property. Some approaches describe this indeterminate state as a *superposition* of spin-up and spin-down, since the wave-function could be interpreted as representing the particle with both properties simultaneously. However, whenever a *measurement* of spin-x is performed, the electron will always be found to be in *either* a spin-up or spin-down state. This reduction from a superposition of values to a single value is referred to as a collapse of the wave-function.

At first sight, this collapse might appear to be simply concerned with the transition in our state of knowledge that is prompted by an act of measurement. Before the measurement, we didn't know what state the electron was in, so its state could be represented as a conjunction of probabilities that it was in either x-up or x-down. After the act of measurement, this description of the system shrinks to the determinate value of spin-x obtained by the measuring device. Empirical evidence, however, such as that arising from the Bell tests, has given physicists reason to believe that quantum objects exist in genuine indeterminate states when they are not in interaction with a macro device such as a measuring instrument.

Much of the debate over the course of the decades has concerned the nature of the act of measurement. How can an act of observation prompt an object to assume determinate values of a property that it did not possess unambiguously prior to the measurement? The current discussion has moved away from considering the measurement *act* in itself to have a special significance. Other factors that invariably happen to be present when measurements take place are suggested as prompting the collapse of the wave-function. All of these approaches are interesting in themselves. However, if the causal process under investigation does not consist in particles, then these sophisticated accounts are ultimately directed at resolving a problem that does not exist.

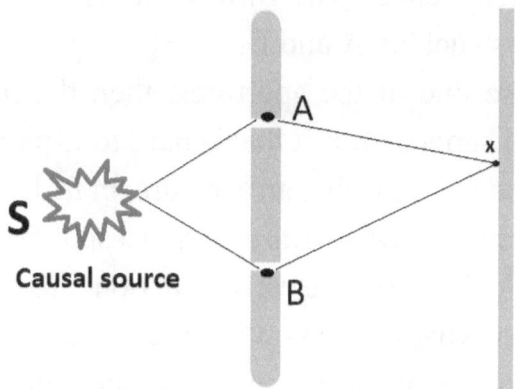

Figure 7.4 A causal process is jointly modified by A and B

Consider again our account of causal processes of this kind (see Figure 7.4). A source object, S, gives rise to a causal process, such as the emission of an impulse of light. From the spatial perspective, this evolving impulse is confronted with a screen that has two apertures in it. The spatial arrangement gives us only a partial viewpoint on how things are at the ontological level. On the level of reality as it is in itself, there *is* a separation of some kind between the source and the screen, but our immediate impression of spatial separation as having an overriding causally insulating significance is to be rejected. The process evolves in the system and comes under the influence of atoms (or groups of atoms) at A and B in each aperture. These *jointly modify* the evolution of the process. The process has a manifestation interval and an associated "length" corresponding to the extent (as measured from the spatial perspective) to which the process will reverberate in the system during that time interval. As discussed in detail in Chapter Three, a process like this will give rise to a classic "interference" pattern when it is modified by two causal players such as A and B.

If we close one of the apertures, then the interference pattern will disappear, a fact that is hard to explain from the particle perspective. Closing an aperture, in fact, has the *very same result as an act of measurement* in a spin experiment. It prompts the "collapse of the wave-function". To see how this is so, consider what happens when we leave both apertures open and turn down the output of causal source S so that it produces only a single impulse at a time. The impulse produces an impact on the screen typical of a particle. But when the impacts are allowed to build up over time, their arrangement forms an interference pattern. Certain areas of the screen are being selected or rejected in a joint fashion by the two apertures together. As soon as an aperture is closed, the interference pattern is interrupted. The quantum mechanical formalism (i.e. the wave-function) that describes the passage of a single particle through the apparatus will use terms that represent both apertures, as if the particle were passing through, or at least being influenced by, both apertures simultaneously. But when an aperture is closed, the mathematical formalism "collapses" to a description of a particle following a simple trajectory through the system.

All of this is very mysterious, and even disconcerting, as long as we believe that there are particles passing through the system. How can a particle be influenced by the state of *both* apertures when it can only pass through one? However, when we describe the situation in terms of a causal process, then it becomes much more natural to think of it as being modified by two players simultaneously. The quantum mechanical formalism describes systems such as that depicted in Figure 7.4 using a wave-function that has terms for *both* apertures (or that at least allows that the particle will be found in positions corresponding to the fact of both apertures being open). The progress of the "particle" through the system is thus mathematically represented as being influenced by the state of the two apertures in unison. And the formalism is right in that respect, even if its terms (such as the position of a particle) does not refer to anything in reality. The process *is* influenced by the state of both apertures, but the process is not a particle.

7.5 The physical basis of the wave-function

One of the principal issues that beset quantum mechanics from the beginning was the character of the wave-function. Does it merely represent our state of knowledge of the system, or is it in some sense "real"? Empirical data such as that arising from Stern-Gerlach-type experiments and from the Bell tests seemed to indicate that the wave-function couldn't be dismissed as a purely mathematical construct that represented our state of ignorance before a measurement was taken: *reality itself* seemed to embody something of the superposition expressed in the formalism.

When we consider the wave-function in the light of the process approach, it becomes clear that it contains certain elements that derive from our state of ignorance of the state of the system. However, it also enshrines features that correspond to real dynamics at play in the evolving causal process. To make this clear, let us return to the thought experiment. Here, there were no particles in the system, but the observer postulated them in response to the empirical data he had gathered. He then came up with a formalism that represented the probability that a given particle would be found in a particular position. To keep the formalism simple, the thought experiment only permitted the particle to be found in one of twenty different locations. The state of being in one of the positions was designated by mutually orthogonal vectors of unit length: A_1, A_2, ... A_{20}. The position of any given magneton could be expressed as a sum of the following sort: $\Psi = x_1 A_1 + x_2 A_2 + ... + x_{20} A_{20}$, where $x_1, x_2 ... x_{20}$ were numbers representing the expansion coefficients of the vector Ψ in the position state space that used A_1, A_2, ... A_{20} as a basis. In the case where the particle was found at spring number 18 on the grid, x_{18} had the value one, and all of the other coefficients had value zero.

Before making an observation, the observer did not know the position of the particle, so he couldn't assign the value one to any coefficient in particular. Statistical data that he had gathered over a long period of time, however, indicated that, in any given interval t, there was an *equal* probability that a particle would be found at any of the twenty positions on the grid. Thus he could assign the following values to the expansion coefficients of vector Ψ: $\Psi = \frac{1}{20}A_1 + \frac{1}{20}A_2 + \ldots + \frac{1}{20}A_{20}$. The *total* of this sum is of course one. When an observation is performed that registers the particle at position 18, the form that Ψ had prior to the observation now "collapses" to: $\Psi = A_{18}$

It is clear that there are many aspects of this wave-function that have no basis in reality at all. While the process is still ongoing, the wave-function describes the system in terms of probabilities that a particle is in a certain position, but there is no particle in the system at all. Thus, interpretations of the wave-function that say things like, "The particle is in a superposition of twenty different states" make the error of attributing too much reality to the expressions used in the mathematical formalism. The formalism may well save the phenomena, but it is inappropriate to look to its terms to try to further our understanding of what is actually happening in the world. In a similar way, Ptolemy's epicycles saved the phenomena but did not tell us anything about the causal mechanism at work in the solar system.

Such a formalism that expresses the state of the system in terms of possible locations of non-existent particles is, of course, a classic example of an empirically adequate theory whose objects fail to refer to authentic structures in reality. Yet there are aspects of the wave-function that capture something of the real dynamics of the system, and it is these aspects that led many in the scientific community to treat the wave-function as if it had some kind of objective reality. Consider a formalism that expresses the position of a particle in a two-path apparatus in the case where both apertures are open. A quantum mechanical formalism of this sort will have terms representing all possible positions of the particle in the system given that both apertures are open. A formalism for the state of the system where one aperture is closed would only have terms related to the open aperture. In this way, the structure of the formalism is *actually determined by the structure of the apparatus itself*. True, the formalism represents this structure in a misleading and inaccurate way. It is couched in terms of properties of non-existent objects. Yet the formalism manages to enshrine the truth that the progress of the causal influence will be affected by the state of both apertures simultaneously.

It is this fact that has ultimately led to the situation where the mathematical formalism is sometimes given a status almost akin to that which would once have been given to the physical object itself. Physicists of a realist persuasion were reluctant to accept that quantum objects existed in indeterminate states when measurements were not taking place. They were inclined to think that the formalism that expressed the state of objects in such terms was merely an empirically convenient construction. But when tests were performed that seemed to show that the object failed to possess the property apart from the experimental conditions necessary to measure that property, then it seemed that the *formalism itself* somehow reflected reality. It appeared to be much more than a mathematical tool for predicting or describing phenomena. The indeterminacy of states expressed by the formalism appeared to be a feature of the quantum object.

It is illegitimate, however, to infer from an *element* of verisimilitude in the formalism to the verisimilitude of the formalism in a much more general sense. The wave-function for a two-path apparatus genuinely reflects reality insofar as it takes into account the *joint influence* of the apertures, but its conjunction of probabilities for the location of the particle have no corollary in real things. This point is particularly relevant for the challenge that quantum mechanics presents for physicists who are committed to a realist view of reality. The realist worldview holds that objects have determinate properties that are possessed autonomously by an object independently of whether those properties are being observed or measured. The capacity of the wave-function to account for the evidence has tended to mitigate against approaches that seek to understand the evolution of the system in terms of particles possessing determinate properties. Interpretations such as that offered by David Bohm try to account for the evidence whilst maintaining autonomous properties like that of position. But many physicists working in practical situations probably have little use for such notions as determinate properties. In one sense, these physicists are already treating the causal event as a process rather than a particle. The state description they work with is in terms of an evolution of causal influence in the system that is affected by all relevant players (such as two open apertures).

The account that we have developed may seem to resonate with the attitude of such physicists who have little concern for the metaphysical reality of the objects that they work with. However, it is likely that there is all the difference in the world between our attitude and theirs. Such physicists may be sceptical or non-committal about the positions and momenta of electrons, for example, leading them to treat the electron as if it were a causal process that evolves in the system (as opposed to a particle that traverses the system with a definite trajectory). But we are neither sceptical nor non-committal about the positions and momenta of electrons. Our approach explicitly argues that electrons do not exist at all, and it explicitly asserts that the *real causal players* in a given system (such as the light source, the two-aperture diaphragm, and the screen) *all* have determinate properties of position and momentum. Thus there is a significant difference about our approach from a realist point of view. Our entire account is in terms of material objects with autonomous properties that exert deterministic influences on each other.

7.6 The completeness of quantum mechanics and indeterminacy

Indeterminacy and indeterminism are distinct terms. One refers to the *state* of a particular object or system (either in itself or with respect to the knowledge of an observer); the other refers to the way that *causal processes unfold*. But the terms have been closely related in the way that the philosophy of quantum mechanics developed from the 1920s onwards. Indeterminacy of properties (or states) refers to the question of whether a quantum object has a determinate property at a given time. If we hold that objects can exist in a condition of indeterminacy with respect to a property like position, for example, then we believe that it makes sense to think of an object as having no definite position at a particular time. Indeterminism, by contrast, refers to the belief that causal processes can evolve in non-deterministic ways. An example would be the claim that an event can unfold in a way that is fundamentally probabilistic.

Collapse interpretations of quantum mechanics relate causal indeterminism to indeterminacy of states with their claim that quantum objects undergo transitions from indeterminate states to determinate states (i.e., at the moment of collapse), and that the physical laws governing the transition are fundamentally non-deterministic or probabilistic. The wave-function expresses the state of the system in terms of probability of states. Properties such as position are expressed in terms of conjunctions of probabilities that the particle can be found in various locations. When the object encounters a measurement device, the wave-function collapses and the object takes on a determinate state. The state assumed by the object during the event of measurement appeared to have a purely probabilistic relation to the state it existed in previously. Once again, much debate has centred on the question of whether this probability expressed our lack of knowledge about the actual state of the system, or whether there was an inherent indeterminism about the manner in which the system evolves. And once again, certain empirical tests seemed to confirm that the indeterminacy of states was a feature of quantum systems prior to measurement; hence the indeterminism of causal evolutions was a feature of reality.

The reader will be aware by now that this question does not even arise from the point of view taken by the process approach. The issue, indeed, refers to the probability that a *fictional* object possesses a particular property. The incapacity on our part to attribute determinate properties to particles that don't exist has no authentic bearing on the real indeterminacy or otherwise of physical reality. Once we dispose of the non-existent particles in the system and view the event in terms of an evolving causal process, then we have absolutely no reason for asserting that real objects exhibit a fundamental indeterminacy of states. On the contrary.

There is a sense, however, in which the probabilistic character of the formalism has a definite basis in reality. But this has no negative consequences at all for determinism, as we shall now see. In our treatment of joint influence patterns (usually described as "interference" patterns), we considered the manner in which a causal impulse reverberates in step with the overall evolution of the system. Each impulse has a manifestation interval (which corresponds to $1/\upsilon$, where υ is the "frequency" of the impulse). At the end of each interval, the evolving causal process has the potential to exert its influence on a target. But this can only occur if appropriate target material is present at that precise moment in the location corresponding to the distance the process has reverberated from its source. On the basis of this model of causal influence, we can speak of the probability that the process will have an *effect* in a particular location at a particular moment in time.

This contrasts with the traditional formulation of the wave-function which dealt with probabilities that a *particle* had a particular position at a particular time.

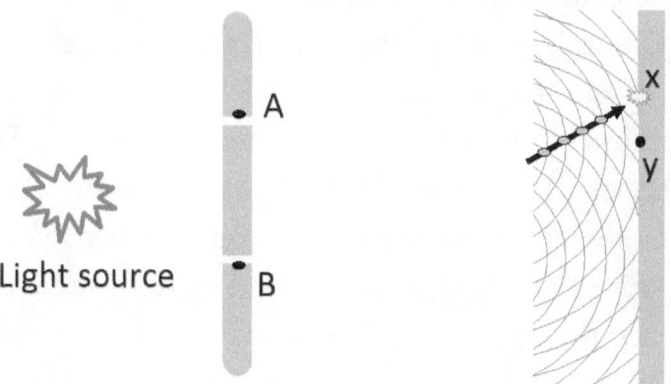

Figure 7.5 Joint influence pattern showing locations of effects, if suitable target matter were present. Before the process eventually has its effect at point x, it evolves through the system and potentially could have effects at the points shown if appropriate material is present. These points do not show us the trajectory of a particle, but the locations at which the process would have manifested itself if it had encountered an object at that position.

Consider Figure 7.5 which depicts a joint-influence situation in terms of the classic wave model. The traditional wave explanation of interference considered the circular wave fronts to represent *physical waves* passing through space. Some species of the quantum mechanical approach tended to see these wave fronts as representing *probabilities* that the particle would be found on one of the points. The process account, by contrast, would describe these wave fronts as representing points in space where the causal influence has the *potential* to give rise to an effect (i.e., a causal change in suitable matter) but this would require appropriate target material to be present there. And of course, even if target material is present on one of those points along the wave front, the process need not necessarily come to completion at that point.

The exact point at which a given causal process will come to completion may depend on many factors that are impossible to know in any complete sense. Take an impulse that is emitted by the light source in Figure 7.5. As it evolves, it comes under the joint influence of atoms or groups of atoms at A and at B. The relative modifying influence prompted by either A or B is simply impossible for a human observer to calculate, based on our current level of instrumentation. These joint influences may prompt the impulse to be deflected upwards or downwards to varying degrees. But if we permit an enormous number of impulses to pass through the apparatus, then we can give an accurate *statistical* description of the outcome of causal processes over an interval of time. We can say that there is a certain percentage probability that any given process will give rise to an effect in the central area of the screen, for example. Descriptions of processes in terms of such probabilities does not entail that any individual process has a fundamental probabilistic character, no more than the use of statistics in thermodynamics to describe the random collisions of molecules entailed that the molecules were not governed by deterministic laws. Science routinely uses probabilities to derive empirical laws in cases where it is impractical to formulate the real deterministic descriptions of a particular phenomenon.

The description of atomic systems in terms of probabilities is perfectly understandable and has no direct philosophical consequences for the question of whether reality itself is governed by deterministic or non-deterministic laws. But we can see why almost the entire scientific community has been led to conclude that indeterminacy prevails at the quantum level. Since the time of Born, the wave-function has been widely interpreted as describing the system in terms of probabilities. Now, if these probabilities were of the kind described in the previous paragraph, then they would have no consequences for determinism; i.e., in that case they would simply derive from the fact that the human observer does not have access to the deterministic factors that govern the evolution of a system (and as a result he is constrained to describe the evolution in terms of probabilities). But the accumulation of experimental evidence since the 1920s seemed to show conclusively that there was more to the wave-function than a simple statistical tool for predicting the final state of a particle that had determinate properties. As we have seen, variations of the Stern-Gerlach experiment indicated that particles did not follow simple trajectories, nor possess intrinsic values of properties such as spin. The superpositions expressed by the formalism for properties like position seemed to correspond to the state of the particles in reality.

The first experimental violations of the Bell inequality in the 1970s and 80s were a further nail in the coffin of determinacy because they helped to corroborate the doctrine that quantum mechanics was "complete". This doctrine stated that the wave-function of any given system is complete insofar as it specifies all of the physical properties of a system. Now the wave-function does not tell us the position, for example, of a particular particle in the system. If the wave-function is complete, this entails that the particle *does not have* a position. The experimental violations of Bell's inequality seemed to demonstrate conclusively that no description of the system in terms of additional parameters (such as the positions of particles) could account for the empirical evidence unless non-local action was possible. These tests, in other words, supported the view that the indeterminacy of states expressed by the wave-function was a fact of nature. And if indeterminacy of *states* is a fact of nature, and if we accept that quantum systems collapse into *determinate* states during measurements, then it follows that causal *processes* evolve in non-deterministic ways.

All of this controversy, and the grave consequences it has for the worldview that reality evolves in an ordered way, seems like so much smoke in a bottle when we eliminate the particles and view the situation from the process standpoint. The wave-function *has* a measure of completeness in the sense that no additional parameters, such as the position of non-existent particles, are necessary to tell us how the process will evolve in the system. In the case of a two-path experiment such as that depicted in Figure 7.2, the evolving process can be statistically described in an adequate manner simply by formulating an expression that takes into account the characteristics of the causal output of the source (in terms of "frequency" and intensity for example) and the structural features of the apparatus (the location and relative configuration of source, apertures and screen). No other parameters can improve the predictive/descriptive capacity of the evolving process because no other relevant parameters are available to us. The trajectories of particles that are not present in the system will certainly not constitute an improvement of the wave-function as we have it.

The wave-function has this curious character of being empirically adequate yet entirely misleading about what is really happening in the system. *It is empirically accurate because it is faithful to the structural aspects of the system.* When physicists formulate the wave-function for a particular experimental set-up, they do so by paying attention to the potency of the source and the overall configuration of the parts of the apparatus. The wave-function is not expressed in terms of determinate properties of the alleged particle that passes through the system, but probable properties of the particle, given that the *system* has such and such structural properties. And it is no surprise that such a formalism is empirically accurate because the way that a process evolves depends predominantly on those same structural aspects of the system. The wave-function thus makes statistical predictions that correspond to the statistical outcome of very many instances of evolving processes in the system.

The empirical adequacy of the wave-function is indisputable, but its mode of expression is enormously deceptive, cast as it is in terms of conjunctions of indeterminate states. The Bell test experiments confirm that this formalism cannot be bettered by a formalism that contains terms for determinate states (unless non-locality is the case). The ever-increasing rigour with which Bell tests are being carried out gives us no reason for doubting this conclusion, but the significance of the conclusion is transformed from the process standpoint. After all, these tests make the unsurprising revelation that a formalism describing the system in terms of the indeterminacy of position of a non-existent particle cannot be bettered by a formalism that includes the position of a non-existent particle. Of course, the Bell tests also underlined the non-local features of quantum systems, but not even this was novel. Bohr had recognized and accepted this non-locality in a quite prophetic way many years previously. We will consider the non-local features of these tests in the next section.

When a measurement is carried out, collapse occurs and the particle is found to have a determinate value of a particular property. This is where determinism is believed to break down and a fundamental roll of the dice is supposed to enter into the causal evolution of systems. Again, this conclusion is based on the deceptive character of the mode of expression of the wave-function. The formalism describes the system prior to the measurement in probabilistic terms, as a conjunction of mutually-exclusive states. As we have seen, the empirical tests seem to demonstrate that this expression has some measure of "completeness". There is simply nothing more to be added regarding the hypothetical particle, such as terms designating determinate properties such as its position. Then the measurement occurs and determinate properties are observed (such as the point of "impact" of the particle on the instrument). A discontinuous transition from indeterminacy of properties to determinate properties would constitute a breakdown in deterministic causality.

The misperception is cleared up once we recognize the real character of this transition. Prior to the measurement, a process is evolving in the system. The process evolves under the joint-influence of the structural aspects of the system, aspects that are indeed expressed adequately in the terms of the formalism, as far as the empirical upshot of these aspects is concerned. When the arrangement of the experimental apparatus creates the conditions for a measurement to happen, the process is coerced into coming to completion: a particular change is induced in a particular target object (such as a flash emitted from an atom on the phosphorescent screen). The target *object* will certainly have a determinate position, and the change induced in the object will have particular characteristics. Thus the transition that is referred to as "the collapse of the wave-function" is *not* a case of a *particle* with an indeterminate position suddenly and discontinuously taking on a particular position. It is a transition from a process that is evolving in response to *all* the causal features of a system (the state of both apertures, the relative configuration of the parts of the apparatus, etc.,) to the culmination of that process in a particular target. Thus the transition is simply that from a process which has multiple causal players involved (it is spread out in a spatiotemporal sense, and indeterminate from our point of view because we can't see its evolution) to the historical event of the process coming to completion in an atom on the screen (which, of course, has a position).

It is paradoxical to think of a point-like particle existing in a state where it manifests some sort of combination of multiple positions at the same time. But it is entirely commonsensical to think of a process as potentially giving rise to an effect in multiple positions, *if* there happens to be target material in those positions at the right time. This multiplicity of *target* positions would constitute a better interpretation of the wave-function than the view that it represents a *particle* in a multiplicity of positions. Once we apply this simple change of interpretation to the wave-function, all notions of a breakdown in causality disappear. The probabilistic transition from the state of having an indeterminate position to the state of having a definite position represents a breakdown in causality. But the transition from the probability that a process will have its effect in one of multiple positions to the fact of having its effect in a single position has no consequences for causality at all.

The indeterminism that is a fundamental part of certain interpretations of quantum mechanics has no genuine empirical foundation. It derives from the adherence to an unsubstantiated assumption with respect to the empirical success of the quantum mechanical formalism. Tests indicated that the wave-function could cater for the data in a way that an account based on determinate properties of particles could not. The unfounded assumption was the supposition that *the particle existed. If* we could be sure that the particle existed, and *if* the wave-function in terms of indeterminate states of the particle could cater for the evidence in a way that a determinate description could not, maybe *then* we could move to the conclusion that the particle genuinely exists in indeterminate states. However, if the particle does *not* exist, then there may well be *other* reasons why an account in terms of indeterminate states of a particle manages to save the empirical data. And, in fact, there are excellent reasons why a formalism in terms of indeterminate states might manage to account for the evidence. The formalism is not pulled out of a hat. It is formulated for a particular laboratory situation in terms of the causal features of the situation, i.e., those very features that determine how a *causal process* will evolve in the system (such as the potency of the causal source and the spatial configuration of the parts of the apparatus). These features determine the structure and content of the formalism. If the formalism is used to represent the position of a particle as it moves through the system, then there will also be a term for position. This might take the shape of an expression that represents an infinite conjunction of positions within the apparatus, as if the particle had an indeterminate position. Now the particular empirical success enjoyed by the formalism *does* seem to require that we maintain the indeterminate character of this term, but, as we have seen, we can reinterpret the meaning of this term to refer to the potential position of the *target atom* of the process.

In this way we account for the particular success of the formalism, *and* the fact that the Bell test indicates that this success depends on the indeterminate nature of the terms used in the formalism. The success of the formalism derives from the fact that it adequately represents the evolution of the particular causal process we are considering. It takes into account the relevant causal/structural features that determine the progress of the influence. And (once we reinterpret the significance of the term usually thought to represent the position of the particle) it contains an expression for the probability that the process will have its effect in a certain location. In the case of spin experiments, the term representing a particular spin of the "electron" is reinterpreted as an expression for the probability that the causal process under investigation will induce an electromagnetic alignment of a particular sort in a *target* atom. Thus we eliminate the claim that particles in transit through the system are in indeterminate or purely probabilistic states. We replace this notion with the claim that a *process* is in evolution through the system, and that this process has a certain probability of inducing a certain state in a target atom. The first claim leads to indeterminism. The second claim is compatible with determinism because it does not envisage probability as being inherent to the process itself. Rather, the process is deterministically influenced by many factors concurrently (the electromagnetic alignment of the ionised atom in the source material, the position of the magnetic deflectors, the presence/absence of barriers or reflectors within the apparatus). The reason that the *final outcome* of the causal process is expressed by the formalism in probabilistic terms is because we simply do not have access to the factors that determine the outcome. The best we can manage is a probabilistic, statistical expression based on the general structural features of the process.

Ironically, the empirical success of quantum mechanics reveals to us a level of order and regularity in nature that is difficult to explain away as a statistical accident on the macro-level. The notion that reality has a fundamental indeterminism involved the illegitimate belief that the wave-function corresponded to reality in way that it does not. Physicists have had their reasons for believing in the "completeness" of the wave-function – for accepting that the indeterminacy of the wave-function corresponded to an indeterminacy in reality. The key to resolving the misperception involves clarifying the way in which the wave-function authentically reflects reality, whilst reinterpreting the probabilistic state of the alleged particle as it flies through the machine in terms of the probability that the process will have a certain result in the target. Once this clarification has been made, we have no grounds for asserting that our knowledge of the quantum world is doomed to being probabilistic or statistical at best. Our "knowledge" of the properties of fictional particles may well be doomed to statistics at best. This king is dead but long live the king for it has absolutely no repercussions for the possibility of improving our understanding of the autonomous and determinate properties of the processes that evolve in our systems.

7.7 Non-Locality

The Bell tests and their fall-out have stimulated controversy for two interrelated reasons. The first is the issue we have just been discussing: the tests indicate that indeterminate states of quantum objects are a feature of reality (as opposed to being due to our ignorance of those states at times when they are not being measured). Indeterminacy was already a feature of the mathematical formalism, but, in the eyes of many people, the Bell test seemed to settle for once and for all the old issue of whether the indeterminacy was ontological or due to ignorance of real states on our part. Secondly, the tests point to a fundamental element of non-locality in the evolution of quantum systems. Again, this non-locality was already implicit in quantum mechanics and empirically verifiable by tests of the Stern-Gerlach type, among others. The Bell test, however, was a much more dramatic and convincing demonstration of non-locality.

Once we accept the process account, the first controversy disappears completely. The second controversy, however, is in need of some elucidation. In what sense do quantum systems defy the constraints of locality? Figure 7.6 shows a schematic representation of a Bell test. Pairs of "particles" (electrons or photons, for example) are made to interact in the source so that they fly apart in opposite directions. The nature of the interaction is such that the particles (if an appropriate measurement is performed) will exhibit opposite values of properties like spin or polarisation. In the case of the electron, when the particle on the left exhibits "up" for x-spin, then the detector on the right will exhibit "down" for the same type of spin. The apparatus incorporates separate mechanisms which randomly change the alignment of the magnetic deflectors at A and at B so that they measure x, y and z spin in a completely haphazard fashion.

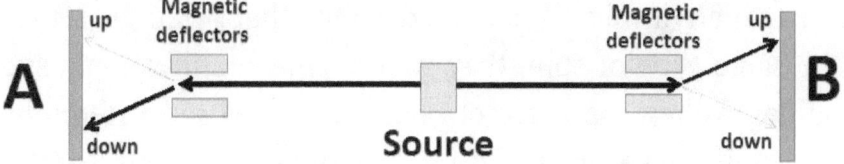

Figure 7.6 The Bell test. *Pairs of "particles" are prepared in the source and they fly apart towards the detectors at A and B. These have equal and opposite values of properties like momentum and spin. The alignment of the deflectors can be changed so that they can measure spin in the x, y or z direction. Statistical analysis of large quantities of empirical data indicates that a measurement of a particular type of spin at A has a non-local influence on the result obtained at B.*

In cases where the deflectors at A and at B happen to measure the *same* type of spin for a particular pair of electrons, it *always* turns out that each electron manifests the opposite value of spin to the other electron in the pair. This is expected since the pair was produced by a process that gives rise to "particles" with opposite values of spin. In cases where the detectors measure different kinds of spin, the values that are produced take on a random character: an "up" value for a measurement of x-spin at A could be accompanied by either a "down" or an "up" value for measurements of y-spin or z-spin at B. This is also expected because we have no reason for believing that "complementary" properties of this kind should show any correlation with each other.

Right away we can see how these two features of *individual* spin measurements ought to lead to noticeable patterns in the *overall* empirical data when the experiment is run many times. When the detectors happen to be set for the same type of spin, they will *always* manifest opposite values. When the detectors happens to be set differently, they will have an *equal* chance of manifesting similar values or opposite values. Therefore, after a large run of tests, there should be conspicuously more instances where both detectors manifest opposite values.

We can put a figure on this disparity very easily. There are nine different combinations of possible settings for the detectors: AxBx, AxBy, AxBz, AyBx, AyBy, AyBz, AzBx, AzBy, AzBz (where AxBx, for example, signifies that the detector at A is set to measure x-spin and that at B is set to measure x-spin). In one third of these possibilities, the detectors are set for the same kind of spin, so they are guaranteed to produce opposite values (by virtue of the process by which each pair is produced). The remaining two thirds of possibilities have an equal chance of producing values that are similar or different (by virtue of the lack of correlations between complementary properties of spin). Therefore one half of these cases (or one third of the entire run of tests) will produce opposite up/down values. The net result of a large number of tests should thus show a pattern where *two thirds* of pairs will produce values that differ from each other and *one third* will produce similar values (either up/up or down/down).

The assumptions upon which we based the above calculation seem sound. It is an empirical fact that a pair of this kind will manifest opposite values for the same measurement of spin. And in cases of pairs where different types of spin are being measured, we have good reason for thinking that the respective measurements by two spatially distant and disconnected detectors should produce values that have no relation to each other. The formalism of quantum mechanics, however, proposes a completely different prediction for an experimental set-up of this sort. It forecasts that half of the runs should produce similar values for "up" and "down" at A and B, and half should produce different values. And when the test is run, the quantum mechanical predictions are vindicated. Ever more rigorous versions of these tests come up with results that are ever more in line with quantum mechanical predictions. These results seem impossible to reconcile with an interpretation of quantum systems in terms of particles that respect the principle of locality.

What is the nature of the quantum mechanical predictions? Why do they differ so much from the expectations of local realism? The process by which the pair of "particles" is produced in experiments like these is mathematically represented by a *single* wave-function. That is not surprising because the function is generated by parameters relevant to the state of the *source* that produced the pair in the first place. And the nature of such a function is that it is unable to provide an independent description of the state of one of the particles on its own. It describes, rather, the state of the system of two particles as a whole, a state that is often referred to as involving "entanglement". When one of the pair encounters the spin detector at A, the resultant influence on the particle is represented in the mathematical formalism in terms of the action of a particular mathematical operator on the wave-function. But the operator must be applied to the wave-function *as a whole*, because there is simply no individual wave-function for the particle at A. This means that, according to the mathematical formalism, there will be instantaneous repercussions for the measurement taken at the other detector. And the empirical results bear out these predictions.

All of this is counter-intuitive and paradoxical as long as we continue to believe that there are particles in the system. The believer in particles has no problem in accepting that electrons or photons can have correlated properties at the moment they separate in the source. But those properties ought to be possessed, nevertheless, by each particle in an *autonomous* way, or so one would be inclined to think. How can an operation carried out at A have instantaneous repercussions for the measurement obtained on the particle at B?

Now let us consider the experiment from the perspective of the process account. According to this view, the causal evolution in operation in the system involves no transmission of particles. It is simple enough to imagine how a process can be produced that will give rise to a "pair" of causal impulses that have opposite effects in their respective target atoms. The processes we have considered so far in this book often involved a pair of genesis-units as source and a pair of genesis-units as target, as was the case in simple ionisation, for example. The Bell test involves creating the circumstances that gives rise to the evolution of a complex causal process that has twin effects in two separate targets. The phenomenon known as "electron emission", we recall, involved a source material reaching a critical level of disequilibrium by virtue of the increased number of ionised genesis-units it contained. This led to a pair of genesis-units in the source emitting two "electrons", giving rise eventually to the ionisation of genesis-units in the target material. The genesis-units in the original pair would have had their B-V poles oriented in opposite directions. Thus the pair of "entangled" electrons would have opposite "spin" in the sense that they would induce effects with *opposite* alignments in their respective target atoms.

The fact that such a process cannot manifest determinate properties typical of a particle (such as position or momentum) in the absence of appropriate target material is something we have already considered in detail. When a target atom is present, then the position or recoil of *that target atom* will become the basis for attributing position or momentum to a fictitious particle. What is not clear, perhaps, is how the twofold process typical of entangled systems can give rise to non-local effects. Consider first of all how two-path experiments manifest the same characteristic.

In Figure 7.7 we reproduce the two-path arrangements that we considered in Chapter Three. In 1 and 2, a source gives rise to causal processes that evolve through the system before having their effect in the target. In 1, both paths through the apparatus are open. We might be tempted to say that the process in scenario 1 took both paths together, whilst the process in scenario 1 took the only path that was open to it. This sort of talk is very typical of the wave-particle duality approach and is entirely misleading. In *both* scenarios the process evolves through the system and is modified by the various causal players that are operative in the apparatus. The process does not take a path, but evolves in such a way that it has the potential to give rise to effects in material objects that (from the spatial perspective) we represent as lying on a particular path. In 2, the evolution of the process is constrained in a way that is not the case in 1 by the presence of the barrier along the lower path.

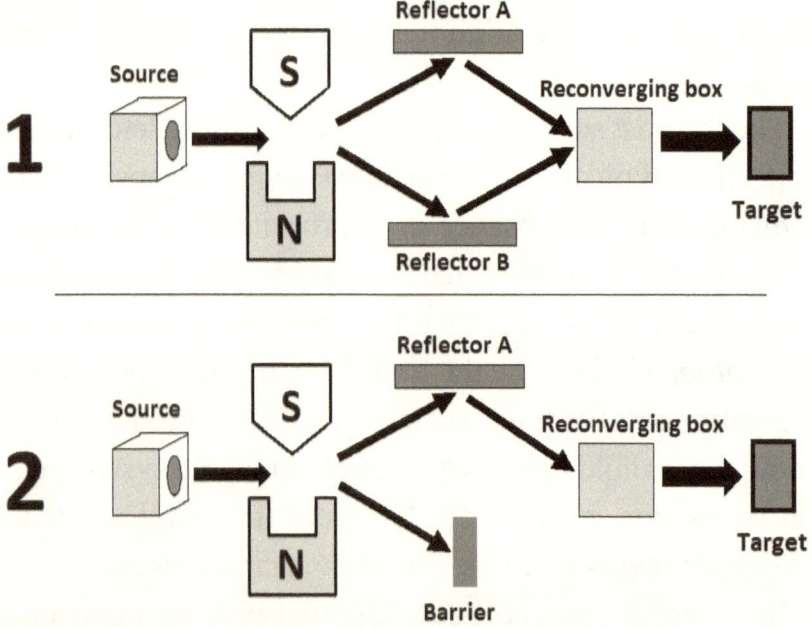

Figure 7.7 Non-locality and two-path experiments

In our consideration of measurements of complementary types of electron spin, we saw how an "open" apparatus such as that depicted in 1 does not disturb the value of complementary spin previously obtained. If the apparatus in 1 measures x-spin, and an "electron" enters the apparatus that had previously manifested y-spin-up, then a subsequent measurement of y-spin on that electron will still yield "up". In the case of apparatus 2, that will not happen. Any electrons that emerge from the apparatus will be found to have scrambled results of their previously recorded values of y-spin. These combinations of results show us that the fact that there is no barrier on the lower path in apparatus 1 has a fundamental influence on the way that the apparatus mediates the process that evolves through it. In other words, the state of *both* paths determines the way that the process unfolds. This conclusion is identical to the conclusion regarding non-locality that derives from the results of the Bell tests, and it was made almost fifty years earlier.

Let us consider in more detail the nature of the non-locality that is a feature of all two-path experiments, a feature that was already operative in the phenomenon demonstrated by Thomas Young at the beginning of the nineteenth century. As the process evolves through apparatus 1, it comes under the influence of the various causal operators in the apparatus. The magnetic deflectors are composed of many genesis-units, but they give rise to combined influences that can be divided into two principal components: an upward influence and a downward influence on any magnetised object or causal process that evolves through them. These influence have tendencies to change in opposite ways the orientation of the B component of any *genesis-unit* that passes through the detectors. Equally, they will tend to change the orientation of the B component of any atom that is a target of a causal *process* that evolves through them.

These combined influences, with their opposite tendencies, are mathematically represented in quantum mechanics by applying an operator to the wave-function that corresponds to the process that is evolving through the system. This jars with common sense if we continue to think of the process as a particle passing through the system. Why should the state of a particle (that can only take one path) be described by a mathematical function that contains terms for both paths together? But for a process, this sort of mathematical representation makes a lot of sense (even if it contains misleading terms that represent mutually exclusive trajectories). A process of this sort just *is* modified concurrently by both components of the deflectors and it continues to be influenced by the state of both paths through the apparatus.

How are we to think of this process? Is it spread out like a wave in such a way that a barrier or a deflector along a certain path influences the way that the wave as a whole propagates? The wave picture is wholly inadequate to represent the evolution of causal processes. A wave, after all, is a spatial phenomenon. A process can only be likened to it in a weak analogical sense because processes evolve in a pre-spatial realm. The eventual *effects* of causal processes generate the spatial perspective, and it gives us an impoverished and indirect viewpoint on the evolution of the processes themselves. For example, in apparatus 2, the evolution of the process in the system at one point encounters the presence of the barrier along the lower path. From this moment on the evolution of the process is modified by virtue of that fact. But we shouldn't think of that as involving some sort of signal being sent from the train of the process that is evolving along the lower path to the train that is evolving along the upper path. In the case of the interference of water waves, there *is* a material transmission of motion from both open apertures that contributes to the overall motion of the wave. But when we consider causal processes in evolution, such a perspective is not helpful. The process is *not* evolving by means of matter in motion. It is evolving in a pre-spatial realm in which reflector A and reflector B don't require signals or ripples in a medium to be passed between them in order for them to jointly modify the evolution of the process.

To recall the account of space that we have presented, we acknowledge that there are ontological divisions between objects, but we caution against assumptions being made on the basis of a naïve acceptance of the phenomenal aspect of spatial separation. The two-path experiments demonstrate that causal processes have a marked non-local character. *This does not mean that they are spread out spatially.* It means that they are jointly influenced by *objects* that appear separated from the spatial perspective. And those objects *do* have an ontological separation of some sort, the magnitude of which is represented by us as their spatial separation in space. But those same objects also have connections with each other (*because* they are mutually guiding the evolution of the same process), connections that are not apprehensible from the spatial perspective, and which have become relevant in this particular experimental set-up. Thus, the set-up of the experimental situation and the process which evolves through it have made these causal players at reflector A and reflector B joint partners in the sort of influence they wield on the evolving process. From the perspective of *this* process evolving in *this* system, it is as if the reflectors at A and at B are causally present to each other. When a barrier is placed at B, no signal needs to pass to A. No non-local influence needs to traverse space instantaneously. The way that the process has evolved has made A and B joint partners in the progress of the influence.

This leads to a distinction within the concept of locality. The particular formulation of locality that is a feature of special relativity states that no signal or material object can travel faster than the speed of light. We will consider this principle in the next chapter, along with the principle that the velocity of light is constant. But a more general statement of the principle of locality (in terms of the impossibility of *instantaneous* action at a distance) is certainly compatible with our model of causal interaction. Different processes evolve at different rates, but the evolution of any process will require a finite time interval. The rate of evolution of the ionising influence that we call the "transmission of an electron" will depend on contingent factors such as the potency of the causal source that prompted the impulse. The greater the potency then the greater the rate of evolution of processes of this kind (accounting for variations in the "kinetic energy" of electrons). The rate of transmission of light does not depend on the potency of the source of light, for reasons that we will consider later. But the salient point is that a finite interval of time is required for the evolution of an impulse in the system, or for an object to change its position in the causal configuration of things.

The sort of non-locality that manifests itself in quantum systems absolutely does *not* involve the transmission of a signal, nor the movement of matter of any sort. In Figure 7.8 we see a version of the Stern-Gerlach experiment where the reflector on path B can be rotated to become a barrier. In 1, the reflector is in place on path B and the process evolves through the apparatus under the joint influence of both reflectors. In 2, we see that the process has two possible outcomes: it can evolve through the apparatus under the influence of reflector A, or it can come to completion by having its effect at barrier B. Consider the case where the reflector at B is transformed into a barrier while the process has *already begun* to evolve through the system. Some elaborations of the old Copenhagen Interpretation made this event sound like an instantaneous and discontinuous transformation from a quantum system that behaved like a wave to one that had particle-like characteristics. The process account provides a more natural and plausible description of the transformation, and the non-locality inherent in the transformation is more easily palatable. If this transformation involved an event at B (the rotation of the reflector) having instantaneous repercussions at A, then it would indeed constitute non-locality of an unsettling kind. We would then be faced with a situation where a causal *process* begins at B and evolves to A instantaneously. This is inconsistent with our model of how processes evolve, the account we have presented of "frequency" and "wavelength", and the whole picture we have defended of the transmission of light.

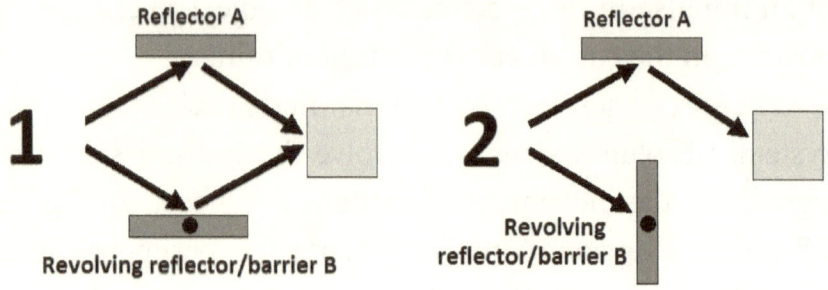

Figure 7.8 Two-path experiment with a revolving reflector/ barrier on the lower path

The kind of transformation prompted by the revolution of the reflector, however, is not to be understood in terms of a new causal process, or sub-process, that modifies the evolution of the original process. Of course the revolution of a reflector *is* a causal event and it *will* give rise to a myriad of individual causal processes emanating from the genesis-units that comprise the reflector. But as far as the *original* causal process under consideration is concerned, the revolution of the reflector does not involve conventional causal interaction of that kind. It represents something much simpler: the modification of the evolution of that process. A and B are causally separated in the sense that an atom in B will require an interval of time to effect a change in an atom at A. But a causal process that has a *separate origin altogether* and comes under the joint influence of A and B in the same moment (because they are equidistant from the source, and sufficient time has elapsed for the process to evolve from the source to them) can be instantaneously modified by a change at A or B. The rotation of the reflector at B does not induce a change in an *atom* at A; it modifies a *process* that has already evolved to A at that moment. This way of looking at things is very much in line with the principle of locality. The process in question evolved from the causal source to A in a manner that fully respected the principle of locality. In the Bell test, the random change in the detector at B seems to have an instantaneous effect on matter at A, but it is really having an instantaneous effect on a *process* which happens to have evolved as far as A in that moment. Did the change at B have an instantaneous influence at A? The answer is that the *source* object had its influence at A, and that influence evolved from the source to A in a timeframe that respected the principle of locality. The alteration at B modified the evolution of the *process* in an instantaneous way.

Thus we can uphold locality in its original form, affirming that no material object can induce an instantaneous change in another object that is separated from it. Causal processes *always* require an interval of time to evolve from one object to another. But a relevant change in a material object will instantaneously influence the *evolution of a causal process* that is already being guided by the object. This kind of non-locality has been a feature of "interference" phenomena since they were first observed hundreds of years ago. It is a feature of Stern-Gerlach experiments and two-path arrangements in general. In fact we can set down a general rule regarding non-locality of this sort. When a causal process is being influenced by multiple causal players, the ongoing evolution of that process will be instantaneously modified by relevant changes in any of those causal players.

7.8 The utility of processes for explaining quantum mysteries in general

If one interprets quantum systems in terms of evolving processes rather than the motion, charge and spin of particles, then many of the traditional obscurities of quantum physics become much more comprehensible. Under the process account, the mysteries of complementarity, indeterminacy of states, indeterminism of causal processes and non-locality all either attain a plausible explanation or disappear completely. These issues have been treated in the previous sections and here we give a brief summary of how they are dealt with using our model of causal interaction. If the reader has already been persuaded by our arguments on this topic so far, then he could skip the next two sections and go directly to Chapter Eight.

The term "complementarity" has been used to refer to incompatible physical *models* of explanation for phenomena (such as "wave" and "particle") as well as to pairs of *properties* that demand conditions for measurement that are mutually exclusive (such as momentum and position). In our scheme, the phenomenon is understood in terms of different experimental arrangements that permit a process to evolve in one of two mutually exclusive ways. In a two-path experiment, we are presented with the option of blocking an aperture or leaving both open. In one case we coerce the process to evolve over one path only ("particle-like" behaviour, single trajectory) whilst in the other we allow the process to evolve over both paths together ("wavelike" behaviour, superimposed trajectories, indeterminacy of position). Such complementarity would be distressing if we thought it involved material reality taking on the form of a wave or a particle in a discontinuous fashion in response to the experimental conditions, but there is no mystery about the evolution of a process being utterly altered by changes in the set-up of the apparatus.

Indeterminacy of states, as we have seen, arises from the erroneous attribution of properties to a fictional particle. When both paths are open, the particle appears to take on an indeterminacy as far as its overall trajectory and particular position at any moment is concerned. But if the causal impulse in the system is a process rather than a particle, then the attribution of positions is simply inappropriate in the first place. The evolution of causal processes is not directly accessible to our sensory apparatus because we are sensitive only to the *effects* produced when a process comes to completion. For that reason there *is* a certain indeterminacy as far as the dynamics of the process is concerned, but this is an indeterminacy only as far as our *knowledge* of what is happening is concerned. The material objects that exist in the world (genesis-units and conglomerations of genesis-units) *all* have determinate properties such as position in the causal configuration of things, mass, electrostatic equilibrium/disequilibrium and relative causal potency in various situations. Empirical data to hand gives us no reason for believing otherwise.

Once we accept that quantum systems do not represent a genuine indeterminacy of states on the ontological level, then we have no reason to assert that there is an indeterminism on the level of causality itself. This indeterminism is postulated on the basis of the belief that quantum systems collapse under certain conditions from a superposition of states to a determinate state. The discontinuous nature of this collapse (and the fact that the outcome of the collapse was believed to be purely probabilistic in nature) seemed to represent a breakdown in causal determinism. From the process perspective, such probabilities are seen to arise from fundamental limitations on the part of the observer, both in the misinterpretation or blatant inappropriateness of the terms that figure in the wave-function, and in the fact that the evolution of processes are hidden from our view and can only be described probabilistically. If the wave-function describes the system in terms of a conjunction of probabilities that the particle is in a certain position, then the measurement event that collapses that wave-function and obtains a definite position for the particle is going to look for all the world like an event that is probabilistic in nature. When the same event is understood in terms of processes, then the probability takes on a different aspect. In this case, a hidden process is in evolution in the system. The positions that appear in the wave-function can now be interpreted as possible positions where the process will manifest itself *if* suitable target material is present. When a measurement is taken, the process is coerced into having its effect on a material target and we simply obtain the position of the target atom. This "collapse" is an epistemic event corresponding to the real physical event of a process with many possibilities coming to completion in a target atom.

Non-locality is perhaps the most counter-intuitive aspect of quantum systems, even though it is already a facet of relatively everyday situations such as the "interference" of light. It is undoubtedly the most potentially fertile feature of modern physics, offering precious insights into the nature of causal processes and the significance of spatial separation. Of all of the physical mysteries that have come to light in our time, this is the one that challenges in the most profound sense our naïve views about space. How are we to react to the Bell test? "Solutions" that talk about instantaneous signals being passed from A to B through a "wormhole" in space try to solve the problem whilst basically holding on to those same naïve views about space. The wormhole is needed only because we attribute a certain ontological character to space itself. Under the process account, space is understood as having no ontological significance in itself; consequently no wormhole is needed. Space is our way of representing a division between objects at the ontological level, a division that implies that a certain interval of time is necessary for causal process to evolve between those objects. We affirm the legitimacy of the principle of locality as far as the *evolution* of causal processes across that gap is concerned. But the causal processes themselves are not "local": indeed, it is from the effects of evolving causal processes that the framework of locality if generated. If an ongoing causal process is being jointly influenced by objects A and B (that are separated from each other), then the *manner* of evolution of the process will be instantly altered by a relevant change at either A or B. This does *not* mean that a change at B (such as the rotation of a reflector so that it becomes a barrier) is capable of having an instantaneous effect at A. A is *not* a target of some sort of impulse from B: it is the *process* that is modified by the alteration at B.

The process may then well go on to make a change at A, but to say that this change at A has been instantaneously induced by a change at B is simply misleading. The process that has changed A has *evolved from the source* (depicted to the left of the apparatus in our diagrams) and the timeframe of the evolution of the impulse from the source to A and B fully respects the principle of locality. The process itself, however, has this characteristic of being instantaneously determined by ontologically separated objects.

7.9 To be or not to be – that is the answer

The notions of wave-particle duality, superposition/collapse, and states of entanglement are all products of an ontological condition (namely, that certain causal processes evolve in the system without the aid of intermediaries and can be modified by multiple players simultaneously) and a particular epistemological attitude (the tendency to reduce natural explanations to matter in motion) working together. We claim that it is *a fact of the matter* that causal activity in quantum systems evolves without the aid of material intermediaries. This fact gives rise to paradoxes only because the causal activity is *wrongly interpreted* to involve material particles in motion, waves or fields. Matter-in-motion cannot spread itself out or shrink in the way that is implied in the notions of superposition and collapse. But we *can* see a way in which the evolution of causal processes is immediately affected by changes in the set-up of the system in which the processes are evolving. In the concluding section of this chapter we will reflect a little on how aspects

On the Revolutions of the Internal Spheres

of causal activity that seem incomprehensible under the particle/wave/field account just disappear when the same causal activity is interpreted in terms of processes.

The observer we presented earlier had a severely limited perceptual apparatus, and this, in part, contributed to the formation of his defective hypothesis that particles were the bearers of causal influence in the thought experiment. We, also, have a severely limited perceptual apparatus, and it is simply not the case that we are in possession of adequate or convincing accounts of the phenomena of magnetism, gravity, or electricity, nor do we know much about what is happening in the atom. But all of these phenomena involve a causal source giving rise to effects in a target object. If a causal *process* has been found to be jointly influenced by two causal players simultaneously, and one of these players is later excluded from the process, then it will be no mystery to find that the causal process is now being influenced *by the other player only.* The new set-up of the apparatus no longer presents the causal process with multiple modifying influences, so the evolution of the process in a form that was wrongly interpreted in terms of "superposition of paths" can no longer arise.

Statements of this sort offer a more correct (though more minimalist) description of the causal dynamics of the thought experiment than that offered by the magneton explanation. When the electromagnet is activated, a causal process is initiated that is mediated by the intervening apparatus (tray with lead balls), and has the potential to produce effects on the grid below. The nature of the effects produced depends on the way that the apparatus is set up. The set-up can allow the causal process to naturally evolve, giving rise to

467

movement in the magnets on the springs, or it can snuff the process out as soon as it begins (magnets clamped rigidly to the frame). Both of these set-ups are mutually exclusive, and naturally give rise to effects that are mutually exclusive (movement or no movement). In either case, *the effects prompt no anti-realist interpretation regarding the nature of the causal process that is active in the system.*

Once the effects are used as evidence for hypothesizing the presence of *particles* in the system, however, then the spectre of anti-realism looms. In quantum systems, when such a "particle" is presented with a new experimental situation in which there are two "paths" of evolution, and the particle appears to take both paths simultaneously, then the ontological condition has been met. If one of the paths is now blocked, then the "particle" will either show up at the barrier, or pass integrally along the open path, causing consternation. How was the particle originally affected by the opening or closing of a path through which it does not pass?

The simple and intuitively natural complementarity of the thought experiment throws light on one of the most controversial aspects of quantum theory. Empirical data from modern experiments seems to show that hypothetical particles cannot be attributed properties in the absence of the appropriate measurement conditions. What bafflement this causes, not to mention soul-searching on the part of physicists of a realist persuasion! But what if the empirical data normally used to quantify the properties of hypothetical particles is really data that arises from natural causal processes in which no such intermediaries are involved? In the thought experiment, one sort of data was based on momentary effects when a natural causal interaction came to a climax or end-point, whilst another sort of data was based on the continuing evolution *of that same causal activity* at the natural level. This was the origin of the brand of complementarity that was manifested in the experiment – *the same causal process* could be either coerced into producing an *instantaneous* effect on an instrument or it could be permitted to act on the instrument over *a period of time*. These options, evidently, were mutually exclusive. The experimenter had the task of setting up the apparatus and, in so doing, excluded not only the "measurement", but also the *genesis* of the other "property". If he set up the apparatus so that it measured a property related to an instantaneous effect, then he excluded the very genesis of the property that was related to the evolution of the process. But once we appreciate that what was being observed was not the properties of *particles* but the manifestations of causal *processes*, then the weirdness disappears from the experimental results. It would be disconcerting to find that a *particle* ceases to have a property when the conditions are not in place to measure it, but it is no surprise to find that a causal process gives rise to different effects depending on whether the apparatus allows the process to evolve in a particular way or not. The contrast between the two mutually-exclusive experimental situations was *not* the contrast between a set-up that measured the momentum of a particle and a set-up that measured the position of a particle, as the standard interpretation would have it. The contrast, rather, was between a set-up that allowed a causal process to induce motion in the target object and a set-up that prevented motion being imparted to the target object.

These considerations from the thought experiment can be generalised to cover all cases of complementarity. In situations where a two-path apparatus is transformed into a single-path apparatus by closing one of the shutters, the new set-up of the system is preventing the evolution of the causal process being jointly modified by both causal players (the atoms in the apertures that "diffract" the evolving impulse). The fundamental point is the same: mutually exclusive set-ups of the system will result in the evolution of the causal influence in mutually-exclusive ways, giving rise to different sorts of empirical data.

The origin of the crisis of understanding in quantum physics originates in faulty hypotheses regarding the *causal players* that are involved in quantum systems. Once the superfluous entities have been removed, then the complementarity of causal effects is explained, whilst superposition/collapse and wave-particle duality simply disappear from our descriptions of atomic systems. The fulfilment of the ontological condition is, by itself, a powerful reason for removing these foreign entities from atomic hypotheses. Empirical data has shown decisively that the causal activity in atomic systems - when presented with multiple modifying influences – tends to be jointly affected by all of them simultaneously. We can be sure (even if three generations of physicists have become progressively more embarrassed to say it unambiguously) that particles cannot take multiple paths at the same time, but we can see a way in which processes can be jointly modified by the state of multiple paths confronting it.

Chapter Eight

THE VELOCITY OF LIGHT AND THE MEANING OF RELATIVITY

"We have first raised dust and then complain we cannot see"

George Berkeley

Overview of this chapter and its principal claims

1. *The constancy of the speed of light.* If light were either a wave or a particle, then its velocity could not be constant across different reference frames. The constancy of its rate of transmission derives from the fact that it is a process that evolves perfectly in step with the overall rate of evolution of the system itself.

2. *Reasons for the constancy.* We discuss the reasons why light evolves at the same rate as the causal evolution of the system. This is in contrast to, say, an electron, which does not have a constant rate of transmission in all circumstances. Light is described as an impulse that resonates with the causal evolution of the system. Its rate of progression is determined by the overall evolution of the causal configuration of things.

3. *The velocity of light as the upper speed limit of the cosmos.* The movement of macro-objects in space is described as a progressive readjustment of causal relations between objects. The speed of objects cannot exceed the velocity of light because it is ultimately

governed by the rate of evolution of cause and effect, which light displays in a pure sense (for reasons that we will consider).

4. *Objects have an absolute motion.* We argue that objects have an absolute motion relative to the causal configuration of things. This is the sum total of all accelerations and decelerations of the object relative to the system since its atoms came into being.

5. *Absolute motion has no particular direction.* The causal configuration cannot be reduced to any privileged inertial frame. Therefore, absolute movement does not have a direction but a *quantity.* The fact that the causal configuration is not aligned with any particular reference frame explains the null result of the Michelson-Morley experiment.

6. *Time dilation.* The claim that objects have an absolute quantity of motion explains the empirically-verified phenomenon of time dilation. We consider how greater causal distances have to be traversed by impulses that are evolving towards a target that has greater absolute motion relative to the system. However, this only slows the evolution of interactions (and the functioning of clocks); *time itself* continues to pass at its universal rate, as the constancy of the speed of light testifies.

7. *The importance of the absolute motion of the target during the last reverberation of the light.* In the crucial moment in which the light arrives at its target, the properties of the target object make their own telling contribution to the causal process that we call the evolution of light. The absolute motion of the target will help to determine the "density" of the causal configuration that the light must traverse, which can lead to time dilation.

8. *The importance of the relative motion of the target during the last reverberation of the light.* The relative motion of inertial frames has no influence whatsoever on the rate of transmission of light impulses because the impulse is dropped into the sea of the causal configuration. The ensuing process does not move *relative to the source* of light, but relative to the system as a whole. However, the potency of the impulse *is* determined in part by the relative motion of source and target, for reasons that we will discuss.

9. *The inadequacy and after-the-fact nature of the spatial trajectories we ascribe to light.* We consider how the evolution of an impulse of light through the system cannot be adequately represented by any spatial trajectory until *after* the impulse has hit its target. While the process of transmission of the impulse is still ongoing, it evolves in different ways relative to different targets, depending on their absolute motion. Down at the level of the causal configuration, this evolution is a simple and unitary thing, but from our perspective we would be forced to represent it in a complex multi-dimensional space, akin to the way quantum mechanics represents the evolution of systems.

10. *Relativity is the saviour of the phenomena par excellence.* We discuss the way in which relativity theory saves the phenomena adequately because it provides correct mathematical transformations for generating one inertial frame from the point of view of another. It tells us how the effects produced by the causal configuration are altered by the relative motion of the observer who experiences the effects. But this does not entail that reality itself is relativistic by nature, nor is it of much use in helping us to penetrate the reality that lies beneath the perspective-generating effects.

8.1 The constancy of the velocity of light

In this chapter we will consider the velocity of light from the point of view of our account of causal interaction. During the nineteenth century, it was widely believed that light was a wave motion and hence required a medium (the luminiferous ether) in order to propagate. When the notion of the ether was dispensed with, the view that light consisted in some sort of wave was retained, but no satisfactory account has since been provided of how a wave can propagate without a medium. Light quite simply cannot be a wave in the sense for which the physical wave model was originally invoked to explain this form of causal impulse. In a similar way, the particle explanation of light is inadequate, failing to explain interference and diffraction phenomena. For much of the twentieth century, the doctrine of wave-particle duality staved off serious examination of the nature of light. The doctrine included the statement that our physical models could never represent reality in any adequate sense. Wave-particle dualism provided a sort of catch-all, hybrid model that could adapt itself to the empirical evidence that arose in different situations without really purporting to represent what was actually happening at the ontological level.

The challenges that will be issued in this chapter to the two traditional models of light might sound overly direct and, perhaps, even naïve. There is a sense in which we are not supposed to take the physical claims of these models too seriously. The way that quantum mechanics corresponds to reality seems to call for a sophisticated and nuanced way of interpreting the physical significance of these models. But what concrete significance can we attribute to a superposition of states, or to a completely probabilistic evolution of a physical system? By contrast, we shall see how the process account, with its commitment to the deterministic evolution of causal influences between objects that have autonomous and (in principle) measurable properties, can explain these apparently counter-intuitive aspects of light in a very natural way.

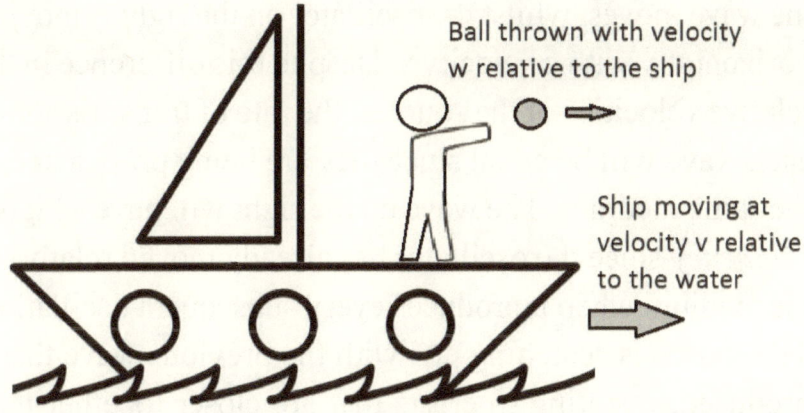

Figure 8.1 The relative velocity of a projectile depends on the relative velocity of the source. *A ship moves at velocity v relative to the sea. The velocity of the ball relative to the water will be v + w*

Consider a ship that is moving at a velocity v relative to the ocean (see Figure 8.1). According to the classical perspective, if someone on board the ship throws a ball in the direction of the ship's movement at velocity w (relative to the ship), then the velocity of the ball relative to the water will be v + w. Traditionally, light was believed to consist in either a stream of particles or a wave motion. If light were a motion of particles, then we would expect the measured velocity of light to be affected by the *relative motion of the light source and the observer*. A light beam shone from the ship in the direction of the ship's motion should move at velocity $c + v$ relative to an observer who is at rest with respect to the ocean.

If, on the other hand, light were a wave motion, we would expect its velocity to be determined by the rest frame of the *medium* in which the wave oscillates. Consider Figure 8.2. The oscillator on the left is at rest (in the horizontal direction) with respect to the medium in which the wave moves, whilst the oscillator on the right is moving horizontally with a velocity v. Despite this difference in the relative velocities of the sources, the rate of transmission of each wave will be equal since they are being propagated in the same medium. The wave on the right will have a higher *frequency* since the oscillator has already moved relative to the medium when it produces every subsequent oscillation. The source is "catching up" with the previous wave that it produced, resulting in crests that are closer together than if the source were motionless with respect to the medium.

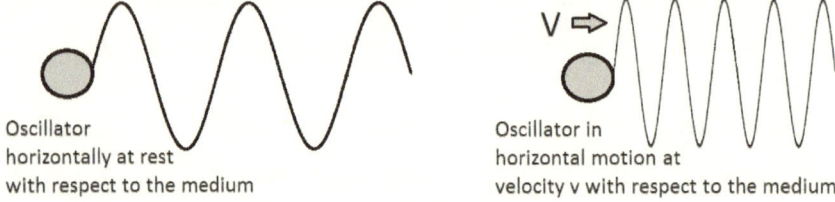

Oscillator
horizontally at rest
with respect to the medium

Oscillator in
horizontal motion at
velocity v with respect to the medium

Figure 8.2 The velocity of a wave through a medium is not affected by the motion of the source

An observer moving through the medium from right to left will measure a higher velocity for the wave motion than an observer who is at rest with respect to the medium. In the case of light, whether we interpret the phenomenon to involve either a wave or a particle, its measured velocity *should* have different values depending on the relative motion of the observer to either the source (if light is a particle) or the medium (if light is a wave). In the case of a particle motion, when the observer is moving towards or away from the source, then the measured velocity of the light should be consequently either greater or lesser. In the case of a wave through a medium, if the observer is not stationary in the medium and moves away from the source, then he should obtain a reduced velocity for the light. When he is converging on the source, he should obtain an increased value.

By the late nineteenth century, the electromagnetic theory had scored enormous successes in unifying the explanations for light, electricity and magnetism. According to the theory, electromagnetic waves oscillated in a medium with theoretical properties that ought to be measurable. The medium was believed to provide an absolute frame of reference for the motion of light. If the earth was moving through the ether from, say, left to right, then the velocity obtained for light should be greater if the light source beyond the earth was pointing right to left. In 1887, Michelson and Morley measured the motion of electromagnetic radiation in perpendicular and opposing directions in order to establish the relative motion of the earth through this hypothetical stationary medium. The failure of the experiment to detect any change in the velocity of light was eventually interpreted as indicating the nonexistence of the ether. The theory of special relativity disposed with the notion of the ether and made the constancy of the speed of light one of the principles upon which the theory was constructed. But the constancy of velocity detected by Michelson and Morley can also be interpreted as an indication that the motion of light does not involve either particles or waves, both of which ought to give rise to different measured velocities depending on the relative motion of the observer.

Let us see how the process account approaches this question. A binary pulsar is composed of a pair of stars, one of which is orbiting the other at a high velocity. Consider Figure 8.3 which shows star A emitting a photon of light towards the observer at two different stages of its orbit around star B. Photon 1 is emitted when A is converging on the observer at an enormous rate, whilst photon 2 is emitted when A is receding. If light were a particle, there should be an enormous differential in the respective velocities of the two photons. In reality, empirical measurements will obtain the same velocity for both photons, contradicting the particle hypothesis, but photon 1 will be found to have a higher frequency. Constancy of velocity with variations in frequency can be accounted for by the wave model if the observer happens to be *stationary* in the medium that carries the motion of light. However, with the orientation of the earth varying constantly as we hurtle around the sun, it is quite clear that we could not be stationary in any hypothetical medium for more than a very brief period of time.

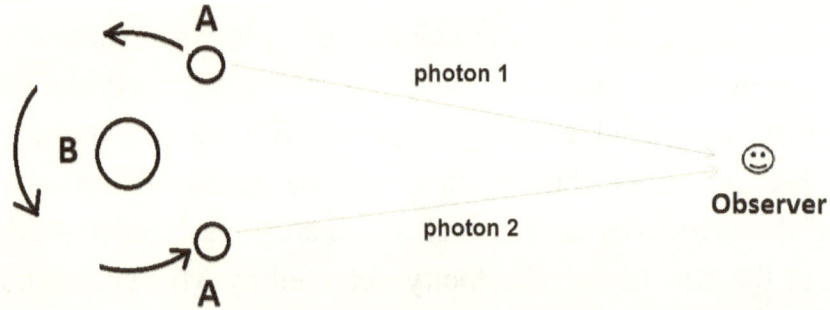

Figure 8.3 Photons of light are emitted as A converges on and then recedes from the observer

The process account that we have developed takes a different perspective on the transmission of light. When photon 1 is emitted, star A is converging on the observer, but this fact has no repercussions for the velocity of the transmission of the influence. At the moment the impulse is emitted, it begins to evolve in the causal configuration of things. This causal configuration does not correspond in a simple way with our spatial representation of the world. As we have discussed earlier, the spatial perspective *has* aspects that are genuine reflections of reality. But reality is not simply *different* to the content of the spatial perspective, it is vastly more *complete*. We generate our perspective from the *effects* of causal processes. They impinge on us by inducing a change in a physical object, or directly stimulating our sensory apparatus. The causal configuration of things is prior to and deeper than these surface effects, and it here at this inaccessible level of reality that photon 1 evolves.

The rate of evolution of the impulse is not affected by the relative velocity of A because the impulse is not some kind of projectile that has been thrown off by A. If it were a projectile, then its velocity *would* take the frame of reference of A as its starting point, but light is a different sort of impulse altogether, as its constancy of velocity demonstrates. Light involves the disassociation and realignment of groups of atoms in electrostatic configurations. This business is significantly different to the emission of a particle. When an object emits a particle, it is shedding part of its own substance. The particle will have mass and a velocity that is determined by the velocity of its source object and the nature of the mechanism within that source object that caused the particle to be emitted. Light, by contrast, does not involve the shedding of part of the substance of the source object (except in cases where atoms in the source object are annihilated). There is no ejection by the source of an autonomous causal entity that has mass and velocity. What is happening instead is something that has much more universal associations. The impulse is "thrown into the ring" of the evolving system in the sense that it begins to evolve with the myriad other impulses that are concurrently progressing through this part of the system. The rate of evolution of the impulse is not in any way influenced by the relative velocity of the source object. Rather it is determined by the overall rate of evolution of the totality of processes in the system itself.

It is in this sense that we can assert that all reference frames are equivalent. They are equivalent insofar as they are all generated in particular observers by the effects emerging from the same world system, and they are all equally *incomplete*: none of them represents the "absolute" reference frame of the causal configuration of things itself. The causal configuration of things is not a *spatial* frame of reference! It is a pre-spatial realm in which the causal connections between objects are evolving. This evolution gives rise to changes in objects and stimulations of our sensory apparatus, generating the spatial viewpoints of various observers. But the causal configuration of things is something much deeper and more comprehensive than any spatial viewpoint. It is in this flux of evolving processes that photons 1 and 2 evolve. Their rates of evolution will be the same and will be measured as equal by any observer, regardless of his particular frame of reference.

8.2 The potency of light depends on relative frame of reference

The *potency* of the causal change induced by the impulse, however, will vary depending on the relative motion of source and target. There is nothing surprising about this and it does not mean that the impulse is some sort of wave or particle, like a ball being released by a fast-running bowler. Even though light is *not* a particle, the potency of the causal change wrought by the light will be greater if the source and target are converging on each other. The fact that photon 1 was emitted when A was hurtling towards the observer will mean that the manifestation interval of the impulse will be shorter (in other words, the measured "frequency" will be higher). As we discussed in Chapter Three, the evolution of light involves a succession of "moments" in which all the evolving impulses in the system interact with each other. During this period of time, the impulse under consideration reverberates away from the source. If suitable target material is present at the right place for the culmination of each reverberation, then the impulse can effect a change in the material. The impulse can only effect a change at the culmination of each reverberation. Until that culminating moment, the impulse is still playing itself out, so to speak, with all of the other impulses in the system. At each successive culminating moment (corresponding to the "crest" of the classic wave model of light) the impulse rears its head from the flux of the system as a whole and has the potential to effect a change in matter.

The evolution of the system is thus envisaged as a pulsating series of moments in which impulses are successively working themselves out in relation to the rest of the evolving processes in the system. Each impulse has a periodic nature: a period of interaction with the rest of the processes in the system (during which the process under consideration is unable to effect a change in matter) and then a culminating point where the process attains the potential to effect a causal change in matter. The more potent the impulse, the shorter will be the interval required for the process to manifest itself by effecting such a causal effect. But the ultimate causal potency of any particular impulse does not have an absolute magnitude. It will be larger or smaller depending on the relative motion of source and target.

Say that photon 1 is emitted at time t (from the perspective of star A) and that it eventually effects a causal change in target T. Say that its manifestation interval is x. During the interval x, the impulse reverberates with the other causal processes in the system. At time $t + x$, and at every subsequent interval $t + nx$ (where n is a natural number), it attains the capacity to effect a causal change in suitable target material. T happens to be present in the right place at one of these moments, which we will call $t + ix$. If we think about it though, we can see there is something incomplete about this account of how photon 1 evolves to T. During each successive interval, x, the photon enters into an interplay with the other evolving processes in the system. The more potent impulses, by virtue of their potency, have priority in the pecking order. They attain the capacity to effect causal change more rapidly and more frequently during the course of their progress through the system. During the interval t

+ *(i-1)x* to *t* + *ix,* photon 1 undergoes its last reverberation before effecting a causal change in T. During this interval, the causal impulse that is photon 1 enters one more time into interplay with the other causal processes in the system, *but now a new causal player has entered the picture.* Target T is converging on the part of the system where the impulse of light is reverberating. It now enters into the mix of the relevant processes that modify the outcome of the impulse known as photon 1.

Any object can be a causal player in a system for a variety of reasons, whether or not it is in relative motion to the object or process under investigation, but we wish to consider the causal import of T simply from the point of view of the fact that it is in *motion relative to the direction in which photon 1 is reverberating.* Let us return again to the moment *t* + *(i-1)x.* At this moment the new interval, *x,* begins. Photon 1 enters into interplay with the relevant evolving processes in the system, and the new causal player, T, enters the scene. The convergence of T on the reverberating influence will mean that the photon will have a higher potency with respect to T than to a target that happened to be receding from the direction of reverberation of the influence. As photon 1 reverberates for the last time during interval *x,* T converges on the part of the system where the impulse is reverberating. We could say that it has "positive kinetic energy" with respect to the evolving impulse. Thus the final causal efficacy of photon 1 does not depend *solely* on the manner in which it was emitted by the source. Recall our treatment of interference and the way in which an impulse of light can be modified by various causal players. This is true also of the *target* of light. In the final interval from $t + (i-1)x$ to $t + ix$, *T itself participates in this causal interplay.* The relative motion of T increases the causal potency of the process, meaning that the manifestation interval of photon 1 is consequently shorter than would have been the case if T had been receding from A.

Now we can see better what was wrong with our original description of the evolution of photon 1. We spoke of manifestation interval x as if it were a fixed magnitude set only by the potency of the impulse as determined by the source of light. It is true that in many cases the source has the principal contribution to make to the potency of a light impulse. That potency is dependent on the magnitude of the electrostatic alignment that has disassociated. But our account of causal interaction allows a process to be modified by all relevant players during its evolution. These players can increase or decrease the ultimate causal efficacy of the impulse, and if that is the case then its manifestation interval will be affected. Think of an "electron" being accelerated by electric or magnetic fields. In this case an evolving process is modified by new causal players and attains greater potency. But the *target* of causal influence is also a genuine player in any interaction. This model of causal evolution does not see the target as a passive recipient of change: it has a crucial role to play in the evolution of the process, but it often becomes relevant only in the very last reverberation interval of the impulse. It is in that interval that the target modifies the process, making it more or less potent depending on what is normally referred to its "kinetic energy" relevant to the direction of reverberation of the impulse.

Imagine a situation where star A emits three impulses of light of the same potency in different directions simultaneously. When we say that they are of the same potency, we mean that they were produced by disassociations of electrostatic alignments of exactly the same magnitudes. One impulse is directed at a target that is stationary with respect to A; the second is directed towards a receding object; and the last impulse has a converging object as its target. Each of these impulses will ultimately undergo different evolutions as a result of the dissimilar modifying influences contributed by their respective targets. During the final reverberation of each impulse (the reverberation in which the impulse finally effects a change in its target), the *target object itself* will become a causal player in the evolving process. The receding target will have a negative kinetic energy with respect to the impulse, resulting in a diminishment of the potency of the causal change wrought in the target. The opposite will be the case for the converging object, whilst the stationary target will have no influence on the potency of the process. Observers present on each target object will obtain correspondingly different values for the "frequency" of the impulse directed towards them.

Clearly, these variations in the measured potency of light with the relative motion of source and target is not relativity by another name. The view we are defending asserts that there *is* an absolute perspective, but it is not spatial by nature. All spatial perspectives are equivalent insofar as they are equally based on *effects that are generated* by the one causal configuration of things. This causal configuration, being the sum total of the causal processes in evolution between all the causal players in the system, is necessarily unitary in nature. All of the causal players are genesis-units and groups of genesis-units and nothing else. The fact that the same causal process *might* have had different consequences if it were directed to a different target does not mean that there is some sort of inherent relativism in causal processes. A receding observer would measure a lower frequency of light than a converging observer would have measured for that same impulse; but this does not mean that the potency of that impulse is fundamentally relative to the observer. In the Newtonian worldview, the impact that a moving billiard ball had on another ball depended on the relative motion of that other ball. If the ball was receding, the impact would be less, but this did not lead to the conclusion that the energy of the ball was fundamentally dependent on the relative motion of the second ball. The first ball simply had its own quantity of motion, but the way that it impacted on the second depended also on the relative motion of the second.

Different values obtained by different observers for frequency, mass, wavelength, duration of time, or other quantities, do not entail that there are no non-relative facts of the matter regarding the objects that populate the world. The constancy of the evolution of light is a characteristic that arises directly from the unitary causal substrate where objects have non-relative autonomous properties. Indeed, this constancy - regardless of the situation of the observer - is a testimony to the existence of such a non-relative pre-spatial realm. The observer generates his own perspective on this substrate from the effects that are induced in his sensory apparatus. The form of his perspective will be influenced by factors that are particular to the observer, such as his motion relative to other objects. Despite this relativism, the absolute nature of the pre-spatial substrate is still discernible in a number of ways, not least in the constancy of the speed of light. Another clue is the simple character of the magnitude of the causal interaction between source and target. Consider again the three impulses emitted by star A. Say that all three have magnitude x (from A's perspective). The first target is stationary with respect to A, so the potency of the causal effect wrought in this target will simply be x. The second target is converging on the reverberating impulse. The energy (let us call it y) that it will contribute to the causal interaction will be related to the kinetic energy of the target genesis-unit. The second photon will thus effect a causal change of potency x + y in the target object. The kinetic energy of the genesis-unit in the receding target will diminish the causal potency of the third impulse. Its magnitude will be x − z, where z represents this diminishment. The potency of each causal interaction depends on the relative circumstances of the target, but the contribution of the source is *always x*, and the relative contributions of the targets depend simply on their kinetic energies relative to the source. And x itself originates in the dissolution of a group of genesis-units of a certain size, a group that is held together by electrostatic bonds of a certain magnitude. No relativity of perspective can alter the facts of the matter of this situation.

We can summarize this section as follows: The rate of evolution of light is always constant - the impulse is released into the flux of the causal configuration of things which determines the speed of evolution. The magnitude of causal change brought about by the impulse, however, will vary, depending on two principal factors. Firstly, the size of the impulse will depend on the *magnitude of the electrostatic alignment* that disassociated to release the impulse in the first place. Secondly, the final effect wrought in the target will depend on the *relative motions of the source and target*. These relative motions contribute to the potency of the interaction, just as a moving ball will hit me harder if I am moving towards it rather than away from it. This is not to identify the light impulse with a particle. It simply acknowledges that the relative motion of source and target is highly relevant for the potency of the causal interactions that occur between them.

8.3 The velocity of light as the upper speed limit of the universe

The theory of relativity holds that the velocity of macro-objects cannot exceed the speed of light. As velocity increases, objects allegedly become heavier (towards a limit of infinity) and more contracted in length (towards a limit of zero), accompanied by a dilation of time (towards the limit where time does not pass at all). The increase of an object's mass towards infinity means that infinite energy would be required to accelerate the object until it attained the velocity of light. This is clearly impossible, making the velocity of light the upper speed limit of the universe.

If our approach is correct and light is not a motion in space at all, then does the rate of its evolution have any consequences for the velocity of material things? Let us step back for a moment and consider what happens when material objects in general change their position in the system. This movement is a representation from the spatial perspective of a *genuine* re-configuration of the causal relationships between objects in the system at the ontological level. Take a system composed of three objects of equal size, A, B and C. Say that their configuration is such that we represent them spatially as being arranged in a line. This spatial representation corresponds to authentic relationships at the ontological level. In some sense, for example, the substance of B stands between that of A and C. Any potential causal influence of A on C will likely be blocked, influenced or mediated by B. By contrast, A will probably be able to influence B without significant interference from C (all of this depends of course on the specific causal properties of A, B and C. If A is highly reactive it may well be able to influence C despite the inhibiting presence of B). Now imagine that a movement occurs and this is represented from the spatial viewpoint in terms of object B going upwards vertically until such point as it forms an equilateral triangle with A and C. In ontological terms this means that the causal configuration of A, B and C has been altered, leading to a situation where they now (potentially) exert *equal* causal influences on each other.

The movement of B at the ontological level no doubt has other significances. But we wish to highlight the simple fact that what we represent as motion from the spatial perspective involves a progressive readjustment of *causal relationships* between objects. When an object moves through the system, it involves successive shifting relationships with all of the objects in its vicinity. Now let us return to the question of the speed limit of the universe. According to our view, the evolution of light through the causal configuration represents the natural rate of evolution of causal influence of this sort through the system. We have good reasons for thinking that the rate of evolution of light (unlike the rate of, say, the impulse we call the "electron") is the purest rate of reverberation of causal influence through the system. The fact that the rate is constant for all observers regardless of their relative motion is a powerful indication that we are dealing with a rate of reverberation of universal significance, the very ticking of the causal clock of the system itself. Other non-material impulses (such as the "electron") have variable rates of motion, demonstrating that their evolution is not simply determined by the rate of causal reverberation of the system, but by additional complicating factors. Light, on the other hand, is a causal impulse of a very simple sort. It involves the disassociation and formation of electrostatic alignments between atoms. The transmission of an "electron", by contrast, involves ionising atoms, a disruption of a genesis-unit's inner state of equilibrium. The evolution of such an impulse through the system will be subject to all sorts of hindering influences that will influence its rate of progression. We can intuitively see how the transmission of light, by contrast, could have the swiftness and universal character that it has.

So much for light. The motion of an *object* has just been described as involving the readjustment of causal relationships between that object and other relevant bodies in the system. It is unthinkable that a material object could move through the system more rapidly than the reverberation of a simple causal impulse like light. A material object is made up of at least one, and usually multiple, genesis-units. The movement of an object involves causal reverberations of an immensely more complicated sort than that involved in the transmission of light. Let us try to elaborate on this a little.

The first point that needs to be discussed is the question of the precise rate at which a causal impulse reverberates through the system. This reverberation is not a movement in space, but it will produce effects in objects that can potentially generate the spatial viewpoint. Instead of thinking of the impulse as reverberating in space, it is more accurate to think of it as reverberating through the complex network of evolving causal relations between objects. Perhaps this approach becomes clearer when we contrast it with the Newtonian conception of space. According to this view, space had an ontologically independent role in the causal interactions between objects. It made perfect sense to think of a spatial universe existing in which there were absolutely no objects at all. Objects had positions and relative velocities in this container-like substrate, and movement simply involved changing position in the container. There is no self-evident reason for denying that an object might be accelerated (if sufficient energy were available) to an arbitrarily high velocity through space of this sort.

According to the causal configuration approach, by contrast, space has no independent existence and cannot be understood except in terms of the causal relationships between objects. When we represent an object as moving in space, it is undergoing a process in which it is continually readjusting its causal relationships with other relevant objects in the system. Spatial voids represent our way of representing causal gaps of some sort. The perceived movement of an object towards another object is the closing of that gap. Any change in an object (such as a change in position in the causal configuration or a change in electrostatic equilibrium) causes reverberations in the system. In other words, those changes have repercussions or ramifications for other objects in the vicinity, or much further afield. But these repercussions are not transmitted *instantaneously* from the causal source to the entire rest of the system. Instead they reverberate through the system and this reverberation has a rate of evolution. Some types of causal change (for example the emission of sound) have rates of reverberation that are determined by the particular characteristics of the empirical situation. Others (such as light) are so simple as to be unaffected by particular empirical factors. Thus their rate is determined by more universal factors.

All of this assumes that the passage of time is something that has a universal character. The system as a whole is envisaged as evolving at a uniform pace. Causal events give rise to their effects according to the dictates of the ticking of a universal clock. When a change occurs in an object, that object generally becomes a source of causal action that operates on other objects in the system. But the effects are not induced in their targets instantaneously. They must work themselves out in the system in reciprocal interaction with other relevant evolving processes. This reciprocal interaction is driven by the consistent evolutionary progression of the system as a whole.

Imagine a large object bobbing with a constant rhythm on the surface of water. Circular ripples emanate out from the object. Now imagine that some smaller objects of different sizes are dropped in the water close to the oscillating object at the centre. These objects will begin to be carried outwards from the central object, but at different rates of progression depending on their size and weight and on the relative magnitude of the waves. Some of the objects will be carried along quickly by the waves whilst others will be hampered by their depth of immersion in the water. The regular ripples of water emanating from the centre represent the ticking clock of the system. The smaller objects represent causal impulses of varying sorts. Some of these impulses will reverberate from the centre more quickly than others, depending on their particular characteristics. But the reverberation of *all* of the impulses is enabled by the oscillation of the water. If the water were still, the objects would remain where they were dropped. In the same way, the system in which we live enables the reverberation of events and *transforms them into causal events*. The system pulsates with a natural rhythm, carrying the repercussions of events through the system. The fact that the system has this natural rhythm, however, does not entail that all causal impulses emanate through the system at the same rate. Just as some of the objects dropped in the water had characteristics that hampered their movement, so some causal impulses (such as the "electron") have traits that hinder their evolution in the system. Other objects, instead, were carried along at the very same rate as the progression of the waves themselves. Light, we believe, has precisely this characteristic. It is a simple form of causal impulse, involving no invasive alterations of the internal equilibrium of genesis-units. It evolves through the system at the pure rate of reverberation of the system's natural clock.

In the next section we will consider the empirical evidence for time dilation and argue that this evidence does not rule out the possibility of absolute time. Right now we wish to return to the issue of the maximum velocity permissible for material objects. When an object moves in the causal configuration of things, it is continually causing reverberations in the system of a causal nature. It is unthinkable that the rate of progression of such an object could exceed the rate of evolution of a simple form of causal reverberation such as light. To see why this is the case, consider Figure 8.4.

Object A is moving towards object B. At the level of things as they are in themselves, A is closing the ontological gap between itself and B. At time t1, A emits a photon in the direction of B. Now let us allow that A is travelling *faster* than the velocity of light. At time t2, A arrives at B but the photon of light is still reverberating in the system and hasn't yet reached B. In the diagram, for convenience, the photon is represented at time t2 as a point in space. Of course, if an impulse of light were still in the process of evolution then it would make no sense to represent it spatially. This is only possible when it has caused a change in an object, thus producing an effect that can potentially generate the spatial perspective.

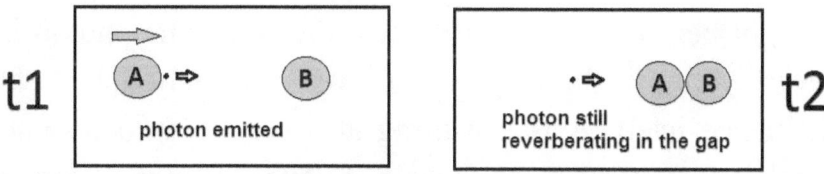

Figure 8.4 The paradoxical consequences of faster than light motion. *At time t1, object A is moving faster than light towards object B and emits a photon of light. At time t2, A reaches B but the light is still reverberating in the original ontological gap. It is unthinkable that an object could close a causal gap faster than the rate of evolution of a simple form of causal influence like light.*

The scenario depicted in Figure 8.4 is evidently paradoxical given our account of how causal processes evolve in reciprocal interaction with each other. Consider the moment (t2) when object A comes in contact with B. At that moment, A becomes a causal player in the evolution of the state of B. Prior to that moment, B will have been typically involved in multiple evolving causal processes. Other players may have been exerting causal influences on B that were playing themselves out and now come to completion in unison at time t2. The impact of A on B will be one of these myriad influences that become relevant in the infinitesimal moment leading up to t2. But A also emitted an impulse of light towards B at a previous moment in time, t1. This causal influence has been reverberating through the system at the purest rate (or so we have reason to believe) at which a causal influence can reverberate. It makes no sense to think that A could have reached B (and is now exerting a causal influence on B) before an impulse that was reverberating in this most rapid fashion. Or to put it another way: At time t1, both A and the photon of light begin a race in the direction of B. The photon is reverberating at the very rate of reverberation of the causal influence itself. A is travelling at a certain velocity and reaches B first, exerting causal influence by contact action on B. But how can A win a race to exert causal influence on B ahead of an impulse that was being carried along at the very rate of reverberation of the system itself?

To return to our analogy of the object bobbing in the water. The ripples emanating outwards represent the pure rate of reverberation of the system. The objects that are dropped at the centre will eventually move to the periphery, but it is impossible that any of these objects could move from the centre to the periphery at a faster rate that the progress of the water waves themselves. It is the water waves, after all, that enable these objects to move outwards. A simple light object might be carried with the same speed as the water waves (analogous to the way an impulse of light is carried along with the reverberation of the system), but no objet that is carried by the waves can exceed the velocity of the waves. In the same way, the natural reverberation of the system is ultimately responsible for the movement of objects through the causal configuration of things. A simple impulse like light can be carried along at the very same rate as this reverberation, but a conglomeration of genesis-units – or even a single genesis-unit – cannot. The movement of macro-objects requires a change in causal relationship with other objects in the system, such that all of the genesis-units composing the object readjust their causal relationships with other relevant objects in the causal configuration in unison. For an object to approach the speed of light signifies that the object is changing its relationship with the rest of the system at almost the rate of change of causal influence itself. Objects cannot move quicker than this rate of evolution since their own movement is itself governed by this rate of evolution.

8.4 Time dilation

It may be true that all spatial frames of reference are equivalent insofar as none of them constitutes a privileged viewpoint on reality at the ontological level. But the scientist can do much more than point out the limitations of the spatial perspective. Reflection on the way in which spatial perspectives in general are generated can aid us in understanding the authentic underlying state of reality. In every spatial description there are aspects that are perspectival and aspects that reveal genuine facts of ontology. Even the perspectival elements themselves are grounded on ontology and can be used to create empirically testable hypotheses about underlying mechanisms. The dilation of time is one of these aspects of our viewpoint that is grounded on ontological facts but manifests itself in a perspectival way, as we shall now see.

According to the theory of relativity, time slows down as objects approach the velocity of light, but the dilation of time should also be detectable for speeds of a much more everyday nature. One of the most celebrated empirical verifications of time dilation was obtained in 1971 when four caesium atomic beam clocks were flown around the world twice. These clocks were then compared with reference timekeepers at the U.S. Naval Observatory and were found to have lost a number of nanoseconds that was consistent with the predictions of relativity. Let us see how such empirically verified dilation of time has aspects that are merely perspectival even though it is grounded in causal relationships at the ontological level.

Clocks in general typically involve oscillation of some sort. The swinging of a pendulum is the classic way of measuring the passage of time, a fixed interval being marked out by each stroke. Atomic clocks function according to much more complex oscillations, but they still involve repeated interactions of a rhythmic sort between atoms. Let us consider a simple model of such a clock. Two groups of atoms, A and B, are set up as depicted in Figure 8.5. A source of energy is attached to A and is turned on briefly. This prompts the atoms in A to produce electromagnetic radiation which passes through the tube to the group at B. The radiation energises the group at B, producing one tick of the clock and causing radiation to be emitted back through the tube. The group of atoms at A are now reenergised by the returning radiation and the cycle begins again.

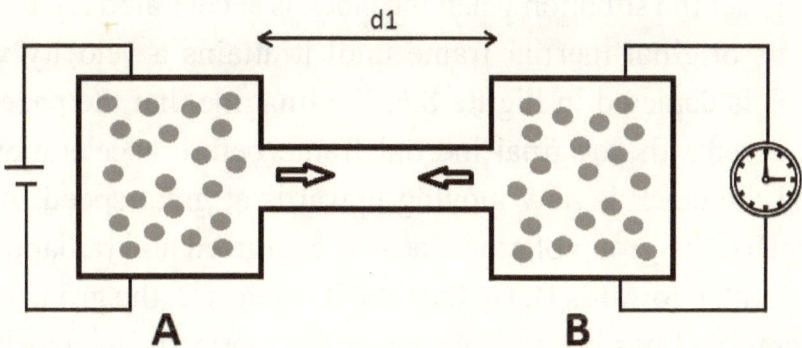

Figure 8.5 Schematic diagram of a simple atomic clock. *The clock is started by switching on the source of energy briefly, energising the atoms in group A. This prompts radiation to pass from A through the tube to the group at B, causing these atoms to be excited, producing one tick of the clock. In turn, the group at B radiates light through the tube, reenergising the group at A, and the cycle begins again.*

Now imagine that this clock is accelerated until it attains a high velocity relative to the frame of reference of an observer. According to the theory of relativity, the timepiece will run slower than a similar one that remains in the original inertial frame. This prediction follows from the principle of the constancy of the velocity of light, and is to be accepted as a brute fact of the world. In this section we wish to argue, by contrast, that *apparent* time dilation arises very naturally from the causal configuration approach to motion. A clock that is moving very rapidly within the causal configuration of things will naturally have its functioning slowed down, even though the *system's* clock will continue to tick at its universal pace.

In Figure 8.5, the distance between the groups of atoms at A and B was represented by the magnitude d1. Now let us consider the situation when the clock is accelerated relative to its original inertial frame until it attains a velocity v. This is depicted in Figure 8.6. We imagine that the paper represents the original inertial frame before acceleration and the clock is now moving upwards at great speed. At time t1, the group of atoms at A is energised and radiation is emitted towards B. By the time it reaches B, the group of atoms at B has been displaced by the motion of the clock. The radiation therefore has to reverberate through a greater portion of the causal configuration of things and will arrive at B later than would have been the case if the clock were not moving at this rate. The intervals between each tick of the clock (t1-t2 and t2-t3) will consequently be longer. This means that the functioning of the clock will be slowed down and it will register a slower passage of time than a clock that remains in the reference frame of the paper. It is clear, however, that this dilation does not constitute a genuine slowing down of time itself. The rapid motion of the clock with respect to the causal configuration means that a larger causal gap must be traversed by the impulse, entailing a greater interval for each oscillation of the clock.

Figure 8.6 is a simplified depiction of what is happening and is misleading in various ways. In Section 8.2 we discussed how light has a different potency depending on the relative motion of source and target, but this difference is only worked out in the final reverberation of the light before it reaches its target. Similarly, any "delay" associated with the transmission of light can only occur during its last reverberation before arriving at its target. If the target has a high quantity of absolute motion then the light will have to traverse a greater portion of the causal configuration during this last reverberation. The increased length of d2 compared to d1 can be thought of as pertaining to this final reverberation only. In Section 8.6 we will discuss this fact further to show that it is the motion of the *target atom only* that is responsible for the apparent dilation of time.

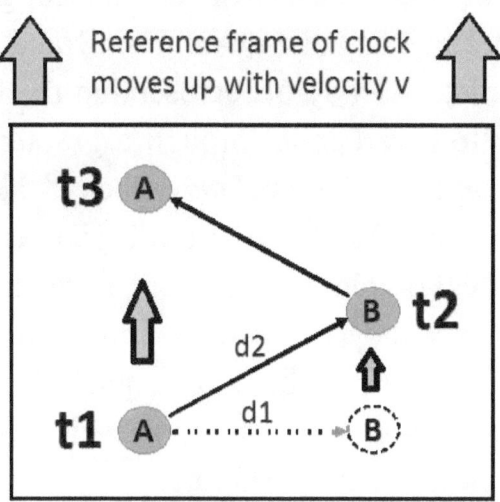

Figure 8.6 Motion through the causal configuration slows the functioning of a clock even though time itself progresses at the same rate. *The group of atoms at B is displaced upwards after the radiation is emitted from A. This means that the radiation has to travel further (distance d2 rather than d1) within the causal configuration before it can produce each tick of the clock. Therefore the interval between each tick is longer, but only because the radiation has a greater causal distance to travel, not because time itself is slowed down. Note that causal distance does not correspond to distance measured from the spatial perspective.*

The approach to time dilation depicted in Figure 8.6 is insufficient as it stands because it is open to the valid rejoinder that experiments of the Michelson-Morley type show that there is *no* lengthening in the interval of time it takes for a photon to close a gap, even if the apparatus is moving at high velocity. The key to approaching this question is that we cease to think of the causal configuration as if it were a privileged *spatial* frame of reference, like a sort of absolute space through which impulses evolve. In these diagrams we represent the ontological gap between A and B in terms of spatial quantities, d1 and d2. But this is merely a pictorial way of representing the fact that the impulse has to reverberate through a *greater portion of the causal configuration of things* when it takes the path represented by d2. *Neither* d1 nor d2 give us a complete sense of the real ontological gap across which the process reverberates, but they do give us a measure (in relative terms) of the greater reverberation required by the impulse when the target atoms are changing their position in the causal configuration at a great rate.

The difference between the causal configuration and a privileged reference frame can be illustrated if we return to the Michelson-Morley experiment that sought to establish the motion of the earth through the stationary ether. Figure 8.7 shows a variation of this type of experiment. A photon of light is emitted at A and is reflected at B, C and D before returning to A. Let us imagine that the stationary ether is represented by the frame of reference of this page, whilst the apparatus in which this photon evolves is moving from left to right at velocity v. When the photon is moving in the same direction as the motion of the apparatus through the ether, it has to traverse a greater distance (d2) through the ether before it arrives at B. But when the photon is returning from C to D, the motion of D through the ether to meet the photon entails that the distance travelled is less. This should result in a measurable difference in the time it takes for the photon to traverse A-B as compared to C-D. Rigorous experiments of this sort, however, have revealed no difference in the intervals at all.

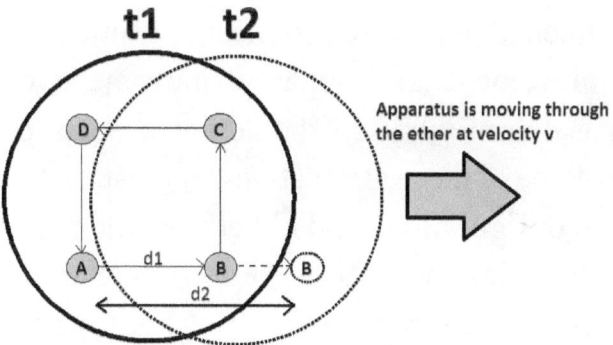

Figure 8.7 Depiction of the Michelson-Morley experiment. *The dark circle shows a photon of light being emitted by A at time t1 and being reflected at B, C, and D, so that it returns to A. But imagine that the entire apparatus is moving from left to right through the stationary ether. This means that the apparatus takes up the position in the ether designated by the broken circle at time t2. The photon has to traverse a distance greater than d1 (namely, d2) to reach the new position of B. When the photon eventually reaches C and begins its journey towards D, the opposite situation will be the case. D will move forward to meet the photon and consequently the distance the light has to traverse will be less than d1.*

If the causal configuration could be likened merely to a privileged spatial frame of reference, then we would expect measurable differences in the time it takes light to traverse what appears to be the same spatial gap from the viewpoint of a particular observer. It *ought* to take longer to cross A-B than C-D. To see why this attitude is deficient, let us consider Figure 8.7 again. The photon takes four different directions before it returns to A. According to the classical account, direction A-B flies *directly in the face* of the oncoming ether, directions B-C and D-A go *across* the ether, whilst C-D is the *same direction* as the motion of the ether relative to the apparatus. If the causal configuration could be likened to a privileged spatial frame of reference, then all movement (and apparent movement) of objects in the causal configuration should correspond to a vector addition of one of these alternatives: against, across or with the causal configuration. And if that were the case, then set-ups like the Michelson-Morley experiment should reveal different intervals for the passage of light over distances that seem equal from a particular point of view.

The causal configuration, however, lies beneath the effects that *generate* the spatial viewpoint. It is not aligned against, across or with any particular direction in space. This great flux of reciprocal causal processes is, in a sense, evolving in all possible directions at once, and for that reason we can legitimately say that it transcends all particular trajectories. We can legitimately think of the causal configuration as being *stationary* with respect to all objects simultaneously. The interesting consequence of this view is that time will be dilated for objects that accelerate and attain high velocities, but experiments of the Michelson-Morley variety will still fail to detect any privileged direction in the causal configuration. Let us examine this curious fact in more detail.

Consider again Figure 8.6. The clock may not be moving through an ether, but it has undergone acceleration and is moving rapidly with respect to the original frame of reference. The process of acceleration places all the atoms composing the apparatus into a relationship of rapidly shifting causal relationships with the other bodies in the original inertial frame. When the clock is now turned on and the group of atoms at A emits a causal impulse towards the atoms at B, the impulse must evolve as impulses always too: in step by step reciprocal interaction with the other dynamics that are concurrently unfolding. Because the clock is moving rapidly in the causal configuration, the atoms in the clock are continually readjusting their relationships with the rest of the system. The causal gap between A and B (in other words, the complexity of the evolution of the impulse) has increased, and the process will take a longer interval of time to be completed.

Now let us return to Figure 8.7. It was felt by those who believed in the ether hypothesis that the photon should take longer to traverse A-B than C-D, because A-B involved crossing a bigger stretch of the ether. From the point of the causal configuration, by contrast, there is no privileged direction in space. The frame of reference of an apparatus situated on earth is *certainly moving* in the causal configuration of things, but this movement does not have a particular direction relative to the configuration itself. When the photon is evolving from A to B, it is moving relative to the causal configuration to the *very same extent* as when it is evolving from C to D. Therefore no experimental apparatus will obtain different intervals for these legs of the journey. From the point of view of our imagination, the task of understanding this situation is made difficult by the fact that we are constrained to visualizing things spatially. If we claim that an object is moving relative to the causal configuration, then we think that this movement must have a particular *direction*. The challenge for our imagination is to refrain from thinking of the system as having a particular orientation to any given frame of reference. This exposes us to the danger of considering that movement and trajectories in space are pure features of our way of organising our experience, with no deep significance in themselves, which is simply not true. But if the causal configuration can somehow be likened to a heaving flux that is moving in all directions at once, then in what sense can the earth (or any reference frame) be said to be moving *at all* relative to such chaos? To clarify this issue we will turn to the question of inertial frames and acceleration.

8.5 Acceleration and the meaning of inertial frames

We have argued that time dilation occurs when a clock is accelerated to a high velocity because such velocity entails constant readjustment of the causal relationships of the atoms in the clock with the rest of the causal configuration. This follows from our understanding of the ontological basis of movement in space. The greater causal gap traversed by an impulse with respect to a moving target (more on this later) takes more time and the functioning of the clock is consequently slowed, but the system's clock continues to tick at its universal pace.

The results of the Michelson-Morley experiment demonstrate that any delay in the transmission of an impulse is not related to the orientation of evolution of the impulse in the causal configuration of things. This might seem an unusual conclusion to derive from the experiment, but it is legitimate. Believers in the ether expected the measured velocity of the photon to depend on the direction in which the source (or intermediate reflector) of the photon was pointed in the ether. When no difference in velocity was detected, it raised the question as to whether the ether existed at all. But we can equally interpret it to support the conclusion that motion delays the evolution of an impulse *regardless of the direction of the impulse* within the moving inertial frame. We can see this more clearly with the aid of Figure 8.8.

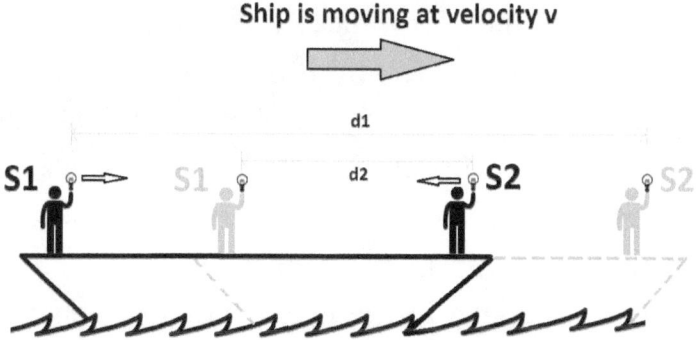

Figure 8.8 Two sources of light emit beams in opposite directions on a moving ship. *The diagram shows the original position of the ship when the impulses were emitted, and the new position of the ship (in broken lines) when the impulses arrive at their targets.*

A ship is moving at velocity v relative to the water. Sources of light (S1 and S2) at either end of the vessel emit light in the direction of each other. By the time the light from S1 reaches S2, the ship has moved a good distance to the right. From the point of view of an observer who is stationary with respect to the water, the light from S1 has to travel distance d1, which is considerably greater than the length of the ship. By contrast, the light from S2 will only have to travel distance d2 (from the perspective of the reference frame of the water), since S1 has converged on the light after it was emitted. Relativity theory recognizes that paradoxes of this sort (the possibilities for setting up maddening variations of these are endless) result from the illegitimate ascription of magnitudes (extension in space, intervals of time) to bodies that inhabit one frame of reference (such as the boat, in this case) from the point of view of a completely different frame of reference (such as the stationary water).

But the point that we want to make is different. If the ether hypothesis was true, and if the water was stationary with respect to the ether, then the photons of light from S1 and S2 would have different distances to travel because of their opposing directions with respect to the relative movement of the ship and the ether. Now of course this author doesn't believe in the ether, but *does* believe in a causal configuration that is "stationary" (in a manner of speaking) with respect to the evolution of light. This might lead some people to think that a moving ship *must* have a trajectory within the causal configuration. And if it had a trajectory, then an impulse of light emitted on board should require a different time to traverse the vessel depending on whether it was going in the same direction as the ship or towards the rear. This is where it is essential to contemplate the non-spatial character of the causal configuration. The impulse of light that evolves from S1 traverses *the exact same amount* of the causal configuration as the impulse from S2. The causal configuration is absolutely neutral as far as trajectories in space are concerned. It is the ground from which trajectories in general are generated. Whatever direction we are going in space has no bearing on our genuine motion relative to the ontological ground of the system.

This is why Michelson-Morley did not detect any difference in interval for the various perpendicular stages of the journey undertaken by their photons. The empirical evidence for time-dilation, on the other hand, shows that motion through the causal configuration *can* lengthen the magnitude of the causal gap that must be crossed by impulses of light. When these empirical results are put together, the picture that emerges is as follows: Impulses of light that reverberate towards atoms in motion sometimes traverse a "denser" causal configuration space (during the last reverberation before arriving at their target). This delays their progress. That is not to say that these impulses are travelling slower than would have been the case if the target atom was not in motion (later we will discuss why it is the motion of the *target* atom, rather than the source, that is responsible for the delay). The rate of reverberation of light through the system remains constant, but the causal configuration space can be made "denser" by the fact that the target is moving. This motion has added causal complexity to the situation and the impulse has consequently more causal configuration "distance" to reverberate through. The *magnitude* of the causal configuration distance, however, does not depend on the direction of motion of the impulse (as would be the case for the ether understanding of the transmission of light). In Figure 8.8, the photon from S1 had to traverse more of the ether because S2 moved away through the ether before the photon arrived. By contrast, the photon from S2 had less of the ether to traverse, because S1 moved through the ether and met the photon while it was on its way. On our understanding (and the Michelson-Morley results can be used in support of our view), the *direction* of motion of the source does not have any such consequences. *The simple fact that the eventual target atom is in motion* will entail that an impulse that evolves to it will have reverberated through a denser causal configuration space. The direction of the motion is not relevant. The impulses from S1 and S2 both traverse the same distance in the causal configuration, even though they are going in opposite directions, because there is no difference in the "density" of the configuration traversed by either.

This leads to the question of the precise nature of the motion that is being referred to here. Up to now, the term "motion" has been used in an ambiguous manner. The universe is composed of objects flying in all directions. Therefore *every* object in the universe is travelling at great speed relative to *some* other object. And the speed of an object relative to one object will be completely different to its speed relative to another. Would clocks on all of these objects show different dilations of time relative to each other? The key to understanding "ground zero" as far as the dilation of time is concerned is the phenomenon of acceleration. The theory of relativity itself, in fact, must appeal to acceleration for resolving questions such as the "twin paradox". According to the paradox, one twin boards a spacecraft and travels around the universe at almost the speed of light. When many years have elapsed on earth, he returns. Time will have slowed down for him, according to the theory, so he will be now much younger than his sibling. The paradox arises from the supposition that all reference frames are equivalent. This would mean that the twin on earth can equally be thought of as having travelled at high speed while the spacecraft remained stationary. Why should time dilate only for the twin on the spacecraft? The answer is that only he undergoes the process of acceleration to achieve his eventual state of high velocity.

It is only when we assert, however, that there is an *ultimate* ontological grounding to which all spatial frames are relative, that we obtain a clear explanation as to why acceleration should slow down the functioning of clocks. From a cosmological point of view, this explanation fits in with the belief that the universe originated in a centralised, concentrated event like the big bang. If all matter has such a single origin, then every material object stands in a certain relation to that historical event as far as its motion is concerned. And they also stand in a definite relationship with each other. In overly simplistic terms, the objects that are now flying outwards at great speed from the original explosive event could (in theory) have their steps retraced to the point and time where their motion began. This "zero moment" represents a state of rest relative to the system as a whole (not to be identified with a frame of reference that is stationary with respect to that moment). The object's present state of motion relative to that original state was achieved with a definite magnitude of acceleration, the sum total of all the accelerations and decelerations that it has undergone since the beginning of time. From this point of view, we think of the object as having an absolute speed which has been achieved by a very definite quantity of acceleration, but (as has been reiterated many times above) we are not to think of this motion as having a definite *direction* with respect to an original absolute frame of reference in the spatial sense.

The challenge of this approach is to be able to imagine that there exists an absolute causal configuration of things, whilst refraining from conceiving of such a system spatially. This permits us to assert that objects can be thought of as having undergone a definite magnitude of acceleration (since the moment of the creation of the world) without saying that the motion of the object stands in a particular orientation to the causal configuration. The system at the ontological level is not spatial by nature. Our motion *does* stand in a certain fixed relation to that ontological substrate from the point of view of the *quantity* of that motion. But because the causal configuration cannot be reduced to a spatial perspective, the *direction* of our motion simply cannot be said to be oriented in a particular way to the ontological grounding of the system. The fact that all motion has a definite *quantity* with respect to the original state of zero movement of the universe means that the functioning of clocks will be slowed by a definite amount for the reasons elaborated in section 8.4. There we argued that the greater the quantity of motion an object has relative to the system, the more complex the causal evolution of impulses transmitted between the parts of such an object. In other words, these impulses will effectively have to traverse a greater "distance" in causal terms. Every object has a history of acceleration that can be retraced back to the big bang. Consequently, the current state of motion of the object (in absolute terms) is a measure of the time interval required for a causal process to evolve between the genesis-units that make up the object.

Objects cannot be said to be moving in a particular *direction* with respect to the causal configuration of things, but it is extremely likely that most objects in the universe *are* moving. The direction of movement of an object only becomes relevant when we start to consider the relationships *between* objects (as opposed to the relationship of an object's direction of movement to the causal configuration as a whole). If an object is moving towards or away from another object, then this fact will influence the potency of any impulses transmitted between them. In Figure 8.8, two light sources on board a ship emitted impulses towards each other. Because they are stationary with respect to each other, the magnitude of these impulses cannot be influenced by their relative motion (since there is none). However, an observer behind the boat who is stationary relative to the water will measure S2's impulse to have a reduced potency (lower frequency). We discussed the relationship between the measured frequency of light and the relative motion of source and target in section 8.2. The conclusion was that the potency of light is influenced by the relative motion of source and target. Whilst there is a component of the potency that depends only on the magnitude of the electrostatic alignment that disassociated to emit the light, the impulse can gain or lose potency depending on the relative motion of source and target.

All of this is perfectly in keeping with the view that there is an ontological grounding for material reality. This grounding (the genesis-units that constitute reality) is responsible for the magnitudes of the causal impulses that are transmitted between objects. Objects have a quantity of motion in relation to this ontological grounding (their "absolute" motion) and that will affect the way that causal processes evolve in them. The *relative* motion of objects will also contribute a component to the *potency* of processes that evolve between those objects, and this will be independent of the absolute motion of those objects relative to the causal configuration. Indeed, a significant portion of the causal processes that evolve in the world (all causal events that operate by contact action, for example) will have a causal dynamics that can be more or less reduced to a pure consideration of relative motion. This observation prompted Galileo to set down his principle of relativity. The version of relativity that appeared in the twentieth century meant the demise of the belief in an absolute frame of reference that had become orthodoxy with the success of electromagnetic theory.

The approach that we are advocating is not a return to an absolute frame of reference but a claim that there exists a common ontological grounding that gives rise to the empirical data which generate frames of reference.

A frame of reference is a particular perspective on the common ontological grounding with respect to the motion of the observer. There is a component in this perspective that is due to the motion of the observer relative to the *system* and a component that is due to his motion relative to other objects. The component that is relative to the system (in other words, his "absolute motion") does not have a particular direction but a *quantity*. To calculate it, we would need to know the entire history of acceleration and deceleration of the physical material constituting the observer since the big bang. This component is responsible for the delay in the functioning of clocks and the extra causal distance in general that must be traversed by impulses that are passed between atoms that have attained this quantity of motion. The *relative* motion between the observer's frame of reference and other objects, by contrast, has no influence on the rate of transmission of impulses, but it does alter their measured potency, giving rise to the phenomena of blueshift and redshift observed in electromagnetic radiation emitted by converging and receding sources respectively.

Do all observers who are stationary with respect to each other have the same magnitude of absolute motion? This question is a good way to reflect on how the causal configuration differs from the classic conception of the ether. Imagine a very long train moving at a hundred miles per hour. A person in the front carriage, X, begins running towards the rear at a hundred miles per hour relative to the train. X is now "at rest" relative to Y, who is standing on the platform, but he probably has a different absolute motion to Y. X underwent the initial acceleration when the train began moving, and then he accelerated again to attain one hundred miles per hour relative to the train. Thus his motion has been increased *twice* with respect to the causal configuration. Most objects, in fact, are in a fairly regular state of ever increasing motion relative to reality. Contrast this with the ether framework. If the platform happens to be in the reference frame of the ether, then the double acceleration undertaken by X actually *returns him* to a state of complete rest relative to the ether.

8.6 It is the motion of a target atom that is responsible for the apparent dilation of time

(As we read the following section it is essential to keep in mind that the quantity of motion of an *object* relative to the causal configuration of things is a completely different matter to the velocity of evolution of an *electromagnetic impulse* through the causal configuration). Section 8.4 took a preliminary look at the apparent slowing of time for moving objects. It was claimed that the time it takes an impulse to traverse a causal gap depends on the genuine quantity of motion of the clock relative to the common ontological grounding of reality. However, there was no discussion as to whether the apparent dilation of time was connected to the motion of the atom that was the source of the impulse or the motion of the target atom. Here it will be sustained that the apparent dilation of time is related to the motion of the target atom only.

Consider the case of the binary pulsar mentioned earlier (see Figure 8.3). Star A is orbiting star B at great velocity and emits photon 1 as it converges on the observer and photon 2 while it is receding. Given the star's circular orbit and its constant process of acceleration and deceleration (due to the dynamic between its own inertia and the gravitational attraction of the other star) there can be little doubt that the quantity of motion possessed by the star relative to the system itself is in constant fluctuation. But all the photons of light emitted evolve towards the observer at the same rate. This conflicts with the observation that impulses passed between the atoms of a moving clock require a longer interval for their transmission. If processes are forced to evolve through a denser causal "space" on account of the movement of an object, then shouldn't the photons that issue from star A require different intervals to reach the observer depending on the quantity of motion A had at the moment of each emission? The solution to this conflict involves once again refraining from the tendency of seeking to apply magnitudes measurable from one frame of reference (that of the observer) to another frame altogether (that of the moving star). An impulse that evolves from the binary pulsar to the eye of an observer has not traversed a denser causal space on account of the motion of the *star*. It has traversed a causal space whose density is determined by the motion of the *observer*. Let us see how this happens in more detail.

The electrostatic alignment of the mini-system of atoms in the star disassociates and the impulse is "dropped" into the causal configuration of things where it begins to evolve. For simplicity, let us imagine that there are only two possibilities for the final evolution of this impulse.

Firstly, there is the possibility that the impulse will give rise to a change in a group of atoms that is already *part of the moving star itself.* Secondly, there is the possibility that the impulse will evolve beyond the star and eventually cause a change in the retina of the observer. Both of these evolutions (and multiple others which we are not considering) actually begin at the same time, but only one of them will actually come to completion. Once it has come to completion, then a suitably-placed observer will be able to say, "The light took this particular path, travelling at this particular speed." But the nature of the causal configuration of things is that processes of this sort have a complexity that cannot be comprehended from the narrow point of view of the spatial perspective. Processes simply evolve from a causal source, being interacted with by other processes that are concurrently in evolution. The impulse emitted by the star reverberates away from its source atoms. During its evolution it may come into reciprocal interaction with another group of atoms in the star itself, or in the eyes of an observer on earth. The final rate of evolution will depend on the quantity of motion of those potential target atoms with respect to the ontological grounding of reality.

As we argued in 8.2, it is only in the ultimate reverberation of the light towards its target that the motion of the target becomes relevant. There are two types of motion that must be distinguished here. Firstly, there is the relative motion of source and target. As discussed earlier, this will affect the *potency* of the light. Secondly, there is the absolute motion of the target with respect to the ontological ground of reality. This will affect the *rate of evolution* of the light during this last reverberation.

In section 8.4 we steered away from this issue in order to develop the approach to inertial frames in a little more detail. When a moving clock has its functioning delayed, it is not the movement of the *source* atoms for each tick that is relevant. When an impulse is dropped into the sea of evolving flux that is reality, it simply begins to evolve concurrently with all the other processes that are in progression. The *relative* movement of source and target will have an influence on the potency of the impulse, but the *absolute* motion of the *source* with respect to the causal configuration will not have any influence on the rate of evolution of the process. However, the motion of the target atom is a different story altogether. The absolute motion of the target will influence the density of causal space that the impulse must traverse during its last reverberation in order to reach its destination. This makes intuitive sense. Once a source drops its impulse in the sea of the causal configuration, its work is done. If the source is destroyed before the impulse reaches its destination (as is often the case when a star emits light that takes millions of years to arrive at a target), then this will have no influence whatsoever on the evolution of the process. It is the *target* that enters into interaction with an evolving impulse. A target that has a high quantity of authentic motion with respect to the system is moving through a denser causal space than an object with less motion. Motion entails constant readjustment of causal relationships between the atoms of the object and the rest of the system.

Why does the last "reverberation" of light before reaching its target have such importance? This is a way of expressing a truth that other writers might be able to describe more clearly. The causal configuration is one. The evolution of light occurs at a constant rate because it is really the rate of evolution of causal change in the entire system itself. But every object in the system has its own individual properties arising from its own particular history. The way that light interacts with each such object will depend on the state of that object - its relative motion relative to the source of light and its absolute motion relative to the system. The "last reverberation" of the light is just a way of referring to the actual moment in which the target object with its individual properties interacts with this evolving causal impulse we call light. Up to this moment, light has evolved to the object through the system in a uniform and constant way, just as it would have evolved to any other target object. Now, however, the *state of the object itself* makes its contribution to how the causal change will finally unfold. It is now that the impulse will have either an increased or diminished effect, and that the timing of its final outcome will be affected by the absolute motion of the target.

8.7 The empirical adequacy of relativity theory

The Newtonian worldview considered space to be an absolute framework within which all causal interactions played themselves out. The recognition that certain properties of objects and events (the quantity of spatial extension, simultaneity, the measured rate of the passage of time) actually depended on the frame of reference of the observer led to the demise of the view that there existed a privileged absolute framework. The perspective from each different inertial frame seemed to be equally valid. But such a swing from absolutism to relativism leaves something vital out of the picture. It is true that causal processes do not play themselves out within some absolute *spatial* framework. However, to conclude from this that all spatial viewpoints have equal validity is only to tell half the story. What *also* needs to be said is that spatial viewpoints in general are woefully *deficient*. The only reason that they are equally valid is because *none of them* provides a comprehensive, non-perspectival description of the causal configuration that produces the effects that generate these relative viewpoints in the first place. If no privileged spatial viewpoint exists, then let us not simply make do with a declaration of the equivalence of viewpoints and the development of mathematical techniques for transforming one into another! These mathematical techniques have proved themselves capable of saving the phenomena with spectacular success. But we can do more than that as far as understanding our world is concerned. We are not bound merely to the effects that generate particular spatial frames of reference. Reflection on the spatial perspective and the way that is generated can reveal much that is non-perspectival about the nature of reality.

Relativity arose in a milieu that favoured a positivist approach to science. Metaphysical frameworks, substances or ethers that could not be verified by hard empirical evidence were frowned upon in many circles. Hypothetical causal mechanisms that evaded empirical probing were considered superfluous. In this environment, it is easy to see how an almost purely kinematic approach to light should have come to the fore. Special relativity does not discuss the nature of the transmission of light and focusses instead on the characteristics of its motion, and on the spatiotemporal characteristics of objects whose velocities approach that of light. But if our spatial viewpoints and all that goes with them (velocity, extension in space, simultaneity) are generated merely from the *effects* of causal processes, then a purely kinematical approach to light and motion will be dramatically incomplete.

Already, in the general theory of relativity, we can see a return to an approach that is no longer content to describe the motion of bodies. The properties of matter (mass, density) determine the properties of space-time, influencing the movement of bodies and the transmission of light in the locality. This converges on our approach in the sense that we seek to understand all physical phenomena in terms of the properties of the matter that produces these phenomena. But we reject the reification of space-time. This is only generated in our perceptual apparatus by the effects of these underlying processes that are evolving between material bodies. Relativity, in fact, is a sophisticated way of calculating transformations between inertial frames. Given, for example, that spacecraft A accelerates relative to an observer until it achieves velocity v, relativity can calculate the time dilation and measurements of lengths or distances obtained by A relative to the observer. There can be no doubt that these sound mathematical techniques *will* save the phenomena. The phenomena, after all, just *are* the product of the effects induced in our sensory apparatus, and these mathematical procedures correctly describe the effects that *would* have been produced if my motion relative to some inertial frame was different than what it is.

It seems fair to say that the description of nature furnished by relativity can stimulate much reflection regarding the non-relativistic substrate that underlies the perspectives generated in different inertial frames. The constancy of the speed of light, ironically, points to a unity and universality in nature that is not belied by the particularity of spatial reference frames. Such constancy is a telling indication that the perspectival is not the last word. Even perspectival elements, such as the measured dilation of time, can be reinterpreted as pointing to a universal world order that generates these perspectival data when the observer has genuine motion relative to the underlying reality. After all, many people accept the common origin of all material bodies in a single event like the big bang. This would have given rise to a unitary world order, no matter how diverse reality might have become in its subsequent development. In theory, it should be possible to describe all objects and events in terms of the way they have unfolded from that moment of common origin. The *current* unfolding of the universe is what we are terming the "causal configuration of things".

The quantum mechanical formalism provides an empirically adequate account of material phenomena because it takes into account the multiple causal players in a system (such as the joint influence of two apertures).

This empirical accuracy can be achieved even if the terms of the formalism (such as positions attributed to particles) do not actually refer to anything in nature. Relativity rightly recognizes the constancy of the speed of light and of the influence that mass and acceleration exert on the evolution of causal processes. Like quantum mechanics, its terms (space-time and its curvature) need not actually refer to anything in reality for the predictions of the theory to be empirically adequate. The theory is fundamentally oriented to *cataloguing what an observer will perceive* in another inertial frame from the point of view of his own. And the fact that it achieves this aim does not mean that the theory is "complete" in any more profound a sense as the "completion" of the formalism of quantum mechanics. Science still has a duty to probe the reality beneath the phenomena of inertial frames and it cannot be content to resign itself to the sentiment that one perspective is as good as another. Rational reflection can aid us in reaching the reality beneath the appearances, a reality that operates according to universal laws and which is governed by a uniform evolution of time.

Bats generate a primitive representation of their environment from the ultrasound signals that they emit. Sounds are emitted by the bat and the delays in the returning echoes indicate the positions of objects in the vicinity. A bat that is moving faster will encounter the returning echoes more quickly, meaning that his "picture" of the world will be consequently different. Imagine a very intelligent bat who develops a mathematical formalism that will calculate what any observer using echolocation would experience of a particular object depending on their relative velocity towards that object. The perfect capacity of this formalism to save the phenomena would not be a measure of the completeness with which the formalism describes the nature of the world. Rather, it would be a measure of the competency of the formalism in describing what moving bats using echolocation will experience of the world.

8.8 The type of relativity that the process approach espouses

The cosmological view defended here is one that seemed to have lost much ground during the twentieth century, even though there can be little doubt that it has continued to guide the practical reasoning of a great number of professional scientists. This is the view that there is a unitary reality that underlies all particular perspectives; underlying facts of the matter that are fundamentally governed by universal laws which unfold in deterministic ways. Despite the solidity and unity of this underlying reality, the perspective generated by the system in an observer's sensory apparatus will be influenced by the situation of the observer relative to other objects. And not only this: the genuine motion of an observer relative to the *system* will influence the way that causal processes change the atoms in the reference frame of the observer, including the atoms that make up the subject himself. The various relativistic and non-relativistic aspects of motion can be summarized as follows:

i) *Bodies have an absolute motion relative to the system.*
   Objects have a genuine motion with respect to the

system. This is the sum total of all accelerations and decelerations of the atoms of the object relative to the system since they came into being. Any change in the current state of motion of an object with respect to the system requires work to be done in order to accelerate or decelerate the body. All acceleration involves a readjustment of the genuine absolute motion of a body relative to the system.

ii) *The causal configuration is not associated with a privileged orientation in space.* If absolute space existed, then a ball thrown from a moving ship would either go against, across, or with the privileged frame of reference. In our account, all genuine motion is relative to the system, but it does not have a direction. Therefore if a ship has genuine motion relative to the system, it will make no difference whether a source on board emits light in the direction of motion of the ship or backwards. The impulse will not have greater or lesser motion with respect to the system because the ship's genuine motion is not associated with a direction in *space*. This is a coherent interpretation of the negative result of the Michelson-Morley experiment.

iii) *The magnitude of measured spatial extension depends on the frame of reference of the observer.* This is a valuable insight from relativity theory. The spatial magnitudes that we observe depend on our frame of reference. This is not surprising given that the way that the causal configuration will generate our viewpoint on any object will be influenced by our motion relative to that object. It is therefore illegitimate to attribute magnitudes to an object in one frame of reference from the results of observations on that object carried out from another frame altogether.

iv) *The dilation of time is perspectival even though processes are delayed.* Processes take a longer time to evolve towards objects that are in genuine motion relative to the system. This is because the movement of the target entails that it is evolving in a denser causal space, delaying the progression of the impulse to the target during its final reverberation. But this does not entail that time itself is slowed down.

v) *The magnitude of potency of an impulse is influenced by the relative motion of source and target.* The potency of an impulse of light is determined in the first place by the magnitude of the electrostatic alignment that disassociates to create the impulse. This magnitude depends on the autonomous properties of genesis-units and is non-perspectival. However, the potency of an impulse will be diminished or increased (experienced as blueshift or redshift) when the sources and recipients of light are in relative motion.

Concluding Remarks

"It is the function of science to discover the existence of a general reign of order in nature and to find the causes governing this order"

Dmitri Mendeleev

To some, a work of this sort might seem unduly ambitious. As if it were not enough to present an entirely new model of the constitution of matter, we have also offered a different picture of the transmission of light, denied the very existence of the electron and the entire zoo of subatomic particles, outlined a new approach to electric charge and magnetism, attempted to place the structure of the atomic table and the nature of chemical bonds on a new physical foundation, and more besides. But the edifice of physics that has been constructed in the past one hundred years is a house of cards standing on foundations that are disordered and problematic. It is not possible to address these issues without the entire house seeming to come down around us.

As stated in the introduction, however, the figures in physics can still be retained. The number that Moseley identified with the quantity of protons in the nucleus can be reinterpreted to represent the number of proactive genesis-units that are fused in an atom. The valence of atoms may not have anything to do with the electrons in the outer shell, but we have seen how it may correspond to a different physical feature of the atom that determines its bonding behaviour. The terms in the quantum mechanical formalism may not refer to anything in nature, but the formalism still works and can be placed on a sounder philosophical basis if we interpret its probabilistic expression of the position of a hypothetical particle (for example) as the probability that a process will have its effect in a particular position. The notions of "frequency" and "wavelength" may be inappropriate ways of describing a non-wavelike phenomenon, but their magnitudes still have a physical basis in the authentic properties of evolving processes. All of the hard-earned formulae and empirical patterns discovered by science constitute a valid contribution to the description of the world. It is the physical interpretation of their terms that needs to be transformed and put into order. The renewal of science can only be carried out by professional physicists working with hard empirical data. With the quality of experimental evidence at hand, all that is needed is that they keep their minds attuned upwards to the order manifested by the heavens.

The project of this book was instigated in the beginning, many years ago, by simple reflection on the mystery of the two-path experiment. The author looked for a particle, but could discern nothing there, just like the little boy who could see that the emperor was without clothes. The ambition of this book is not entirely misplaced, for it is the ambition of a child who instinctively knows that the world is an orderly place, not governed by chaos, not the product of the unfolding of the probabilistic movement of "particles" that simply cannot be particles.

This conviction led to the realization that all of the anti-realist elements of quantum physics (indeterminacy of states, complementarity, superposition, collapse, breakdown in causality) could be accounted for in terms of evolving *processes*. A consideration of processes challenges us to look at the way that the spatial perspective is generated, how it has its origin in the effects produced when a causal influence prompts a change in a material object. The mysteries of complementarity, wave-particle-duality and indeterminism all fall before this key. As processes evolve, they can be jointly modified by multiple causal players (such as the open state of two apertures), giving rise to wave-like characteristics. If only one modifying causal player is involved (one path open), then the process will evolve along that path, simulating the trajectory of a particle. When processes come to completion (the act of measurement) they always have their effect in individual target atoms, producing localized effects typical of a particle. In short, processes evolving under the joint influence of multiple causal players produce patterns typical of a wave; those evolving under a single influence produce patterns typical of a particle; but their ultimate upshot is always to produce changes in individual target atoms. Thus wave-particle duality and complementarity are explained by the being or not being of mutually-exclusive evolutionary conditions confronting a process. To be or not to be, that is the answer.

The shortcomings of this book will be evident to professional scientists, perhaps to almost everyone. It is not certain, however, that the conceptual content offered here has any greater deficiencies than that contained in current elements of theory such as "spin", "entanglement", and a host of others. Nor is it obvious that the simple picture presented of matter is more objectionable than the contemporary picture of the atom. It is the search for simplicity and order that must characterise science. The dramatic simplicity of the picture outlined by Copernicus describing the revolutions of the celestial spheres was eventually taken as a hallmark of its truth. The first thing we ask of a hypothesis is that it save the phenomena in a coherent and plausible manner. Have we done that? Only the scientific community can answer. Then the question can be posed as to whether or not the order and simplicity of our account is the order and simplicity of truth.

Appendix

REDUCTIONISM AND SCIENTIFIC METHOD

"Give me matter and motion, and I will construct the universe"

Rene Descartes

Summary

From the seventeenth century onwards, natural philosophers began to reduce the set of physical principles that were considered necessary to describe the behaviour of an object. Eventually, the properties of position and momentum became the heart of a dominant programme in physics that we call the "matter-in-motion project". The aim of this project was to reduce all material phenomena to a description of matter in motion regulated by basic principles such as the principle of contact action. Today the project continues in a more nuanced form. Particles, waves and fields are all attempts to account for material phenomena in terms of the transmission of an impulse by means of an intermediary or through an intermediary. The success of the project in bringing a large range of causal interactions under its umbrella has only served to further convince scientists that all phenomena are reducible to dynamics of this sort. But the project is excessively reductionistic. Genuine features of the evolution of causal processes are left to one side in the effort to reduce all mechanisms to one particular school. We consider the reasons why the project has been so successful in the description of the behaviour of macro-objects. We examine why success of this sort should not be taken as entailing that reality itself is limited to causal mechanisms that operate according to these principles only. The ironic thing about the matter-in-motion project is that it arises from an attempt to provide a simple and coherent account of how material phenomena originate. However, when it is applied to mechanisms that *do not* operate according to these simple principles, then it ends up constructing a fantastic and ad-hoc description of their genesis.

A.1 The reductionism of seventeenth-century science

During the seventeenth century, natural philosophers turned their attention to the autonomous physical properties of objects themselves, and away from the anthropomorphic and teleological considerations that had dominated earlier attempts at understanding celestial and terrestrial phenomena. As the scientific revolution advanced, there was a progressive and dramatic narrowing of the set of physical principles that were considered essential for a complete description of the behaviour of objects. Unlike the natural philosophy of previous ages, it was now believed that a material object could be fully described in terms of a few definitional properties like position and momentum. If we leave aside for the moment the question of the nature of gravitation, the causal efficacy of material objects could now be reduced to a single dominant principle - the principle of *contact action*. Objects exerted causal influence by imparting motion to other objects through direct substantial contact, or by means of intermediaries like waves or particles that guaranteed such contact.

The narrow set of definitional properties of objects, plus the principle of contact action, became the heart of a dominant programme in science that has never since been displaced from centre stage. This programme can be accurately labelled "the matter-in-motion project". During the seventeenth century, the motion of matter was invoked to explain such diverse phenomena as heat (in terms of vibration of particles) and gravitation (in terms of the effects of "vortices" in the medium that permeated space). Action-at-a-distance treatments of electricity and magnetism were tolerated temporarily until the nineteenth century, when accounts of fields and lines of force (in terms of stresses in a material medium) were offered as attempts to provide mechanical contact of sorts between cause and effect. As time wore on, only four phenomena were considered to offer significant resistance to the matter-in-motion project: gravity, electromagnetism, and the strong and weak nuclear forces. Today, the project is still in full swing in a physics that is dominated by the analysis of particle collisions (the motion of matter taken to an extreme) and the search for hypothetical particles that will ultimately reduce these four "forces" to matter in motion.

This dominant project, with its minimalist set of mechanical principles, is in fact *reductionistic* in a sense that increases the predictive/descriptive expediency of its mathematical component, but mitigates against its capacity to express the truth about material reality. Later in this appendix we will consider how a theoretical framework can increase its power to predict/describe natural phenomena by the very act of *ignoring* genuine dynamical components in nature. We will claim that the matter-in-motion project has systematically disregarded certain modes of causal interaction in its attempt to explain the world, and it has done so to its own detriment, as the conceptual confusion in atomic physics testifies.

Before considering what the project ignores, we will first consider what it prioritizes. The priority of the matter-in-motion project has been to explain natural phenomena in terms of causal mechanisms that operate according to the principle of contact action.

A.2 The centrality of the principle of contact action

The Aristotelian dictum that an object cannot act where it is not present has reverberated through science until the present day. By the middle of the seventeenth century a mechanical version of this dictum had not just been established at the heart of physics, but had become the dominant principle by which natural philosophers fully expected to explain all material phenomena. The mechanical version of the principle is intuitively plausible. The idea is that an object cannot have causal influence on another object unless the objects are in physical contact with each other and thus in a position to exert causal influence by means of this contact. This principle became one of the pillars of physics from the seventeenth century onwards, and it tried to reduce the nature of causal influence exclusively to *motion*. Objects were believed to influence each other by imparting motion to each other, and this required substantial contact. A principle of the conservation of motion (or momentum) became the most fundamental law governing the causal interaction of objects with each other. In the influential Cartesian scheme, the world was construed as a collection of particles in motion, and all natural phenomena including gravitation were believed to arise from the collateral effects of the collisions, governed exclusively by the law of conservation of motion.

There is an aspect of the mechanical version of the principle of contact action that is easy to overlook, and that is the view of the nature of space that underlies the principle. There is an implicit assumption that position in space *causally insulates* objects that are located in different regions of space. This belief in the causal insulation entailed by spatial separation was more clearly evident in the Aristotelian form of the dictum. But why are we inclined to think that spatial separation entails causal insulation?

In the "absolutist' conception of space that became dominant in classical mechanics during the eighteenth century, space is a container that has a sort of ontological priority to the objects that are scattered within it, as if it were a kind of substance in itself. It would make sense, in this conception, for a completely *empty* space to exist with no material objects inside it. Objects receive two of their three principal definitional predicates from this absolute framework – position and velocity relative to the structure of space (the third fundamental property of material objects in this conception is mass).

It is easy to appreciate how action at a distance is anathema to such a conception of space. Objects have absolute positions in the container at any given moment, and their movement relative to the framework ultimately determines which other objects they will interact with. This approach implicitly assumes that the position occupied by an object at a given moment causally insulates the object from any other object that does not occupy, or seek to occupy, the same position in that moment. Causal influence between spatially separated objects thus necessitates motion of one object towards the other (resulting in impact), or the emission of causal intermediaries, such as particles or waves, that will bear contact action between the objects.

The law of universal gravitation, as originally formulated, caused considerable controversy in the late seventeenth century because it was felt that it alluded to an "occult" force acting at a distance. This controversy demonstrates the insistence with which natural philosophers sought explanations that described the genesis of phenomena in terms of a causal story that respected the principle of contact action. The way in which the scientists of the eighteenth and nineteenth centuries became accustomed to speaking of gravitation as if it *were* a force acting at a distance is anomalous in the natural science of that period. When magnetic and electric phenomena were brought under inductive-empirical laws of the same kind as the inverse-square law for gravitation, physicists *continued* to seek a way to account for these phenomena in mechanical terms. The dissatisfaction of the scientific community with causal "explanations" of the action-at-a-distance sort (or *lack* of explanation, rather) is attested by the concerted attempts in the nineteenth century to reduce electromagnetism to a contact-action account. The notions of lines of force and the electromagnetic field were invoked in an effort to explain electromagnetic phenomena in terms of purely mechanical stresses and strains in a material medium. The relentless effort to couch physical explanations in contact-action terms continues today unabated as the physics community seeks to account for all natural phenomena in terms of the interaction of subatomic particles and the properties of fields.

This project – the modern face of the ancient matter-in-motion project – contains a number of unwarranted assumptions about the significance of contact action and the kind of causal insulation that is entailed by spatial location. Elsewhere in this book we have discussed how our spatial framework gives us a very partial perspective on the true causal configuration of the system. Any attempt to reduce the causal configuration to a spatial description will only succeed up to a certain point. Moreover, classical mechanics is reductionist in its inability to distinguish between the "classical" properties of an atomic unit and what we might call its "natural" properties. The classical properties of an atom include its position, mass and motion relative to other atoms. Such properties are universal features of all matter regardless of the type of substance that the atomic unit is an instance of. Along with these classical properties, atomic units also possess other properties that arise from the nature of their substance. The most important natural property is the polarity between the B and V components that constitutes matter itself. This property in turn has given rise to certain natural properties of the entire *system*, such as its tendency to maintain or increase electrostatic equilibrium. Physicists during the twentieth century have tried to account for these natural properties in terms of the *classical properties of more fundamental particles* constituting atoms, such as electrons, protons and neutrons (and an entire zoo of other particles).

These efforts to reduce the behaviour of atoms and the properties of systems to a classical account of fundamental particles moving in space have foundered. Atomic phenomena (not to mention gravitation) have resisted being subsumed under the reductionistic terms of the programme. Given that the framework cannot even save the empirical evidence coherently, it seems right and proper to reject the reductionistic bias of the framework, introducing instead the notion of natural properties that are irreducible to their classical counterparts.

The four "forces" mentioned earlier, that sit uneasily in the classical framework alongside purely mechanical laws, can all be interpreted as manifestations of these non-classical properties of physical systems. The failure to reduce these forces to a classical account can be taken as a philosophical justification for our introduction of a distinction between classical properties and natural properties that are irreducible to classical treatment. Similarly, the paradoxical behaviour of atomic objects in modern experiments, and phenomena such as the unintelligible *superposition* of classical properties and *collapse* of the wave-function, can also be read as providing justification for the introduction of our distinction.

During the twentieth century, the failure of the classical project led the pioneers of the standard interpretation of quantum mechanics to renounce the possibility of a spatiotemporal account of atomic interactions. But this renunciation was not accompanied by a presentation of an alternative causal dynamics for the atomic realm. Instead, the properties of atomic objects were understood in positivist or anti-realist ways, and causality itself was considered to break down at the atomic level.

Our approach, by contrast, interprets the failure of the classical project in much less catastrophic terms. As we see it, the project should *not* be seen as something that was performing admirably and then broke down inexplicably in twentieth century atomic experiments. The limitations of the classical framework *have been manifest since the seventeenth century* in its failure to account for gravitational, electrical and magnetic phenomena in contact-action terms. What twentieth century experimental results have indicated in a more conclusive way is that the project to reduce all physical phenomena to a classical description is doomed to failure. It is in the light of this *experimentally corroborated* irreducibility of physical phenomena to a classical account that we claim to have ample empirical justification for introducing a distinction between classical properties and the natural properties that derive from the B-V polarity in matter.

A.3 An analogy to illustrate how the natural properties of objects can often be reduced to their classical counterparts

Take the case of a macro-object composed of a large number of smaller objects. The nature of the bond that enables the parts to compose the larger object need not concern us. It is a simple fact of nature that when smaller units are bound together to compose a larger object, then the *characteristic* properties of each unit tend to have a *reduced* role in the behaviour of the newly-formed macro object. The properties that come to dominate in an account of the behaviour of the macro-object are *universal* properties of matter like its momentum and position, i.e., the properties that figure in the classical mechanical description of its behaviour.

Say that macro-object Y is composed of a collection of very many grains of sand, multiple particles of clay and a large number of droplets of water. When these elements are combined to form Y, the result is a large durable spherical object. When Y comes in contact with another similar macro-object, Z, the individual properties of the sand, water and clay that compose Y and Z will be largely irrelevant in describing their interaction, but the overall mass, relative position and velocity of the macro-objects will dominate the description. Thus the dynamics of the interaction between Y and Z can be *reduced* to the dynamics of the *combined* classical properties of the individual units making up the objects (D_C), whilst the dynamics that arises from the particular properties of the constituent parts (D_N) can effectively be ignored.

This rather crude analogy cannot be taken as a true demonstration of the difference between natural and classical properties. However it does succeed in showing how a form of reductionism can be extremely useful in accounting for the behaviour of macro-objects. Sand, water and clay have particular properties that derive from the nature of their substances, but they also possess the universal properties of material objects, such as position, mass and velocity. When these objects are combined to form a larger object, their particular properties tend to drop out of the description of the behaviour of the larger object. Once combined into a new unit, brute classical properties like mass and velocity can be summed together to form very significant quantities that dominate the dynamics of the larger object, whilst the particular properties of water (fluidity), for example, are not readily "summed" to the particular properties of clay (powder-likeness). In fact, in this case, such particular properties tend to negate each other's influence and do not figure at all in the description of the behaviour of the composite object.

This ability of classical mechanics to conveniently reduce the dynamics of Y and Z to D_C, ignoring the particular dynamics of the individual parts, D_N, can be explained in different terms as follows: The various parts of each macro-object have a combination of natural properties and classical properties. Natural properties derive from the particular nature of the individual unit in question (here we disregard gravitation for the moment) and may not be shared by other substances that compose the macro-object. Classical properties, instead, are universal features of *all* material objects. This universality means that at the level of the macro-object, Y, it is right and proper to speak of Y having a certain mass (the combined masses of all its various parts) but it makes no sense to speak of Y having a "combination of fluidity and powder-likeness". In this rather artificial example, thus, each individual unit has an amalgamation of natural and classical properties, but the macro-object can accurately enough be said to *exhibit* classical properties only. The set of properties that the object *exhibits*, however, does not exhaust the set of properties that are actually *possessed* by the object.

In contrast to the dynamics of macro-objects, the behaviour of *atomic* objects will be influenced by their individual natural properties in a relatively significant way. Take a system consisting of two atomic units a and b in isolation. It is reasonable to suggest that the dynamics that arises from the particular natural properties of a and b (D_N) is likely to have as significant an influence on the overall dynamics of a and b as the dynamics that arises from their classical properties, such as momentum and position (D_C). After all, a and b are individual atomic units in isolation and their masses will be relatively insignificant, allowing their other properties to have a more appreciable influence on the overall dynamics of the system. But now consider the case where a is part of a macro-object, A, composed of millions of atomic units, and b is part of a similarly large macro-object, B. In this case, the influence of the natural properties of *a* and *b* (as well as of the natural properties of all of the other individual atomic units composing *A* and *B*) will be insignificant in comparison to the *combined* influence of the universal or classical properties of those units, such as momentum and position. Universal properties tend to work together, resulting in the dominance of these properties at the macro-level (magnetised macro-objects, such as bar magnets, are an exception to this general rule. In these cases, the natural magnetic properties of the individual units composing the bar magnet are aligned and work together to produce magnetic influences at the macro-level). The fact that the natural properties of the individual units will tend to drop out of a description of the observable behaviour of the composite object does *not* mean, however, that natural properties *are not dominant* at the atomic level. Historically, the reductionistic project of classical mechanics in describing/predicting the *external* behaviour of macro-scale objects had such success that it became natural to think that its principles could be used to describe the *internal* workings of the atom as well. As a result, there has been a general failure to recognize that the natural properties operative at this level are irreducible to classical properties.

A.4 A more detailed look at why classical principles are so effective in describing the behaviour of macro-objects

The example given earlier of the macro-object made of sand, water and clay illustrated the way in which the dynamics of a macro-object can often be described accurately without reference to the natural dynamics of its individual parts. Why is classical mechanics so effective in accounting for the behaviour of macro-objects in purely classical terms? The dynamics of classical interactions (collisions, blocking influences, interplay of positions and momenta - in short a dynamics based on contact action) tends to come to the fore when we are considering large-scale bodies, and the natural dynamics of an object's constituent atoms can effectively be disregarded at this level.

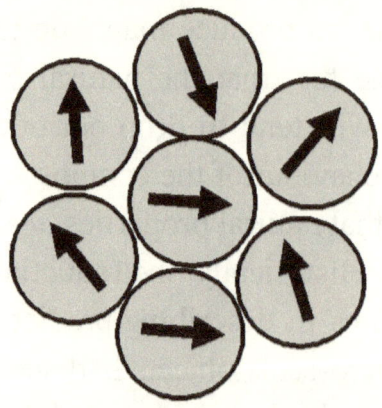

Figure A.1 The influence of natural properties in a body composed of many atoms. The natural properties of each atom tend not to work in unison to produce macro-level effects (if we disregard gravitational phenomena for the moment, and rare phenomena at the macro-level such as magnetism and radioactivity). In fact, the natural causal influences of individual atoms tend to cancel each other out.

Consider the object in Figure A.1 composed of a large number of atoms a_1, a_2, a_3, ..., a_n, each with their own natural polarity. These properties will have influence in a very reduced sphere, because the causal potential of the polarity in any atom (given its insignificance in relation to the world as a whole) will be extremely limited. When the number of atoms composing the macro-object becomes very great, the B-V polarities tend not to be aligned. Thus, macro-objects in general exhibit the collective summation of all the universal properties of their individual atoms, whereas their natural B-V polarities are not cumulative in this way (except when the object is magnetised), and tend not to influence the observable behaviour of the body on the macro-level. If these polarities do not influence the observable behaviour of the object, then they simply will not figure in the dynamical account that physicists formulate to describe that behaviour.

Figure A.2 illustrates the way that the universal properties of individual atoms combine together to produce significant classical effects at the macro-level. The very fact that the atoms are bound together in an individual body means that their universal properties *are* in harmony with each other. Each atom tends to have the same momentum in the same direction as the macro-object as a whole, and the proximity of each individual unit to each other effectively means that they operate as a single body with a unitary position. When such a body comes in contact with another body, we can effectively disregard the natural properties of the constituent atoms and describe the dynamics of the interaction purely in terms of the classical properties of the macro-object. The most important effects of classical properties involve the *impartation of momentum* through contact action, and thus the transmission of momentum through substantial contact became the linchpin of the dynamics of classical mechanics.

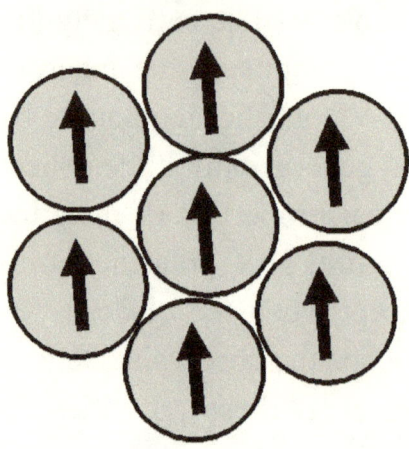

Figure A.2 The combined influence of universal properties in a body composed of many atoms. *The very fact that the atoms are combined together to form a body ensures that they have the same classical properties of (approximate) position and momentum in a given direction. Thus these properties have dominant influence on the macro-level and become the corner-stone of classical mechanics and its attempt to reduce all phenomena to a matter-in-motion description*

When the founders of quantum mechanics left aside large-scale phenomena and focussed their attention on the dynamics of atomic interactions, they were guilty of an understandable, but major, oversight. In the macro-world where billions of diverse atoms are gathered together into objects, it is highly expedient to disregard the natural properties of individual atoms and focus on the overall classical properties of macro-objects. But when these pioneers turned to consider the dynamics of *single* atoms, they continued effectively to disregard their natural properties! This disregard is expressed in the *projection* of the properties of large scale conglomerations of atoms into their account of subatomic mechanisms. It was a case of classical mechanics falling victim to its own success. A reductionistic dynamics was so effective in accounting for crude, contact-based, interaction between macro-objects, that scientists began to believe that the principles of this dynamics could explain *everything*. This belief was carried over to the world of atoms, despite the mounting evidence that elements had unique properties that were in some sense irreducible to classical properties.

The history of twentieth century physics is the story of the unravelling of this belief, coupled with a failure on the part of physicists to appreciate why classical dynamics does not work on the atomic level. It cannot work because it actively *disregards* fundamental properties of the constituent units of the world. Such a disregard may be just what is needed for a workable description of the dynamics of objects on the *macro-level*, but it is clearly inappropriate when we actually turn to consider those *natural properties themselves*. There are *good reasons* why classical dynamics is successful on the macro-level, and one of these reasons is *not*, as we have tried to indicate, that basic causal mechanisms *actually operate* according to a purely classical dynamics.

Some people might think that the most controversial aspect of the argument that we wish to defend lies in the claim that there exists a natural dynamics between objects that is irreducible to the classical framework. But the challenge of defending such an apparently controversial assertion is not as daunting as might appear at first sight - the entire weight of "strange" evidence from quantum physics can be invoked in support of this claim, not to mention the everyday poorly understood phenomena of gravitation, electricity and magnetism.

A.5 Two methods in science

The historical tension between the *hypothetico-deductive method* of science and the *empirical-inductive method* can be explained on the basis of our distinction between classical and natural properties. Natural philosophers using the hypothetico-deductive method have sought to formulate explanations of natural phenomena in terms of a causal story that operates according to the principles of classical mechanics, with the principle of contact action at the heart of the framework. In the case of gravity, electricity and magnetism (as well as a host of atomic phenomena), they have failed to come up with comprehensive and coherent explanations in classical terms. The best that can be achieved for such phenomena are *empirical-inductive laws* like the universal law of gravitation, and similar inverse-square laws for electricity and magnetism. Such laws save the phenomena but do not explain them. The distinction between classical and natural properties neatly explains this dichotomy in science. The hypothetico-deductive method is capable of offering explanations in classical terms for phenomena that *actually operate* according to the principles of classical mechanics (restricted sense of the term), such as the tectonic plate account of seismic phenomena. But in the case of natural phenomena (such as gravity and electromagnetism), which operate according to natural principles that are irreducible to the narrow classical account, the best that has so far been accomplished is an empirical-inductive law that saves the phenomena.

For more than a hundred years now, a concerted effort has been made to understand the interior structure and workings of the atom. Rigorous experimental techniques and sophisticated mathematical tools have been brought to bear on the operation. Vast quantities of empirical data have been gathered. How do scientists use this data to advance our understanding of the atom? How is the interpretation of this data shaped by elements of theory that already have been accepted into the canon of physics?

The method used to convert raw data into atomic theory could be described as a combination of the hypothetico-deductive approach and the empirical-inductive approach. The hypothetico-deductive method examines the empirical data and tries to infer a physical structure that would conceivably give rise to such data. Once a causal story has been formulated that can account for the genesis of the empirical evidence, the method tries to deduce other empirically-testable consequences from the causal story (hence the name hypothetico-*deductive*). If these empirical consequences are observed, then the causal hypothesis is considered to be *corroborated* by the new evidence. On the other hand, the discovery of empirical evidence that is incompatible with the hypothetical causal story may be interpreted as *falsifying* a hypothesis. As various philosophers of science have emphasized, the scientific community is sometimes reluctant to discard a hypothesis that has been contradicted by new empirical evidence. This reluctance may derive from the lack of an alternative hypothesis to explain the phenomenon in a simple and elegant way, or it may be because the falsification of a given hypothesis would result in drastic consequences for related components of the dominant theoretical framework. When contradictory evidence appears that is hard to accept, the standard method of maintaining the old hypothesis is to introduce an ad-hoc supposition that keeps the original causal story in place, but "explains" why the story does not have the expected consequences in this particular empirical situation.

In the nineteenth century it was believed that all of matter and space was permeated by an invisible medium, the luminiferous ether. This medium was hypothesized as being absolutely necessary to explain how electromagnetic waves passed through apparently empty space. As mentioned in Chapter Eight, the "ether wind" produced as the earth moved through the stationary medium should have measurable effects in either slowing or hastening the relative velocity of light as experienced by an observer on earth. When no "wind" was detected by the Michelson-Morley experiment of 1887, the hypothetical medium was not immediately discarded. G.F. Fitzgerald's claim (developed by Lorentz) that the length of objects was contracted along their line of motion attempted to preserve the hypothesis by "explaining" how the ether might evade empirical detection. This example has been well-chronicled by historians of science and is the standard illustration of an ad-hoc supposition introduced in order to maintain a deeply-entrenched hypothesis.

The *empirical-inductive method* is markedly different to the hypothetico-deductive approach. It correlates phenomena and formulates law-like regularities that often take a complex mathematical form. The inverse-square laws of gravitation and electrostatic interaction associate properties of objects (such as mass or electric charge) with their mutual disposition to move towards/away from each other. No causal mechanism or explanatory story for the disposition towards movement is given. Some people, like Berkeley, believed that the law of universal gravitation, which dealt only with effects (and not with generating causes) *was incapable* of offering an explanation in any meaningful sense of the word.

A generalisation can be made about the diverse goals and scope of the hypothetico-deductive and the empirical-inductive approaches. The hypothetico-deductive method is oriented radically towards uncovering the *causal genesis* of a phenomenon. This involves efforts to *individuate* the components of the causal story that give rise to the empirical data, to *visualize* (in the case of mechanisms that cannot be observed directly) how these components are *configured* relative to each other, and to describe the *mechanical principles* that govern the interactions between the components that eventually give rise to the observed phenomena.

Copernicus' heliocentric model lacked a description of the causal mechanism for the movement of the celestial spheres. Yet his scheme had dramatic explanatory power *in comparison* to rival hypotheses. It explained *that what we see* in the skies is *a result of the fact* that the planets go around the sun, even if we don't know *why* they go around the sun. As such it remains a prime example of the hypothetico-deductive method in operation. In effect, it posed the hypothetical question: if the planets go around the sun in simple orbits, then can it be deduced that we would see the celestial phenomena that we do see (epicycles, retrograde motions of planets)? The answer is *yes*. Any hypothesis that offers *reasons* for what we observe is explanatory in some limited sense, even if it does not contain a full-blown story of the causal genesis of what is observed.

If the hypothetico-deductive method is aimed primarily at *explanation*, the empirical-inductive method devotes itself to *correlation of empirical regularities into mathematically formalisms*. This allows prediction in a precise fashion of the outcome of carefully controlled situations, and has been fundamental in ushering in the technological age. Often the *terms* used in the empirical-inductive generalisations refer to elements that have been inferred by the hypothetico-deductive method, such as subatomic particles. Thus the empirical-inductive method is often dependent on the hypothetico-deductive method for the structure of its content, and this structure is composed of hypothetical entities or notions such as "electron", "nucleus", "shell", or "strong interaction".

Attempts to *visualize* the structure of the atom, and to understand the causal mechanisms that give rise to atomic phenomena, are instances of the hypothetico-deductive method in operation. Efforts to establish the properties of hypothetical structures within the atom (such as the nucleus), on the other hand, often have to rely on empirical-inductive techniques such as statistical analysis of vast numbers of particle collisions.

Earlier we stated that a hypothesis *may* be discarded if empirical evidence is found that is incompatible with the inferred causal structure. But hypothetical structures that are deeply ingrained into a theoretical framework can often be retained in the face of contradictory evidence if certain ad-hoc assumptions are made. Whenever this happens, the legitimate suspicion arises that the hypothesis is being retained *despite* the evidence, and this can bring the hypothetico-deductive method into disrepute. The aforementioned introduction of the notion of length contraction by G.F. Fitzgerald (to save the idea of the stationary ether) is possibly the most striking instance of an ad-hoc supposition in the history of science. Atomic theory, too, contains ad-hoc features that sustain deep-rooted hypothetical elements that are empirically questionable. If the hypothetico-deductive method is to be used fruitfully in probing the structure of the atom, then it is crucial that we evaluate objectively the manner in which atomic theory has retained elements that have debatable empirical warrant.

The most dramatic example is the planetary model of Niels Bohr. It had intuitive appeal but soon found itself in major difficulties. An electron held by electrostatic forces should result in the emission of electromagnetic radiation as the electron went round on its orbit, leading the particle to lose energy and spiral into the nucleus. But most atoms are stable, so this evidently does not happen. The problem was "resolved" by Bohr with the assertion that electrons can exist in special, previously unheard-of, states of motion called "stationary orbits". What are these stationary states? They are an ad-hoc invention that conveniently saved the planetary model of the atom, and provided an explanation for atomic spectral lines in the process. An electron in such a state would defy the known laws of physics and not spiral into the nucleus. A "solution" of this sort negates the primary virtue of the hypothetico-deductive method. The method seeks to explain phenomena in terms of an underlying causal structure. Once a structure has been hypothesized, empirical conclusions are drawn that should issue from such a causal structure, if reality is indeed as it has been conjectured to be. When empirical consequences are observed that are incompatible with the hypothetical structure, then the physicist would naturally be expected to return to the drawing board and come up with a hypothesis consistent with the empirical findings. This did not happen. Even though we read everywhere that the Bohr model is now "obsolete", its basic structure still underpins the standard account of atomic bonding and the structure of the periodic table.

A.6 The gap between empirical data and atomic theory

Rutherford's gold leaf experiment led to the hypothesis that the atom had a positively-charged nucleus that contained most of the atom's mass. It would be some years before the idea would develop that the nucleus itself was composed (in part) of the positively charged *particle* that came to be known as the "proton." Well before that, physicists began to wonder if the place of an element in the periodic table might be linked to the charge on the nucleus. During the nineteenth century, the particular number assigned to progressively heavier elements in the periodic table seemed to be largely a matter of convention, but by the second decade of the twentieth century physicists were speculating that the key to the periodic table might lie in the positive charge on the nucleus. Henry Moseley's seminal analysis of x-ray spectra was taken to confirm this beyond doubt. Pure samples of various elements were placed in an evacuated glass tube. The causal impulse that is usually interpreted to consist of "electrons" was concentrated on the sample, resulting in the emission of x-ray radiation. This radiation was diffracted through a crystal, allowing the characteristics of its spectrum to be individuated. It was found that there was a strict mathematical relationship between aspects of the emitted spectrum and the atomic number of the element used in the sample. This empirical fact established beyond doubt that there was an objective connection between the atomic number and certain physical features of an atom of that element. It wasn't long before there was a general consensus in the scientific community that the feature in question was the charge on the nucleus. Moseley himself considered his discovery to strongly support the planetary model of the atom, and while it seems right to affirm that his work demonstrates that *some* physical aspect of the atom has a strict relationship to its atomic number, it by no means follows that this aspect must be the arrangement of the atom into a central nucleus surrounded by shells of electrons.

From the middle of the second decade of the twentieth century onwards, physicists began to formulate explanations of molecular bonds in terms of the planetary model of the atom. Given that electrons were believed to inhabit the outer reaches of the atom, it was natural to assume that the bonds between atoms involved interactions primarily between these electrons. As theory developed, various numerical regularities were formulated that generated the number of electrons that were permitted in each shell, and the overall number of shells in a given atom. The electrons in the atom's outer shell were believed to be responsible for the formation of bonds with other atoms, and this same outer shell was considered to determine many of the reactive properties of elements. The stability and inert nature of the noble gases was believed to derive from the fact that each of these elements had an outer shell that was completely filled with electrons. Other elements had outer shells which "lacked" various numbers of electrons, and their bonding patterns with other elements were considered to be driven by a natural tendency to achieve full outer shells.

When it was eventually accepted that the nucleus of atoms was composed of a number of positively charged particles (the proton), the obvious question was how these particles managed to stay together despite their mutually repulsive electrostatic force. In the 1930s (shortly after the hypothesis became accepted that the nucleus of most atoms also contains a particle with no charge - the *neutron*), the notion of a *nuclear force* was systematically developed to account for the supposed stability of the nucleus. This hypothesis has much in common with Bohr's postulation of stationary states. In order to explain a "fact" that defied common sense and had no independent empirical foundation, a completely novel component of material reality was postulated. The theoretical justification for introducing this dramatic new force of nature was the known "fact" of the stability of the nucleus. But the very *existence* of the nucleus was not a known fact: it was a conjecture from empirical evidence that could have been interpreted in other ways, as we have argued earlier.

In order to overcome the force of electrostatic repulsion, and to explain other supposed properties of the nucleus, the nuclear force is believed to be about one hundred times greater than the electrostatic force. This enormous force does not have empirically measurable consequences beyond the nucleus because, we are told, it only makes itself felt at extremely confined distances. Again, the postulation that the force has an incredibly short range is a conjecture that was custom-made to fit the empirical data (or lack of it), not a conclusion based on independent empirical findings.

As the theory has evolved, the nuclear force is now believed to be a residue of the "strong interaction," a force that binds quarks together, and the quarks, in their turn, make up the protons and neutrons in the nucleus. The strong interaction is mediated by another particle called the "gluon." Both the quark and the gluon are hypothetical particles whose properties have a very tenuous connection to the empirical evidence. One of the properties of the quark is called "colour confinement," a characteristic that allegedly makes the quark impossible to isolate and observe directly.

In classical physics, theoretical entities were expected to have empirical warrant. We only have to think of the controversy that erupted when Newton, in his first edition of the *Principia,* seemed to make an allusion to a gravitational force acting at a distance, a force that had no direct empirical warrant apart from its effects. In our day, few qualms are expressed about the lack of empirical warrant for the quark. Part of the theory of the quark, in fact, is that no direct empirical evidence for its existence is possible. And entities whose properties and states cannot be observed directly can be attributed very unusual characteristics indeed. Quarks and gluons are believed to exhibit strange properties such as "colour charge," a triple-valued type of charge completely unrelated to electrostatic charge. As with the notion of electron shells, such properties only express numerical or logical relationships between components of theory that could equally be described in alternative physical terms without invoking the hypothetical particles.

This is only a small flavour of the difficult theoretical area that is contemporary nuclear physics. The problem is *not* that the theoretical properties of hypothetical entities like the quark and the gluon are completely cut adrift from the empirical data. On the contrary, their properties are carefully formulated and honed to fit the greatest mass of empirical data ever accumulated in the history of science. But there is a formidable gap between the empirical data that supposedly provides justification for the properties we attribute to these entities and the theoretical entities themselves. To make the leap from the evidence to the theory requires remarkable feats of human imagination, ingenuity and mathematical expertise.

At this point it may be useful to state a general principle that can be a rule of thumb for evaluating the epistemic content of a theory. The more closely a theory can be tied to the empirical data that gave rise to the theory, then the less likely it is that the theory has extraneous or fictitious elements. The most celebrated example from the history of science of the gap that can prevail between empirical evidence and theoretical entities is provided by the aforementioned case of the luminiferous ether. Following Young's discovery that light exhibits interference behaviour, it was concluded that light must involve the propagation of waves in space. Waves require a medium, so physicists returned to the idea put forward a century earlier by Christiaan Huygens that all of space was permeated by an invisible material that permits the propagation of light. This medium became known as the luminiferous ether. But in order for this medium to facilitate the transmission of light, yet not hinder the movement of objects, it was necessary to attribute to it all kinds of amazing (and contradictory) properties. It had to be fluid-like in order to fill all of space, yet with incredible rigidity to support the high frequencies of electromagnetic radiation. It had to be massless and without viscosity to permit the motions of the celestial spheres, as well as being transparent and incompressible.

It was the *inadequacy of our understanding of light* that led us to the hypothesis of a medium with fantastic properties that would permit light to behave as it does.

Similarly, an inadequate understanding of the atom is leading us to hypothesize nuclear particles with fantastic properties that permit the atom to have the properties that it is observed to have. It is not that the quark and the gluon (and the rest of the zoo of particles) do not have a relation to the empirical evidence. They form a genuine bridge between empirical evidence and a way of looking at the atom that is simply wrong. The theoretical bridge has a certain validity because it provides a comprehensive account of how such theoretical entities, *if they existed*, would give rise to the empirical evidence. In the same way, theorists of the nineteenth century worked out theoretical properties of the ether that, *if it existed*, would have permitted the transmission of light and the unobstructed movement of the planets.

Our general rule of thumb gauges the epistemic content of a theory by calculating how far that theory stands from the empirical evidence. In the case of the ether, light was believed to be a wave motion in space. This was the first hypothesis. On the basis of this hypothesis, we demanded an ether that would facilitate the transmission of the wave motion through space. So the ether is already two steps removed from the evidence, and an essential theoretical bridge between it and the evidence is the hypothesis that light is a particular kind of wave motion in space that requires a material medium of some sort.

As theoretical entities go, the ether was not far removed from the apparently compelling evidence that light was the kind of wave motion that required a medium. In the case of nuclear entities like the gluon, we are already many steps from the evidence. One of the foundational hypotheses was Rutherford's postulation of the positively charged nucleus. Then came the hypothesis that the positive charge is composed of individual elementary particles that constitute the nucleus. This required the postulation of a new kind of force that overcame the mutual repulsion of like charges. Later, quarks and gluons were hypothesized to account for the way in which this "strong interaction" is generated, and for the peculiarly potent, yet restricted, properties that it is alleged to bear.

It is becoming progressively more difficult to return to basics in atomic physics. Our interpretation of the empirical data is mediated to us by layer upon layer of theory, a heavy crust that has been deposited in a haphazard fashion over the past century. The larger the gap between the empirical evidence and the theoretical entities that we hold dear, then the greater the amount of theoretical huffing and puffing that is required to make the leap from experimental data to full-blown theory. Top heavy theoretical baggage may be a *good indicator that our theoretical entities do not reflect the structure of reality.* This was true of the account of the solar system presented in the *Almagest*, and it is true of contemporary atomic and nuclear physics. The theoretical acrobatics of modern physics, its mathematical opaqueness, the Pandora's box of fantastic entities that we call the nucleus, the incredible properties that grate against the intuition – these seem the hallmarks of a framework that has been custom-designed to make the onerous leap from experimental data to purely fictitious entities.

www.ingramcontent.com/pod-product-compliance
Lightning Source LLC
Chambersburg PA
CBHW020718180526
45163CB00001B/18